Future City

Volume 10

Future City Description

As of 2008, for the first time in human history, half of the world's population now live in cities. And with concerns about issues such as climate change, energy supply and environmental health receiving increasing political attention, interest in the sustainable development of our future cities has grown dramatically.

Yet despite a wealth of literature on green architecture, evidence-based design and sustainable planning, only a fraction of the current literature successfully integrates the necessary theory and practice from across the full range of relevant disciplines.

Springer's *Future City* series combines expertise from designers, and from natural and social scientists, to discuss the wide range of issues facing the architects, planners, developers and inhabitants of the world's future cities. Its aim is to encourage the integration of ecological theory into the aesthetic, social and practical realities of contemporary urban development.

More information about this series at http://www.springer.com/series/8178

Sigrun Kabisch • Florian Koch
Erik Gawel • Annegret Haase • Sonja Knapp
Kerstin Krellenberg • Jaime Nivala
Andreas Zehnsdorf
Editors

Urban Transformations

Sustainable Urban Development Through Resource Efficiency, Quality of Life and Resilience

Springer

Editors
Sigrun Kabisch
Department of Urban and Environmental Sociology
Helmholtz Centre for Environmental Research-UFZ
Leipzig, Germany

Florian Koch
Department of Urban and Environmental Sociology
Helmholtz Centre for Environmental Research-UFZ
Leipzig, Germany

Erik Gawel
Department of Economics
Helmholtz Centre for Environmental Research-UFZ
Leipzig, Germany

Annegret Haase
Department of Urban and Environmental Sociology
Helmholtz Centre for Environmental Research-UFZ
Leipzig, Germany

Sonja Knapp
Department of Community Ecology
Helmholtz Centre for Environmental Research-UFZ
Halle (Saale), Germany

Kerstin Krellenberg
Department of Urban and Environmental Sociology
Helmholtz Centre for Environmental Research-UFZ
Leipzig, Germany

Jaime Nivala
Centre for Environmental Biotechnology
Helmholtz Centre for Environmental Research-UFZ
Leipzig, Germany

Andreas Zehnsdorf
Centre for Environmental Biotechnology
Helmholtz Centre for Environmental Research-UFZ
Leipzig, Germany

ISSN 1876-0899 ISSN 1876-0880 (electronic)
Future City
ISBN 978-3-319-72894-0 (PB)
ISBN 978-3-319-59323-4 (HB) ISBN 978-3-319-59324-1 (eBook)
DOI 10.1007/978-3-319-59324-1

Library of Congress Control Number: 2017953426

Printed on acid-free paper

This Springer imprint is published by Springer Nature
The registered company is Springer International Publishing AG
The registered company address is: Gewerbestrasse 11, 6330 Cham, Switzerland

Foreword

This volume is dedicated to the topic "Urban Transformations: Sustainable Urban Development Through Resource Efficiency, Quality of Life and Resilience". After the Habitat III conference and the adoption of the New Urban Agenda, the question of how to tackle the challenges in terms of transforming our cities in a sustainable direction is more urgent than ever. We need to think about appropriate instruments, strategies and mechanisms and have to consider adequate governance patterns, existing power constellations and the inclusion of the civil society.

Our responsibility, as scientists, is to provide an interdisciplinary framework for understanding urban transformations that can serve as a reference not only for professionals in many of the disciplines involved in this volume but also for those involved in policymaking. This has to be done in a way that political action can become the result of the interaction of research, policy and citizenship and the counterpart of the taxes paid by citizens that serve to support research. This has never been an easy task. Thus, this book responds to the challenge of providing a reference tool for researchers, practitioners and policymakers and a stimulus to strengthen the link between research and policy that contributes to reinforcing the idea of transforming cities into more sustainable places to live, within the general European strategy for reaching a more sustainable lifestyle in Europe in the horizon of 2050.

The need to change towards a more sustainable lifestyle is a priority for the European Union. This has generated a number of European research programmes around the Horizon 2020 research framework programme; these aim at emphasizing the need to transform cities, at improving the quality of life by making the management of resources more efficient and sustainable and at promoting resilience in the way cities deal with the new demands. The problem of urban contexts urgently requires a change in our way of managing and using natural resources. This involves two types of changes: On the one hand, there is a need to activate lifestyle changes and support initiatives for social innovation that challenge existing patterns and propose alternatives. On the other hand, we need to transform our economies and generate urban environments that support wellbeing and social cohesion in a sustainable direction. Both require an empirically grounded understanding of the complex interactions among economic, social, cultural, political and technological factors that influence sustainability and urban transformations towards a greener future. The European Commission states that changes to a more sustainable Europe need to address the demand side as well as to re-evaluate economic and growth models. This goes hand in hand with efficient changes in governance systems, to enable necessary paradigm shifts.

The 21 contributions of this volume demonstrate convincingly projects and pathways for urban transformations towards sustainability. They give evidence that scientifically based solutions and recommendations are needed in order to achieve a more sustainable world, as specified in the Sustainable Development Goals. The contributions of this volume provide conceptual framings, problem descriptions and explanations, data sets, modelling approaches and case studies, as well as solution-oriented research results, and they deal with diverse topics related to urban development. Starting with conceptual considerations about the current state of the scientific discussion on urban transformations, the authors provide their own conceptual frame to define urban transformations. They focus on resource efficiency, quality of life and resilience as key dimensions of urban transformations. Some of the contributions analyse one of these dimensions in depth, whilst the majority of articles stress the interrelatedness and the interdependencies of the three dimensions, in order to capture the complexity of urban development. The contributions refer to the initially introduced conceptual frame of urban transformations and reflect upon the authors' own position as well as formulating particular and, sometimes, divergent opinions. Thus, the urban transformation concept provides an umbrella that is able to invite different disciplinary origins to exchange and collaborate. This is the prerequisite for interdisciplinary dialogues and approaches that produce the added scientific value needed to capture complex, dynamic and occasionally wicked problem constellations. The contributions of this volume give excellent evidence of a variety of interactions, exchanges and cooperation, in particular when they cross the borders between the social and natural sciences. Furthermore, the close collaboration with practitioners and stakeholders, which is characteristic of transdisciplinary research, guarantees the relevance and the impact of research results for societal practice. In order to ensure this focus on practical relevance, governance issues, including institutional frameworks and power constellations, are an important component of the research itself, and they invite us to acknowledge that applying research into political practice involves many players, with numerous social, political and urban dynamics, all of which need to be taken into consideration.

Improving the welfare and quality of life of citizens must go hand in hand with the more efficient management of urban resources, mitigation of the impact of risks and design of more urban resilience. For this reason, we need a practical engagement, considering that political action should be rooted in knowledge based not only on different sciences but also on a variety of actors, underlining the need to consider the contribution that practitioners, policymakers and civil society – in its many forms of organizing and manifesting itself – make to the generation of the knowledge base. Thus, the transfer of research results to social and urban transformations and to environmental policy, as well as the use and exploitation of public resources, requires a comprehensive strategy and not just focusing on a single source of obtaining knowledge. Therefore, a new paradigm of knowledge development is needed, starting from methodologies that integrate citizens, users, politicians and stakeholders within a comprehensive system of co-generation of knowledge. This refers to reconnecting science with its primary objective, which is the systematic generation of well-structured knowledge through observation, reasoning and experimentation in disciplines working in cooperation and attempting to explain the functioning of urban phenomena via theories and systems.

The co-production of knowledge, incorporating the knowledge of citizens, policymakers and stakeholders, involves a change in the configuration of politics towards a conceptually different approach, aimed at decision-making and at new styles of governance in the urban context, one that incorporates social innovation as a key element. One of the most important characteristics of this new approach is that it involves the coming together of actors who have differing opinions, agendas, languages and expectations. It also demands the creation of a suitable space for innovation and social participation, where social innovation can emerge as a result of the creative processes arising from the encounters between the various actors. Finally, it demands a consensus on governance models, discarding those that are guided by criteria of experience, in favour of models that are committed to the implementation of participatory mechanisms for building up community policies.

Creating new spaces in which disciplines and actors can establish common terms of reference and a shared language thus becomes absolutely necessary. But perhaps it is even more important, in addition to the shared language, to share the idea that social and environmental policy, based on knowledge of the social sciences, can contribute to a better society. This aspect has not always led to agreement between scientists and policymakers. Nevertheless, it is a matter that affects the prestige of the environmental sciences in general and social sciences in particular, if they are to be considered as the base for decision-making.

This volume would not have been possible without the specific research environment provided by the Helmholtz Centre for Environmental Research, with its broad disciplinary background and its integrative research strategy. This volume delivers a convincing example of what interdisciplinary, future-oriented and relevant urban research should look like.

From my perspective as president of the International Association of People-Environment Studies (IAPS), which places emphasis on strongly fostering interdisciplinary exchange, particularly concerning sustainable urban development, I appreciate the scientific contributions documented in this volume. They provide a sound basis for further scientific exchange about urban transformations towards sustainability.

President of the International Association of People-Environment Studies (IAPS)
A Coruña, Spain

Ricardo Garcia-Mira

Contents

Contributors

Sohail Ahmad Researcher at the Mercator Research Institute on Global Commons and Climate Change (MCC), Berlin

Thomas Arndt Federal Agency for Nature Conversation, Division II 4.1 Landscape planning, spatial planning and urban areas

Ellen Banzhaf Researcher at the Department of Urban and Environmental Sociology at the Helmholtz Centre for Environmental Research-UFZ, Leipzig

Robert Bartz Researcher at the Department of Ecology, Technical University of Berlin, Berlin, Germany

Norman Bedtke Researcher at the Department of Economics, Helmholtz Centre for Environmental Research-UFZ, Leipzig

Chloe Begg Researcher at the Department of Urban and Environmental Sociology, Helmholtz Centre for Environmental Research-UFZ, Leipzig

Sebastian Bohnet Regional Manager at Delitzscher Land, Former researcher at the Department of Bioenergy Systems, Deutsches Biomasseforschungszentrum gemeinnützige GmbH (DBFZ) in Leipzig

Miriam Brenck Researcher at the Department of Economics, Helmholtz Centre for Environmental Research-UFZ, Leipzig

Francisco de la Barrera Researcher at the Institute of Geography and the Centre for Urban Sustainable Development, Pontificia Universidad Católica de Chile

Ulrich Franck Researcher at the Department Core Facility Studies, Helmholtz Centre for Environmental Research-UFZ, Leipzig

Erik Gawel Head of the Department of Economics at the Helmholtz Centre for Environmental Research-UFZ, Leipzig, Professor of Economics, especially Environmental Institutional Economics, and Director of the Institute for Infrastructure and Resources Management at the University of Leipzig

Annegret Haase Researcher at the Department of Urban and Environmental Sociology, Helmholtz Centre for Environmental Research-UFZ, Leipzig

Dagmar Haase Professor for Landscape Ecology at the Humboldt University of Berlin

Bernd Hansjürgens Head of the Department of Economics at the Helmholtz Centre for Environmental Research-UFZ, Leipzig, Professor for Economics, especially Environmental Economics at Martin-Luther University Halle-Wittenberg

Daniel Hertel Researcher at the Department of Urban and Environmental Sociology, Helmholtz Centre for Environmental Research-UFZ, Leipzig

Sigrun Kabisch Head of the Department of Urban and Environmental Sociology at the Helmholtz Centre for Environmental Research-UFZ, Leipzig, Honorary Professor for Urban Geography at the University of Leipzig

Annegret Kindler Researcher at the Department of Urban and Environmental Sociology, Helmholtz Centre for Environmental Research-UFZ, Leipzig

Christian Klassert Researcher at the Chair of Economics/Institutional Environmental Economics, University of Leipzig; Guest Researcher at the Department of Economics, Helmholtz Centre for Environmental Research - UFZ

Bernd Klauer Researcher at the Department of Economics, Helmholtz Centre for Environmental Research-UFZ, Leipzig, Honorary Professor for Sustainability and Water Resources Management at the University of Leipzig

Heinz-Josef Klimeczek Senate Administration for Urban Development in Berlin

Stefan Klotz Head of the Department Community Ecology, Helmholtz Centre for Environmental Research-UFZ, Halle

Sonja Knapp Researcher at the Department Community Ecology, Helmholtz Centre for Environmental Research-UFZ, Halle

Florian Koch Researcher at the Department of Urban and Environmental Sociology, Helmholtz Centre for Environmental Research-UFZ, Leipzig

Ingo Kowarik Professor for Ecosystem Science and Plant Ecology, Technical University Berlin, Germany

Kerstin Krellenberg Researcher at the Department of Urban and Environmental Sociology, Helmholtz Centre for Environmental Research-UFZ, Leipzig

Christian Kuhlicke Researcher at the Department of Urban and Environmental Sociology, Helmholtz Centre for Environmental Research-UFZ, Leipzig

Jan Ladiges Institute for Geosciences and Geography, Martin-Luther-University Halle-Wittenberg

Volker Lenz Head of the Department Thermo-chemical Conversion, Deutsches Biomasseforschungszentrum gemeinnützige GmbH (DBFZ) in Leipzig

Volker Meyer Researcher at the Department of Economics, Helmholtz Centre for Environmental Research-UFZ, Leipzig

Abdelrhman Mohamdeen Researcher at the Department of Urban and Environmental Sociology, Helmholtz Centre for Environmental Research-UFZ, Leipzig

Roland A. Müller Head of the Centre for Environmental Biotechnology at the Helmholtz Centre for Environmental Research-UFZ, Leipzig, Honorary Professor for Integrated Water Resources Management (IWRM) at Leipzig University of Applied Sciences

Jaime Nivala Researcher at the Centre for Environmental Biotechnology, Helmholtz Centre for Environmental Research-UFZ, Leipzig

Moritz Reese Researcher at the Department Environmental and Planning Law, Helmholtz Centre for Environmental Research-UFZ, Leipzig

Sonia M. Reyes-Paecke Assistant Professor, Department of Ecosystems and Environment. Faculty of Agronomy and Forestry Engineering Pontificia Universidad Católica de Chile.

Dieter Rink Deputy Head of the Department of Urban and Environmental Sociology at the Helmholtz Centre for Environmental Research-UFZ, Leipzig, Honorary Professor for Urban Sociology at the University of Leipzig

Cornelia Rönsch Researcher at the Department of Bioenergy Systems, Deutsches Biomasseforschungszentrum gemeinnützige GmbH (DBFZ) in Leipzig

Kay Schaubach Project Manager at the Department of Bioenergy Systems, Deutsches Biomasseforschungszentrum gemeinnützige GmbH (DBFZ) in Leipzig

Uwe Schlink Researcher at the Department of Urban and Environmental Sociology, Helmholtz Centre for Environmental Research-UFZ, Leipzig, Honorary Professor at the Institute of Meteorology, University of Leipzig

Nina Schwarz Deputy Head of the Department of Computational Landscape Ecology, Helmholtz Centre for Environmental Research-UFZ, Leipzig

Katja Sigel Researcher at the Department of Economics, Helmholtz Centre for Environmental Research-UFZ, Leipzig

Daniela Thrän Head of the Department of Bioenergy at the Helmholtz Centre for Environmental Research-UFZ, Leipzig, Head of the Department Bioenergy Systems at Deutsches Biomasseforschungszentrum gemeinnützige GmbH (DBFZ) in Leipzig, Professor for Bioenergy Systems, University of Leipzig

Maximilian Ueberham Researcher at the Department of Urban and Environmental Sociology, Helmholtz Centre for Environmental Research-UFZ, Leipzig

Manfred van Afferden Deputy Head of the Centre for Environmental Biotechnology, Helmholtz Centre for Environmental Research-UFZ, Leipzig

Juliane Welz Researcher at the Department of Urban and Environmental Sociology, Helmholtz Centre for Environmental Research-UFZ, Leipzig

Manuel Wolff Researcher at the Department of Urban and Environmental Sociology, Helmholtz Centre for Environmental Research-UFZ, Leipzig

Andreas Zehnsdorf Deputy Head of the Centre for Environmental Biotechnology at the Helmholtz Centre for Environmental Research-UFZ, Leipzig, Honorary Professor for Bioprocess Technology at the University of Cooperative Education of Riesa

Introduction: Urban Transformations – Sustainable Urban Development Through Resource Efficiency, Quality of Life and Resilience

Sigrun Kabisch, Florian Koch, Erik Gawel, Annegret Haase, Sonja Knapp, Kerstin Krellenberg, and Andreas Zehnsdorf

Motivation

"Cities are the future" is the title of the introduction to the special issue of SCIENCE, published in 2016, that stresses the planetary challenges and chances of ongoing global urbanization. It underlines that "the decisions we make today about how we build and live in cities will affect generations to come" (Winnigton et al. 2016, p. 905). Also in 2016, the German Advisory Council on Global Change (WBGU) in the flagship report "Humanity on the move: Unlocking the transformative power of cities" (WBGU 2016) points to urban development as one of the major challenges of present and future societal development. All this emphasizes that, at present, cities are ubiquitous, with the majority of mankind now living in urban areas. And, urbanization, as a mega-trend, is ongoing. In 2050, more than two thirds of the

S. Kabisch • F. Koch • A. Haase • K. Krellenberg
Department of Urban and Environmental Sociology, Helmholtz Centre for Environmental Research-UFZ, Leipzig, Germany
e-mail: sigrun.kabisch@ufz.de; florian.koch@ufz.de; annegret.haase@ufz.de; kerstin.krellenberg@ufz.de

E. Gawel
Department of Economics, Helmholtz Centre for Environmental Research-UFZ, Leipzig, Germany
e-mail: erik.gawel@ufz.de

S. Knapp
Department Community Ecology, Helmholtz Centre for Environmental Research-UFZ, Halle (Saale), Germany
e-mail: sonja.knapp@ufz.de

A. Zehnsdorf
Centre for Environmental Biotechnology, Helmholtz Centre for Environmental Research-UFZ, Leipzig, Germany
e-mail: andreas.zehnsdorf@ufz.de

global population will live in cities, with particularly high population growth in African and South-Asian cities (UN 2015b).

The major question connected with this urbanization process is how to organize and provide living conditions in cities that enable a dignified life for the current inhabitants as well as for newcomers and future generations. This includes the efficient provision and fair distribution of resources to ensure an acceptable quality of life for all urban residents within a resilient urban environment. To reach this target, fundamental changes towards sustainability – called, in this volume, "urban transformations"[1] – are necessary. By applying a research approach that recognizes the complex and adaptable nature of urban areas, knowledge and solutions can be produced that really respond to current and urgent urban challenges.

Cities, on the one hand, demand, use, and, in many cases, also overuse natural resources such as land, water, energy, or biodiversity. Seto et al. (2011, p. 1) point out that the conversion of land surface to urban uses has to be classified as "one of the most irreversible human impacts on the global biosphere". Cities also increasingly grab natural resources that are far from their original territory, as the debate on urban land teleconnections (Seto et al. 2012) shows. Thus, urban development impacts on the development of resource use in many parts of the world. Furthermore, cities produce 75% of the global carbon emissions, and use 60% of the residential water (Grimm et al. 2008, p. 756). 90% of the global economic power (GDP) and 65% of the global energy consumption are concentrated in urban areas (Solecki et al. 2013).

On the other hand, cities are places where most innovations are created, and where people of diverse backgrounds and experiences come together, bringing in capacities and skills to produce ideas for more sustainable development. Moreover, cities mean a concentration and compaction of people and infrastructure that also has positive impacts on the environment, such as minimizing land consumption and sprawl, supporting short travel distances, making efficient use of water, energy and traffic infrastructure, as well as providing access to services such as school and health facilities. Thus, living in the so-called urban age is nothing to worry about per se. As several urban scholars have highlighted, cities can be seen as places of hope and change (e.g. Glaeser 2011). They play a key role in dealing with the challenges of global climate change (Rosenzweig et al. 2015), as arrival cities that raise hopes of a better life (Saunders 2011), or as economic engines and places where innovations occur (Sassen 2012).

However, there is also a consensus that, overall, urban developments are often not sustainable in terms of resource use, social equity and inclusion, economic welfare, and environmental justice. Accordingly, recently approved programs and declarations place the need for ambitious structural transformations to reach sustainability at the center stage of decisions. The 2015 United Nations Sustainable

[1] In science and politics, the term "urban transformations" is given very different meanings in different contexts and different disciplines. Our definition is presented in the following sub-chapter: *Our approach and understanding of urban transformations*. Nevertheless, some contributions in the volume deviate from this definition and describe their specific handling of this concept. Similarly, there is also no clear-cut definition of urban areas or cities. Within this volume, we consider large urban agglomerations, such as megacities, as well as smaller cities, as cities themselves but also include their relationships with the peri-urban hinterland.

Development Summit marked a turning point with its declaration "Transforming our world – the 2030 agenda for sustainable development". Within the 17 Sustainable Development Goals (SDGs), Goal 11, "make cities and human settlements inclusive, safe, resilient and sustainable", places emphasis on urban areas (UN 2015a). This way, and for the first time, a UN declaration assigns cities a major role as places of human habitat. Besides Goal 11, other goals are also closely connected with urban life, such as Goal 1, "end poverty", Goal 3, "ensure healthy lives", Goal 6, "provide clean water and sanitation", Goal 9, "build resilience infrastructure", Goal 13, "take urgent action to combat climate change", or Goal 16, "promote peaceful and inclusive societies"; they overlap and cumulate in urban areas. Even goals that, at first glance, seem less connected to urban life, such as Goal 15, "halt biodiversity loss", are concerned: Urbanization threatens global biodiversity but "Cities can reconcile […] biodiversity by creating environments that are ecologically sustainable" (CBD 2012, p. 2). This variety of goals and the respective targets that specify each goal mark the complex and diverse character of urban structures and challenges that require innovative, tailored, and adapted approaches and pathways to reach the goals. To take action, it is essential to consider the particular urban context, including the starting conditions, pressing problems, and governance structures, to define clear and realistic aims and pathways.

Sustainability Through "Urban Transformations"

A review of the scientific literature yields evidence that urban transformations, understood as fundamental, partly radical, multi-dimensional and, in many cases, non-linear alterations in and of cities, are necessary to reach sustainability targets (e.g., UN Habitat 2016, WBGU 2016, Pickett et al. 2013, McCormick et al. 2013). In order to achieve sustainable development in cities, various authors describe the process with the term "sustainability transformations" (e.g. Olsson et al. 2014). Thus, the terms "sustainability" and "transformations" are both used in close interconnection. Sustainability can be considered as a general societal vision. It has been concretized within the Sustainable Development Goals (SDGs) to encourage concrete actions based on binding agreements and to make the progress measurable. Thus, urban transformations are a response to global trends affecting urban areas; these include climate change, resource scarcity, economic globalization, demographic change, social polarization, or increasing migration streams.

However, urban transformations towards sustainability represent processes and not an endpoint (Kabisch and Kuhlicke 2014), in which stakeholders and decision-makers develop and execute innovative solutions in line with the SDGs in their local context. This encompasses examples of collective and individual alterations in behavior, economic and power relationships, or technological innovations. Often, transformations are a combination of technical, organizational, economic, institutional, socio-cultural, and political changes towards a sustainable way of life (McCormick et al. 2016, van den Bergh et al. 2011, Markard et al. 2012, Forrest and Wiek 2014).

Moreover, in the practice of urban development, selected sectors or themes, which have their own complexity, become a focus. Examples include infrastructure transformations (Bolton and Foxon 2015), energy transformations (Moloney and Horne 2015), or governance transformations at city district levels (Mahzouni 2015).

Due to the variety of trends and their specific local impacts in cities, different transformation paths towards sustainability are necessary. This means that cities do not follow one single or clear-cut pathway. Rather, multiple urban transformation processes take place and can interfere with each other at the same time, and in different forms, shapes, and temporal steps. They may pursue intermediate and final goals, according to the various stakeholders and interests involved, as well as in terms of available resources. Accordingly, it has been acknowledged that we should speak of 'urban transformations' in the plural (Rink et al. 2015; Sommer and Welzer 2014; see also Pelling 2012; O'Brien 2012; Pickett et al. 2013; Park et al. 2012; Castán Broto and Bulkeley 2013; McCormick et al. 2013 for more general descriptions).

In the light of these manifold challenges, urban transformations towards comprehensive sustainability can only succeed if innovative scientific solutions and governance approaches are developed and implemented; governance approaches, in particular, have to critically consider existing institutional, political, and power contexts with respect to their capacities to facilitate, enable, or hinder transformations.

Thereby, local key actors and city leaders push actions and programs for innovative solutions that are steps and components on the way towards urban sustainability.

Our Approach and Understanding of Urban Transformations

The point of origin of our research on urban transformations presented in this volume stems from our socio-environmental science background. Social, natural and legal scientists as well as economists, urban and landscape planners, modeling experts, and engineers are engaged in investigations of commonly defined research topics. This multi- and interdisciplinary perspective is needed in order to tackle the complexity of urban transformations towards sustainability. In our research program, we identify three transformation dimensions that help us to tame the exuberant concept of urban sustainability and that operationalize different sustainability aims. The dimensions are: **resource efficiency, quality of life,** and **resilience.** They have been selected as the major foci of an integrated socio-environmental approach. Thus, there is a need for defining the content of each dimension but, additionally, and what is important, is the specification of the interplay and the interdependence of the three dimensions within the urban context.

These dimensions include a normative and an analytical connotation as well. Therefore, they are used in order to develop a coherent research perspective throughout this volume, including conceptual debates and empirical evidence.

Concerning **resource efficiency**, the analysis and evaluation of urban resource use in close relation to technical infrastructure services, institutional framings, and governmental reflections are addressed. In particular, unprecedented land consumption, changing urban energy services, challenges of water quality and water scarcity, deficient sanitation, air pollution, and biodiversity losses are in focus. With regard to **quality of life**, we investigate living conditions of different socio-demographic groups of residents with respect to environmental justice, health and well-being, as well as social segregation. These components are impacted by urban nature, with its biological diversity and related ecosystem services, housing conditions, and infrastructure facilities. **Resilience** encompasses, above all, but not exclusively, the challenges of climate change and hazardous weather events. We introduce research results concerning risk assessments and adaptation response measures and address, in particular, the governance and planning perspective.

In the following, these three dimensions are introduced in more detail, according to the contributions in the volume that refer to them.

Resource Efficiency

The increasing competition for scarce land, affordable energy, or sufficient drinking water raises the question about how to use the available resources in an "efficient" manner and avoid any kind of wastefulness. Generally, resource efficiency is an attempt to create or produce the same or an even better output with less input (EC 2011). The higher the efficiency of a system is, the better its input-output relationship (Fang et al. 2013). This notion of mere technical efficiency has to be recharged by considerations of "economic efficiency", which also takes into account the opportunity costs of technically efficient alternatives. Thus, economic efficiency is value-driven, whilst technical efficiency might be misleading if only simple input-output relationships are used. For example, energy-saving lamps may be technically efficient, but they might not be the economically most efficient way to reduce greenhouse gases (Mennel and Sturm 2009). To make it even more complicated: technical and economic efficiency have to be addressed due to their effects on social equity and livability, too.

With regard to the increase of resource efficiency, a simple juxtaposition of behavioral changes (e.g., lifestyle or consumer patterns), on the one hand, and technological innovations, on the other hand, are not sufficient. For instance, attempts to decrease residential water consumption require not only technological advances (Lee et al. 2013) but also the awareness and acceptance of such innovations, suitable tariff structures, as well as the economic efficiency and affordability of water services (Klassert et al. 2015).

Similarly, the transformation of urban infrastructure systems, most of which were designed under the premise of long-lasting, stable demographic, political, and economic conditions, needs to take into account the wider societal context (Kiparsky et al. 2013). Inflexible, pipe-dependent water supply and disposal systems are increasingly exposed to changing conditions such as climate change or alteration in population numbers; these influence the number of users, as well as increasing the requirements regarding efficiency. Hence, there is a need for adaptive and flexible facilities that can deal with variable and possibly unknown requirements and context conditions. These processes, however, question many established practices and routines, as well as institutionalized decision and policy-making processes in the context of urban transformations.

Further examples of the efficient use of resources are retrofitted buildings, to reduce energy consumption (Vandevyvere and Nevens 2015), and the maintenance and enlargement of urban green spaces (Haase et al. 2014), to provide cost-efficient benefits from the delivered ecosystem services, such as reducing air pollution and urban heat island effect. Currently, ongoing urban land use and surface sealing are limiting the existence and the functionality of ecosystem services; e.g., green spaces are reduced by housing construction and infrastructure enlargement. Based on estimates that indicate a growth of urban areas worldwide by 1.5 million km^2 by 2030, in comparison to 0.7 million km^2 in 2001 (Seto et al. 2011, p. 8), the pressure on natural resources will increase.

In our understanding, the efficiency of resource use in urban areas, particularly with regard to land, energy, water, and biodiversity, can and must be improved. It is vital to pay attention to the socio-economic and organizational-institutional contexts of urban development, to technological advances and innovations, to environmental impacts as well as to governance processes in a broader sense. It is a transformation area that, in general, provides major prerequisites for simultaneously improving both urban quality of life and a city's resilience.

Quality of Life

Quality of life is a broad term that "can mean many things to many people" (EEA 2009, p. 6). Thus, quality of life is a multidimensional concept that includes issues ranging from material living conditions such as housing to personal and social well-being, as well as environmental circumstances (Garcia Diez 2015). It involves immaterial as well as material, subjective as well as objective, individual as well as collective elements of welfare, satisfaction, and happiness (Marans and Stimson 2011). Therefore, the availability of and access to resources is a core prerequisite of urban quality of life. Whilst quality of life has increased in many areas of life and for many social groups, it has been decreasing, e.g., with respect to lifestyle-related diseases, and social inequalities also persist (EEA 2009).

In this volume, the focus is on quality of life from a social-environmental perspective. In this sense, the access to green infrastructure, "the network of natural and semi-natural areas, features and green spaces in rural and urban, and terrestrial, freshwater, coastal and marine areas" (Naumann et al. 2011, p. 1), its biodiversity and ecosystem services is basic to quality of life. People who live in urban areas with a high share of green spaces, for example, have both lower mental distress and greater well-being than people in areas with smaller shares (White et al. 2013). Similarly, evidence suggests that high biodiversity positively affects human health (Keesing and Ostfeld 2015) – but access to biodiversity is not distributed equally in society (Strohbach et al. 2009). Equal access to a clean environment and equal protection from environmental harm (e.g., noise, pollution, heat stress, lack of green space) for everyone, independent of any differentiating features of socio-economic status, in terms of environmental justice (Schwarte and Adebowale 2007), is thus vital to quality of life.

Subjective indicators of quality of life, such as individual perceptions of living conditions and/or the environmental situation, are at least equally important as the more classical objective indicators such as concentrations of environmental pollutants and other technical measures of environmental burdens (Marans and Stimson 2011). Therefore, to obtain differentiated insights into the quality of life of urban inhabitants, it is necessary to differentiate between social, demographic, lifestyle, and socio-economic groups with regard to their expectations and perceptions, and between different scales, from the entire city to the neighborhood. In order to provide a high quality of life for everyone, it is necessary to combine technical innovations, social innovations, and ecological interests with participatory governance modes, by involving citizens (Nevens et al. 2013; de Flander et al. 2014).

Consequently, quality of life can only be achieved if a) access to natural resources and protection from environmental harm in and between cities are more equally distributed and b) this distribution process is designed in a democratic, transparent, and integrative way that refers to the concepts of procedural justice and justice as recognition, thus giving equal access to information and participation in decision-making (Walker 2012).

Resilience

The resilience concept points towards the question of how to deal with rapid, mostly unanticipated, and therefore surprising alterations, as well as shocks or crises (Kuhlicke 2013). Resilience is often defined as a system's capacity to adapt or respond to singular, unique, and, most often, radically surprising events and/or processes (Mykhnenko 2016; Olsson et al. 2014). The resilience concept used in the context of this volume refers to the "[t]he capacity of a social-ecological system to cope with a hazardous event or disturbance, responding or reorganizing in ways that maintain its essential function, identity, and structure, while also maintaining the capacity for adaptation, learning, and transformation" (IPCC 2014, p. 3). In this light, adaptation is seen as "incremental adjustments that preserve systems integrity when conditions change" (Pelling et al. 2015, p. 116). The overall orientation of the discourse is to consider how the concept of resilience might be useful for urban planners or management agencies in order to enhance, build, or develop capacities in communities, urban areas, and entire systems, to be prepared for and to cope with unexpected, quite often threatening events or developments.

Cities are particularly exposed to such alterations, e.g., in terms of natural hazards and climate-related extreme weather events, due to the large number of people and assets located in risk-prone areas, but also in terms of other disturbances, such as terrorist attacks or financial crises, which can affect every city with significant and far-reaching consequences. Making cities more resilient means, therefore strengthening a system against turbulences and constructing functions and structures that are flexible and elastic in terms of less vulnerability during times of crisis (Revi et al. 2014). Therefore, increasing the resilience of a city is seen as crucial element of urban transformations (Schneidewind and Scheck 2013).

Resilience is, thus, not only a concrete and observable capacity that is, in a normative sense, desirable, but also a field that has to take into account how actors make sense of their capacity to deal with rapid changes (Kuhlicke 2013). New forms of governance targeting the creation of 'preventive resilience' are needed, in order to implement more resilient strategies and behavior. In a broader understanding, resilience also showcases the future orientation of urban transformations. A city's resilience towards climate change, for example, may require deep structural changes in the urban economy and renewed forms of governance. In this context, other concepts, such as adaptation, need to be considered and win-win situations have to be evaluated. The crucial, although political question is, which structural elements need to be changed?

Synopsis

All three transformation dimensions – resource efficiency, quality of life and resilience – are interrelated and influence each other. Accordingly, synergies and trade-offs, but also risks, feedbacks, and conflicts between the three transformation dimensions need to be addressed.

For example, new technologies promoting higher resource efficiency are not necessarily resilient, socially sustainable and enhancing for the quality of life at the same time, as the example of new and highly innovative sanitation systems shows. Although they make use of resources from wastewater flows, and provide wastewater reuse (as drinking water), they often lack acceptance because people feel restricted in their quality of life, due to the loss of comfort. Another example is related to the ongoing pressure on urban land and land-use changes. More compact urban structures, to restrict urban sprawl, impact on a variety of urban functions because they involve land sealing. They limit inner-urban ecosystem services such as recreation or retention capacity, in case of heavy rainfalls (Banzhaf et al. 2017).

To analyze and evaluate the interrelations between the three transformation dimensions, city-specific economic, political, social, or cultural characteristics need to be considered. This is undispensable in order to generate context specific options for policy recommendations and practice solutions.

Purpose of the Volume

The focus of this volume lies on the concept of urban transformations as an *umbrella term* because of its normative as well as analytical power: Whilst the normative dimension refers to the concept of (multi-dimensional) sustainability as an overarching goal or direction of change, the analytical dimension focuses on understanding how change is occurring, framed, shaped, negotiated, implemented and interpreted. This is based on empirical studies and includes indicators and a well-defined methodological design. Within this perspective, the contributions of this volume critically address synergies, conflicts, trade-offs, and limits, as well as scopes of action of urban transformations. They reflect on urban dynamics, drivers and triggers, and discuss governance options and decision-making processes, highlighting the transformation dimensions resource efficiency, quality of life, and resilience, as well as their interactions.

This volume provides both theoretical contributions related to the concept of urban transformations and its interpretations, as well as corresponding empirical research contributions. Our empirical research is dedicated to a range of specific topics, e.g., land-use change and its consequences for ecosystem services, "the direct and indirect contributions of ecosystems to human well-being" (TEEB 2010, p. 33) and quality of life, governance of urban infrastructures, or urban resilience and adaptation towards climate change. We provide scientifically based recommendations and solutions for the particular challenges. Furthermore, we connect the different approaches, point out interdependencies and define the relevance of the contribution for the sustainability targets. Particular attention is given to governance by exploring decision-making, participative approaches, and power relationships. This perspective is decisive for a true understanding of how and if, at all, knowledge can be implemented into practice. It also allows an explicit view on related trade-offs (between different interests, actors, or policy fields) as well as existing conflicts and failures.

A variety of geographical scales are addressed by the contributions, ranging from global to regional and local, down to the neighborhood scale. The empirical case studies were mainly carried out in Germany. Some examples are located abroad (Chile, Jordan, India), acknowledging their specific cultural contexts. Thus, the volume reaches beyond a merely additive description of single case studies. It provides evidence for the additional value of close collaborations between social, natural, and technical scientists, all of whom are engaged in efforts directed at sustainable urban development. Thereby, the volume demonstrates that urban transformations are multi-facetted processes that depend on local contexts, and particularly on the actors who determine the governance pattern. What stands out is the interdisciplinary collaboration in many of the contributions. Subsequently, we pursue an integrated research approach combining different disciplines, concepts, and methods.

The volume consists of four parts, framed by an introductory chapter and a final reflection. Each part has a thematic focus, nevertheless strong interrelations between the parts exist. **Part I** evolves along a variety of conceptual debates about sustainable urban transformations in different disciplines and topics. **Part II** is dedicated to the improved use of urban resources and infrastructures by different governance options. **Part III** sheds light on quality of life, especially with respect to air quality, thermal conditions, green spaces, and ecosystem services. **Part IV** deals with urban risks and resilience, both conceptually and empirically, in a variety of urban contexts.

References

Banzhaf E, Kabisch S, Knapp S, Rink D, Wolff M, Kindler A (2017) Integrated research on landuse changes in the face of urban transformations. An analytic framework for further studies. Land Use Policy 60:403–407

Bolton R, Foxon TJ (2015) Infrastructure transformation as a socio-technical process – Implications for the governance of energy distribution networks in the UK. Technol Forecast Soc Chang 90(B):538–550

Castán Broto V, Bulkeley H (2013) A survey of urban climate change experiments in 100 cities. Glob Environ Chang 23(1):92–102

CBD – Secretariat of the Convention on Biological Diversity (2012) Cities and biodiversity outlook. CBD, Montreal

de Flander K, Hahne U, Kegler H, Lang D, Lucas R, Schneidewind U et al (2014) Resilience and real-life laboratories as key concepts for urban transition research. Gaia-Ecological Perspect Sci Soc 23(3):284–286

EC – European Commission (2011) Roadmap to a Resource Efficient Europe. COM:571

EEA – European Environment Agency (2009) Ensuring quality of life in Europe's cities and towns. Tackling the environmental challenges driven by European and global change. EEA, Copenhagen

Fang C, Guan X, Lu S, Zhou M, Deng Y (2013) Input–output efficiency of urban agglomeration in China: an application of Data Envelopment Analysis (DEA). Urban Stud 50(13):2766–2790

Forrest N, Wiek A (2014) Learning from success — toward evidence-informed sustainability transitions in communities. Environ Innov Soc Trans 12:66–88. doi:10.1016/j.eist.2014.01.003

Garcia Diez S (2015) Indikatoren zur Lebensqualität. Vorschläge der europäischen Expertengruppe und ausgewählte nationale Initiativen. Statistisches Bundesamt, Wiesbaden

Glaeser E (2011) The Triumph of the city. Penguin, New York

Grimm N, Faethl SH, Golubiewski NE, Redmann CL, Wu J, Bai X, Briggs JM (2008) Global change and the ecology of cities. Science 319(5864):756–760

Haase D, Larondelle N, Andersson E, Artmann M, Borgström S, Breuste J, Kabisch N (2014) A quantitative review of urban ecosystem service assessments: concepts, models, and implementation. Ambio 43(4):413–433

IPCC – Intergovernmental Panel on Climate Change (2014) Climate change 2014: impacts, adaptation, and vulnerability. Part A: global and sectoral aspects. Contribution of working Group II to the Fifth Assessment Report of the Intergovernmental Panel on Climate Change. Cambridge University Press, Cambridge/New York

Kabisch S, Kuhlicke C (2014) Urban transformations and the idea of resource efficiency, quality of life and resilience. Built Environ 40(4):475–485

Keesing F, Ostfeld RS (2015) Is biodiversity good for your health? Science 349(6245):235–236

Kiparsky M, Sedlak DE, Thompson BH, Truffer B (2013) The innovation deficit in Urban water: the need for an integrated perspective on institutions, organizations, and technology. Environ Eng Sci 30(8):395–408

Klassert C, Sigel K, Gawel E, Klauer B (2015) Modeling residential water consumption in Amman: The role of intermittency, storage, and pricing for piped and tanker water. Water 7(7):3643–3670

Kuhlicke C (2013) Resilience: a capacity and a myth: findings from an in – depth case study in disaster management research. Nat Hazards 67(1):61–67

Lee M, Tansel B, Balbin M (2013) Urban sustainability incentives for residential water conversation: adoption of multiple high efficiency appliances. Water Resour Manag 27(7):2531–2540

Mahzouni A (2015) The 'Policy Mix' for sustainable urban transition: The city district of Hammarby Sjostad in Stockholm. Environ Policy Gov 25(4):288–302

Marans RW, Stimson RJ (eds) (2011) Investigating quality of urban life. Theory, methods, and empirical research. Springer, Dordrecht

Markard J, Raven R, Truffer B (2012) Sustainability transitions: An emerging field of research and its prospects. Res Policy 41(6). doi:10.1016/j.respol.2012.02.013
McCormick K, Anderberg S, Coenen L, Neij L (2013) Advancing sustainable urban transformation. J Clean Prod 50:1–11
McCormick K, Neij L, Mont O, Ryan C, Rodhe H, Orsato R (2016) Advancing sustainable solutions: an interdisciplinary and collaborative research agenda. J Clean Prod 123:1–4
Mennel T, Sturm B (2009) Energieeffizienz – eine neue Aufgabe staatlicher Regulierung? Zeitschrift für Wirtschaftspolitik 58(1):3–35
Moloney S, Horne R (2015) Low carbon urban transitioning: From local experimentation to urban transformation? Sustainability 7(3):2437–2453
Mykhnenko V (2016) Resilience: a right-wingers' ploy? In: Springer S, Birch K, MacLeavy J (eds) The handvolume of neoliberalism. Routledge, Abingdon\New York, pp 190–206
Naumann S, Davis M, Kaphengst T, Pieterse, M, Rayment M (2011) Design, implementation and cost elements of Green Infrastructure projects. Final report to the European Commission, DG Environment, Contract no. 070307/2010/577182/ETU/F.1
Nevens F, Frantzeskaki N, Gorissen L, Loorbach D (2013) Urban Transition Labs: co-creating transformative action for sustainable cities. J Clean Prod 50:111–122
O'Brien KL (2012) Global environmental change (2): from adaptation to deliberate transformation. Prog Hum Geogr 36(5):667–676
Olsson P, Galaz V, Boonstra WJ (2014) Sustainability transformations: a resilience perspective. Ecol Soc 19(4). doi:10.5751/ES-06799-190401
Park SE, Marshall NA, Jakku E, Dowd AM, Howden SM, Mendham E et al (2012) Informing adaptation responses to climate change through theories of transformation. Glob Environ Chang 22(1):115–126
Pelling M (2012) Resilience and transformation. In: Pelling M, Manuel-Navarrete D, Redclift M (eds) Climate change and the crisis of capitalism, Studies in human geography. Routledge, London\New York, pp 51–65
Pelling M, O'Brien K, Matyas D (2015) Adaptation and transformation. Clim Chang 133:113–127
Pickett ST, Cadenasso ML, McGrath B (eds) (2013) Resilience in ecology and urban design: linking theory and practice for sustainable cities. Springer, Dordrecht
Revi A, Satterthwaite D, Aragón-Durand F, Corfee-Morlot J, Kiunsi RBR, Pelling M et al (2014) Towards transformative adaptation in cities: the IPCC's fifth assessment. Environ Urban 26(1):11–28
Rink D, Banzhaf E, Kabisch S, Krellenberg K (2015) Von der„Großen Transformation" zu urbanen Transformationen. Zum WBGU-Hauptgutachten Welt im Wandel. GAIA – Ecol Perspect Sci Soc 24(1), 21–25
Rosenzweig C, Solecki W, Romero-Lankao P, Mehrotra S, Dhakal S, Bowman T, Ali Ibrahim S (2015) ARC3.2 summary for city leaders. Urban Climate Change Research Network. Columbia University, New York
Sassen S (2012) Cities in a world economy. Pine Forge Press, Thousand Oaks
Saunders D (2011) Arrival cities. Random House, New York
Schneidewind U, Scheck H (2013) Die Stadt als "Reallabor" für Systeminnovationen. In: Rückert-John J (ed) Soziale Innovationen und Nachhaltigkeit. Springer VS, Wiesbaden, pp 229–248
Schwarte C, Adebowale M (2007) Environmental justice and race equality in the European Union. Capacity Global, London
Seto KC, Fragkias M, Günneralp B, Reilly MK (2011) A meta-analysis of global urban land expansion. PLoS ONE 6(8): e23777. doi:10.1371/journal.pone.0023777, 1–9
Seto KC, Reenberg A, Boone CG, Fragkias M, Haase D, Langanke T et al (2012) Urban land teleconnections and sustainability. Proc Natl Acad Sci 109(20):7687–7692
Solecki W, Seto KC, Marcotullio PJ (2013) It's time for an urbanization science. Environment 55(1):12–17
Sommer B, Welzer H (2014) Transformationsdesign. Wege in eine zukunftsfähige Moderne. Oekom, München

Strohbach MW, Haase D, Kabisch N (2009) Birds and the city: urban biodiversity, land use, and socioeconomics. Ecol Soc 14(2):31

TEEB – The Economics of Ecosystems and Biodiversity (2010) The economics of ecosystems and biodiversity: mainstreaming the economics of nature: a synthesis of the approach, conclusions and recommendations of TEEB.. http://www.teebweb.org. Accessed 11 Jan 2017

UN Habitat (2016) New urban agenda. Quito Declaration on Sustainable Cities and Human Settlements for All. https://habitat3.org/the-new-urban-agenda. Accessed 11 Jan 2017

UN – United Nations (2015a) Resolution adopted by the General Assembly on 25 September 2015: Transforming our world: the 2030 Agenda for Sustainable Development. http://www.un.org/ga/search/view_doc.asp?symbol=A/RES/70/1&Lang=E. Accessed 11 Jan 2017

UN – United Nations (2015b) World population prospects: The 2015 revision, methodology of the United Nations Population Estimates and Projections. ESA/P/WP.242

Vandevyvere H, Nevens F (2015) Lost in transition or geared for the s-curve? An analysis of Flemish transition trajectories with a focus on energy use and buildings. Sustainability 7(3):2415–2436

Van den Bergh JC, Truffer B, Kallis G (2011) Environmental innovation and societal transitions: Introduction and overview. Environ Innov Soc Tran 1(1):1–23

Walker G (2012) Environmental justice: concepts, evidence and politics. Routledge, London

WBGU – German Advisory Council on Global Change (2016) Humanity on the move: unlocking the transformative power of cities. WBGU, Berlin

White MP, Alcock I, Wheeler BW, Depledge MH (2013) Would you be happier living in a greener urban area? A fixed-effects analysis of panel data. Psychol Sci 24(6):920–928

Winnigton NS, Fahrenkamp-Uppenbrink J, Malakoff D (2016) Cities are the future. Introduction to special issue Urban Planet. Science 352(6288):904–905

Part I
Conceptual Approaches of Sustainable Urban Transformations

Outline

Sigrun Kabisch and Annegret Haase

In this first part of the volume, five conceptual approaches to the issue of urban transformations will be introduced. The contributions provide evidence for the diverse ways of urban transformations and show the added value of interdisciplinary discourses. Paying attention to the transformation dimensions resource efficiency, quality of life and resilience, their interrelations, trade-offs and constraints are addressed. All papers start with a theoretical discussion of the scientific state of the art of their topics and illustrate their argumentation by highly topical empirical cases.

Based on belief that cities are key players in the process of transformation towards more sustainability, the first paper by **Rink et al.** contrasts the main features of "transition management" and the idea of a "great transformation"; these are two systematically elaborated conceptual approaches. To illustrate their particularities, constraints, as well as overlaps, the topics of the post-fossil city and the resilient city are scrutinized. The authors stress the prior importance of governance and the need for new governance modes to overcome the still existing resistance against urban transformations. **Bedtke and Gawel** combine transition theories and the economic theory of institutional change by using the example of urban infrastructure transformations. They point out that infrastructure transformations consist of a technological shift as well as a decisive change of the institutional and legal framework. Hence, the authors take an institutional viewpoint to explain the remarkable inertia within the insufficiently flexible and non-sustainable urban water sector. They conclude that there is a need for an integrated transformation approach of the urban system that is based on adaptively efficient institutions. **Gawel and Kuhlicke** raise the topic of the efficiency-equity trade-off in the urban context. Equity and resource efficiency may turn out to be competing guiding principles for sustainability-oriented transformation processes. The authors shed light on the theoretical background of these normative principles and their conceptual relationship, and also discuss the empirical examples of the flood-risk management and water service pricing. They provide evidence for the challenges facing urban transformations when aiming at addressing resource efficiency and equity concerns simultaneously. They plead for overcoming the, to date, usually fragmented theoretical discourses as well as decision-making solely based on one of these principles, and stress the benefit of interdisciplinary cooperation on this topic being aware of the potential trade-offs. **Haase et al.** deal with the connection between crises and sustainability transformations and elaborate the influence of crises on the progress and feasibility of change, both in theoretical terms and with the help of empirical examples such as urban water infrastructure crisis or the nexus between financial crises and urban austerity policies. They argue that there is no straightforward relationship between crises and transformations; crises can support or hinder transformations. They conclude that a more systematic and tailored consideration of the crisis factor is important but needs dispassion and balance. **Koch and Ahmad** provide an approach to measure the progress of the achievement of Sustainable Development Goals (SDGs) adopted by the UN in 2015. They shed light on Goal no. 11, which is dedicated to inclusive, safe, resilient, and sustainable cities. By using the example of German and Indian cities, they specify the underlying seven targets related to appropriate indicators as well as the relevancy and the availability of data. They conclude that Goal no. 11 can be read as a global guideline for the minimum standards a city should fulfill in terms of sustainability.

By and large, all the papers in this part of the volume show the wide range of conceptual ideas that are under discussion with respect to urban transformations towards more sustainability. The papers invite readers to reflect on and critically discuss the described approaches.

Exploring the Extent, Selected Topics and Governance Modes of Urban Sustainability Transformations

Dieter Rink, Sigrun Kabisch, Florian Koch, and Kerstin Krellenberg

1 Introduction

With cities acknowledged as key players for active involvement towards more sustainability (Parnell 2016), the city – or the urban scale – is seen as a, or even *the*, decisive field that can make transformations towards sustainability work (WBGU 2011; UBA 2015b). And, there is "general agreement that effective and integrated solutions can only be found and efficiently implemented through cities and urban areas" (McCormick et al. 2013, p. 2). Cities are, today, assigned a role as pioneers or frontrunners of sustainability transformations (Wolfram 2014; Wolfram and Frantzeskaki 2016). Nevertheless, it still needs to be seen whether these aspirations and expectations about the role of cities is only born out of desperation about slow progress toward sustainability or if they can actually become the real focal point of transformation. The increasingly important notion of "transformation" in policy and research, when talking about changes that need to be undertaken, provokes the question about possible differences between the terms transformation and sustainable development. In this regard, Brand (2016, p. 24) has argued that the term transformation is "more radical and more attractive than the term sustainable development" and stresses both the urgent need for action and the "how to come to" the implementation of sustainability. For Brand (2016, p. 24), transformation is the "new critical orthodoxy", characterized by a radical problem diagnosis, promising far-reaching change, but also involving a relatively incremental understanding of the processes and steps of social change, in order to cope with the problems.

D. Rink (✉) • S. Kabisch • F. Koch • K. Krellenberg
Department of Urban and Environmental Sociology, Helmholtz Centre for Environmental Research-UFZ, Leipzig, Germany
e-mail: dieter.rink@ufz.de; sigrun.kabisch@ufz.de; florian.koch@ufz.de; kerstin.krellenberg@ufz.de

S. Kabisch et al. (eds.), *Urban Transformations*, Future City 10,
DOI 10.1007/978-3-319-59324-1_1

Loorbach concludes that "Sustainable Development itself has become part of the problem" because it has become part of the established regime and has primarily served to make it a little less unsustainable (Loorbach 2014, p. 32). According to this ambitious target, we explore three aspects of urban transformations: *extent, topics, and governance* (UBA 2015b).

In order to shed light on the *extent* to which cities are expected to change, we have selected two conceptual approaches, Transition Management and the Great Transformation. Both approaches are prominent in the current debate about the societal change needed in order to respond to the challenges of sustainable urban development. The *topics*, which are empirically-based examples of urban transformations, are framed by the climate change debate: namely, the post-fossil and the resilient city. They have been selected because of the growing pressure that climate change exerts on cities and, vice versa, for cities to adapt to and deal with climate change (McCormick et al. 2013; O'Brien 2015). In the discussion on how to turn desirable transformations into real-life environments, *governance*, in terms of local politics and participation, is decisive.

The chapter is structured according to the three aspects. In order to contribute to the broader debate on urban transformations and to open the floor for further discussion, it addresses extents and topics of urban transformations, and looks at the governance of urban transformations in terms of actors, networks, and policy paradigms. Various approaches are presented and discussed. By illustrating and critically reflecting the state of the art and identifying challenges, we distillate pressing questions for further research related to the emerging field of urban transformations towards sustainability. The chapter's aim is to provide a conceptual background for transformation on the urban scale.

2 Extent of Urban Transformations: Transition Management and the Great Transformation

The extent of transformation describes the degree to which an existing system is expected and able to change. It can range from stepwise incremental alterations to change of the complete system. This refers to recent science and policy discussions that have evolved around pathways, power relations, mechanisms, and instruments needed for achieving substantial progress towards sustainability. Conceptual approaches on this issue include, amongst others, transition approaches, socio-ecological transformations, sustainability pathways, transformative adaptation (Patterson et al. 2016), and the concept of the Great Transformation (WBGU 2011).

In order to exemplify this, the two approaches Transition Management and the Great Transformation are introduced. We are aware that the terms transition and transformation are broadly used within other discourses and disciplines, but to pursue our goal, we have made a specific selection. We have chosen these two approaches because they are elaborated systematically and have a prominent position in national and international academic, public and political debates. Furthermore, the concept of the Great Transformation, first introduced by Polanyi (1971) and followed up by the WBGU (2011), is partly related to the concept of Transition Management. Nevertheless, central terms such as Great Transformation are used in divergent, partly contradictory ways and the usage of the terms transformation and transition is not at all clear cut or distinct. Within the literature, both terms often describe the same circumstance: fundamental change of societal relationship to nature. Thus, others, such as Brand (2016) or Kabisch and Kuhlicke (2014), conceptualize only transformation as a fundamental change and consider transition more as a gradual or step-wise change that is, or can be, part of a fundamental transformation. Overall, the adoption of both the Great Transformation and the Transition Management approach marks a turning point within the sustainability debate.

2.1 Transition Management

Transition Management (TM) has its disciplinary background in economic theory, particularly in innovation research. It is embedded in the transition research that was elaborated at the Dutch Research Institute For Transitions (DRIFT) in Rotterdam. It is the aim of the concept or paradigm to explain the shift from one regime to another and not only from one system, product, or technique to another. Nevertheless, TM considers incremental changes in so-called niches of markets as the starting point for paradigm change. Niches are defined as specific markets or application domains in which radical innovations can emerge without being included in the selection process of the prevailing regime (Grin et al. 2010; Kemp et al. 1998; see also the contribution of Bedtke and Gawel, "Linking Transition Theories with Theories of Institutions – Implications for Sustainable Urban Infrastructures Between Flexibility and Stability", in this volume). Based on this concept, the Multi-Level-Perspective (Geels 2002) was developed. This refers to "landscape" as the overarching socio-technical setting that includes institutions and the market on the macro level, "regime" as the dominant practices, rules and technologies on the meso level, and "niche" as the place or area where space for radical innovation and experimentation is provided on the micro level (Fig. 1). Each level has its own set of actors, comprising parties, politicians, and organizations from government, enterprises, associations, and unions from the market, as well as civil society actors such as NGOs, initiatives, and projects, all interacting in different ways. The main goal of this heuristic model is to explain a regime shift in order to obtain knowledge on how transition can be systematically initiated, promoted, and steered. The idea is that niches will emerge, grow, and spread and destabilize the level of the old regime. TM attempts to actively support this process with a variety of activities: strategic, tactical, and operational. This should be primarily achieved by engaging a wide range of actors and stakeholders from different fields and over multiple levels. In this regard,

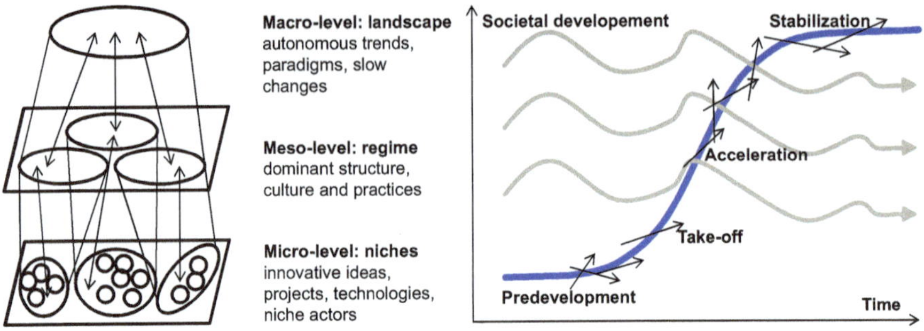

Fig. 1 Phase model of transition (Source: Grin et al. 2010, p. 131)

TM highlights the role of experimentation and learning, so that innovation can emerge and diffuse efficiently (Loorbach 2010). "Transition management is theoretically a form of meta-governance: creating conditions under which the actions of autonomous agents somehow add up to contribute to a bigger whole" (Loorbach 2014, p. 67). To describe the transition, a "multi-phase concept" has been elaborated with a so-called predevelopment phase, a take-off with innovation, then an acceleration followed by the stabilization of a new regime (Grin et al. 2010) (Fig. 1). Because there is no automatism in this phase model, the alternatives of a lock-in, a backlash, or a system breakdown, if the take-off is not followed by a successful acceleration and a subsequent stabilization phase, have been introduced (ibid.).

The TM approach or paradigm has gained its conceptual framework by analyzing expired transitions (Grin et al. 2010). Transition research, so far, has focused predominantly on socio-technical transitions, in particular, in domains such as water infrastructure, energy, waste, food, mobility, care, and housing (Loorbach 2014). Recently, TM has been used to explain the transition to sustainability, to elaborate a new regime, and to steer this from its starting point. Loorbach has distinguished here a so-called "New Transformation" of "the Great Transformation". The New Transformation is defined as "a fundamental shift in power-relations and -structures: a reconfiguration of the basic socio-economic regimes that cuts across our society and societal domains" (Loorbach 2014, p. 53). With this conceptualization, he links TM with the concept of the Great Transformation by giving it a specific interpretation (see Sect. 2).

In urban settings, TM approaches have been used to frame technological and socio-technological changes related to infrastructure (Geels 2011), in addition, the application of the Multi-Level-Perspective in cities encounters critical issues such as mobility or energy production (Hodson and Marvin 2011). More recently, the TM approach has been connected with the so-called Urban Labs or Living Labs. These are environments where "an episode of transformation is realized and, as such, Urban Labs are facilitated sites for creating (social) innovation and within which social change agents can initiate or inflict urban sustainability" (Nevens et al. 2013, p. 115). These labs are often related to climate change issues and try, for example, to show pathways to low carbon societies or CO_2-neutral urban structures (ibid.). Thus, these labs are seen as niches from which innovation can emerge and spread. Even though the idea is that these innovations will be incubators, in order to lead to more comprehensive forms of sustainability transformation, the scale at which urban sustainability transformations begins is local and the extent of change focuses on specific technological/ socio-technological sectors. In contrast to TM, with its attention to niches, the Great Transformation deals with large and radical changes.

2.2 *The Great Transformation*

Within recent decades, the highly appreciated study, "The Great Transformation" (first edition 1944), by the economist Karl Polanyi has become a leading paradigm for approaches that aim at large transformations. Here, "transformation" is conceptualized as an overarching category of a fundamental social change. According to Polanyi, transformation targets the entire society or the entire societal system in its relation to the environment as an object. Transformations are considered within processes of long-term societal change. Following this understanding, until today, only two transformations can be named: 1. the transformation from nomadism to settledness – the so-called "neolithic revolution", and 2. the transformation from the agrarian to the industrial society, the so-called "industrial revolution" (Polanyi 1971; UBA 2015a, b). Both transformations started without a clear aim or distinct state, and the contours of the new society only gradually emerged in the course of the transformation. Processes can be recognized in periods before and after the transformation; they are determined by the intensity of (gradual) changes. Compared to phases of relatively stability the duration of a Great Transformation is relatively short, but it can vary between some years and a couple of decades. The duration of a transformation often correlates with the regular regeneration or replacement cycle of the related object or process (UBA 2015b).

Two fundamental applications of the term or the concept of the Great Transformation can be distinguished: (1) a strategic, and (2) an analytical one. The strategic use of the term and concept of the Great Transformation (which can also be called its "normative" use) can be found within documents of a programmatic nature such as declarations, charters, or "flagship reports". Within the last 10–15 years, international but also national organizations and institutions, in particular, have elaborated and published these kinds of strategic or programmatic texts. Starting with the report of the Stockholm Institute in 2002 (Raskin et al. 2002), similar positioning papers have been published by the New Economics Foundation (NEF 2010), the United Nations Economic Commission for Europe, in cooperation with the United Nations Development Program (UNECE/UNDP 2012), the World Business Council on Sustainable Development (WBCSD 2010) or – as already mentioned – at the national level in Germany, by the German Advisory Council on Global Change (WBGU 2011; 2016). Despite the fact that a number of documents are lacking a definition of the used transformation concept, the term

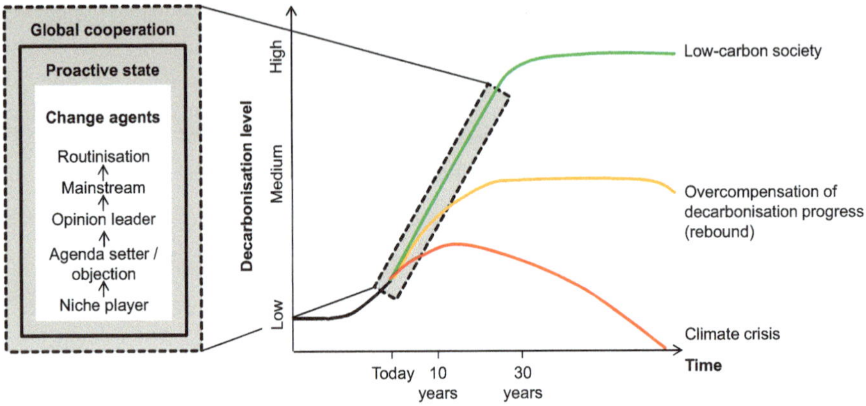

Fig. 2 Temporal Dynamic of Transformation (WBGU 2011, p. 7)

"transformation" (or "transition") has been introduced into the political sphere and has become well known.

We are focusing on the WBGU, which has elaborated the concept of a Great Transformation systematically in several reports (WBGU 2011, 2014, 2016). In its first so-called "flagship report" of the WBGU (2011) "World in Transition – A Social Contract for Sustainability", a new, third transformation, the transformation from the industrial to the sustainable society was introduced. This "new" Great Transformation was conceptualized as an intended and directed, planned and steered process. For this Great Transformation, a relatively short time period of about 30–40 years was assigned. The decisive turning points have to be regulated within the next 10 years, as a prerequisite for fundamental changes to become reality within the following 30 years, and in order to complete the Great Transformation until the year 2050 (WBGU 2011). Based on the phase-model of the Transition Management concept, the WBGU has elaborated a model of the Great Transformation for a sustainable society (see Fig. 2).

It is argued that, only by adhering this timeframe, the transformation from a fossil to a low-carbon or sustainable society can be ensured in due time, and a global climate crisis be averted (ibid). With regard to the role of cities, the WBGU (2011) has stressed four points: (1) The required transformation of inefficient urban infrastructures in industrialized countries; (2) The climate-friendly design of newly emerging cities in urbanizing countries, particularly in Asia; (3) The overcoming of poverty in urban areas in poor, developing countries, particularly in Africa and South Asia (through slum clearance, capacity building, etc.); and, (4) The strengthening of capacities to adapt to climate change.

In 2016, the WBGU published its most recent report: "Humanity on the move: Unlocking the transformative power of cities", which applies the Great Transformation concept to the urban scale. Here, transformation as a "fundamental social change" has gained momentum. Eight activity fields are identified in which action is required in order to create sustainable urban areas, including urban health, land-use, urban mobility and transport, urban design (WBGU 2016). In addition, the WBGU has illustrated its concept for different types of cities in the developed world, such as Copenhagen and the German Ruhr-Area (WBGU 2016, pp. 249–264, 279–297), and for cities in developing countries, such as Cairo and Kigali (ibid., pp. 233–249; 297–307). The discussion on urban transformation is linked to international agendas where cities are identified as places that should become more sustainable, not only in order to create a more sustainable habitat for urban dwellers, but also to achieve sustainability on a global scale and mitigate climate change. In other words, cities are perceived as hot spots for driving environmental change at multiple scales (Grimm et al. 2008, see also SDGs adopted by the UN 2015, especially SDG No. 11). In addition, other programmatic papers, for example, the UN Habitat III's New Urban Agenda (UN Habitat 2016, p. 2) state that "there is a need to take advantage of the opportunities of urbanization as an engine of sustained and inclusive economic growth, social and cultural development, and environmental protection, and of its potential contributions to the achievement of transformative and sustainable development". Even without mentioning the Great Transformation, this UN agenda contains a wide range of so-called transformative commitments, including infrastructure, public space, and the urban way of life.

Thus, both the WBGU and also UN Habitat acknowledge the need for a stepwise approach to implement urban transformations and their adjustment to each city's specific context, in order to achieve fundamental change. This approach goes far beyond the discussed Multi-Level Perspective and the niche management ideas of the TM concept. Despite the fact that the terms transition and transformation are not clearly differentiated, both concepts possess special characteristics. Whereas Great Transformation aims to conceptualize a fundamental global change within a distinct time frame, Transition Management does not specify the extent of change (Table 1).

Table 1 Overview of main features of "Transition Management" and "Great Transformation"

	Transition management	Great transformation
Theoretical background	Economy, innovation research	Political economy
Goals	Transitions in socio-technical systems	Fundamental societal change
Characteristics	Managed single or coupled transitions	One fundamental transformation, consisting of a variety of coupled transitions
Mechanisms	Niche innovations taken up into broader, pathdependent regimes	Globally steered change to a new society-nature-relationship
Trajectory of change	Emergent, triggered	Triggered, steered
Time frame	No exact or explicit time frame(s)	Within the next 30–35 years

It is argued that urban transformations from a Great Transformation perspective will only succeed if different topics are combined, a long-term perspective for the transformation process to a low-carbon society is chosen, and a large extent of changes is envisaged (Rink et al. 2015). At least two topics of urban transformations stand out, as a recent literature review has confirmed: the post-fossil and the resilient city (de Jong et al. 2015). These will be discussed in the following section.

3 Topics of Urban Transformations

Similar to the extent of transformations, the topics that are the objects of transformations are also not clearly defined. Urban transformations with a variety of targets can be identified or conceptualized. Nevertheless, focusing on the topics in the context of climate change these can be broadly divided into two main categories, which are considered as potential solutions for more sustainable urban development: the post-fossil and the resilient city (Bulkely et al. 2011; Geels 2012).

3.1 The Post-fossil City

Overarching consensus exists that cities are the main source of greenhouse gas emissions and that ongoing urbanization can lead to an increase of these emissions if they are not steered to a climate compatible degree. This calls for substantial and fast reductions of greenhouse gas emissions (Müller et al. 2013). The main challenge is to organize and steer the transformation from a city with high energy consumption, based on primary fossil energy sources, towards a city with lower energy consumption, based on renewable energy sources. This transformation has to be carried out within a manageable period of time, often within a defined time frame of 30–50 years (WBGU 2011). Different terms are used to describe this transformation, such as „Low Carbon City"(Chan et al. 2013), „Post-Oil City" or "Post-fossil City" (OEKOM 2011), or the "Climate Neutral City" (Vandevyvere and Nevens 2015). From our point of view, the term post-fossil city most precisely addresses the transformation approach that links to attempts to mitigate climate change in cities. The aim of making a specific city CO_2-neutral is connected to deep and far reaching changes that will affect nearly all municipal policy fields (Libbe 2013). Technological, economic, and social innovations in terms of energy and heat production, energy consumption, energy-related retrofitting of buildings, and mobility behavior form the basis of this approach. However, the conversion of the energy base of the city is a comprehensive municipal task involving political steering, planning, and economic budgeting. In addition, this needs to fit into national and international energy networks, because cities have to cooperate within overarching governance networks. CO_2-neutrality is strongly linked to the transportation sector because improved mobility patterns and innovative traffic modes can substantially contribute to more resource efficiency in terms of energy demand. Furthermore, the fundamental transformation of a city's energy system also has many social implications (Renn 2015). Questions of affordability and acceptability arise, which are related to people's behavior, consumption patterns, and life styles, and need to be addressed in order to increase the quality of life of the citizens rather than avoiding the failure of the intended transformation. This is connected to the statement that we need "cities for people" and not "cities for cars" (Gehl 2010).

There is a small number of cities of various European countries including Germany, the so-called "100% communities" (Marinakis et al. 2015), that have set the target to become CO_2-neutral in order to address the energy and climate goals of the European Union for 2030. They have sought the most appropriate organizational, technical, and financial means to reduce consumption and produce renewable energy. Thus, "until now hardly any city is proceeding towards this transformation in a structured way" (Libbe 2013, p. 214), mainly because there is still uncertainty about the mechanisms to be used to effect such a fundamental shift (Bulkeley et al. 2011). The involved stakeholders, such as utility and housing companies, landlords, urban policy and planning entities, tenants' associations, and consumer protection organizations, are confronted with uncertainties and risks with regard to future energy needs, security of energy supply, cost effectiveness, energy savings, and climate protection effects. It is still unclear what can be actually achieved in this regard, and how this will impact on the social compatibility of the transformation to the post-fossil city. The diversity of existing options and the expectations about the impact of technological and social innovations enable a quick conversion to energy savings and renewable energies. Thus, changing the whole energy system calls for strategic decisions regarding investments and new infrastructure systems (Libbe 2013). Here, the Achilles heel of the transformation to the post-fossil city can be identified, in developed countries, as centralized energy systems with, often, long-term obligations for, in part, extensive financial resources. Changing these systems thus can be considered a truly fundamental transformation (see Bedtke and Gawel, "Linking Transition Theories with Theories of Institutions – Implications for Sustainable Urban Infrastructures Between Flexibility and Stability", in this volume who discuss this point with the example of water infrastructure). On the contrary, in developing and industrializing countries existing energy systems are still developing, targeting, first of all, the connection all inhabitants to the system, in order to cover basic needs. Here, new energy systems based on renewables may have the opportunity to be more easily implementable, particularly if affordability is guaranteed.

In some frontrunner countries such as Germany or the Netherlands, the transformation to the post-fossil city is embedded within the national policy of energy transformation, with probably the German "Energiewende" being the most well known example. On the national level, energy transformation policies represent the main instrument to implement the targets of climate policy. Thus, these policies also need to address other targets, particularly to secure energy supply and cost effectiveness. This means that policies of an energy transformation on the local, regional, and national level are connected with a number of various issues and interests. Furthermore, transformation, as a whole and in its components, is particularly vulnerable to crisis and failures; a "high immanent potential for conflict" is inherent

(Renn 2015, p. 79). In addition, the transformation is confronted with a time trap because decisions are often made under time pressure, which does not allow a fully *ex ante* assessment of the possible implications they may have. Consequently, a new governance approach is needed to follow a robust concept for accompanying and steering the process that will allow for better control of the transformation of complex systems within a relatively short time (ibid., p. 69). This holds particularly true for the conversion from centralized to decentralized energy systems.

3.2 The Resilient City

Unlike the post-fossil city, the transformation to the resilient city in the context of climate change targets fundamental changes in a city that is able to adapt to, or to respond to, singular, unique and most unexpected events in terms of shocks or crisis (see Kuhlicke et al., "Resilience, Adaptation and Transformation: Conceptual and Empirical Insights from Two Case Studies in Germany and Chile", in this volume). The overall orientation is, hence, to enhance, build, or develop the capacities of actors or systems to cope with threatening hazardous events such as flooding, heat, or storms and, often, their occurrence at the same place (e.g. Kuhlicke 2010; Childers et al. 2015; Carter et al. 2015). To be resilient means to bounce back efficiently and to restore the status quo *ex ante* after a shock or crisis and, even more, to be able to change, learn, and transform, and hence to radically alter structures and functions simultaneously (IPCC – 2014). Evidence from various case studies has shown that the adaptation activities many cities are undertaking in terms of climate change are still far away from being called a transformation, although cites are, due to the concentration of people, enterprises, and infrastructure, particularly vulnerable to hazardous events. All over the world there are cities located on rivers or in low-elevation coastal zones, which make them per se more vulnerable to flooding. This means that, in the long run, many harbor cities will be increasingly threatened by the sea level rise (UN Habitat 2016).

Thus, aiming at urban transformations towards a resilient city, those should be concerned with the less visible roots of urban vulnerability, such as social, cultural, economic, and political factors that often overlap and interact on different spatial and temporal scales. This is in accordance with O'Brien (2012), who argues that a broader and more holistic approach to adaptation involves viewing the vulnerability context from different spatial and temporal perspectives. Adaptation hereby refers to incremental adjustments that preserve systems' integrity when conditions change (Pelling et al. 2015). Admittedly, urban vulnerability to climate-related hazards varies in degree. And, it is, in particular, the interrelation between the spatial-structural conditions of an urban area exposed to a discrete and identifiable event in nature (or in society), together with the underlying susceptibility and response profile of its inhabitants, that shapes the degree of vulnerability (Krellenberg et al. 2016). Thus, the way of tackling the challenges also very much depends on the institutional capacities. Hence, vulnerabilities are neither felt nor distributed equally between and within cities. In this context, the adaptive capacity of the city to climate change refers to a broad set of resources (skills, competences, and social relations). This includes the degree to which local authorities have integrated (or are in the process of doing so) climate change considerations into their long-term planning and development processes, in order to be prepared and to react in an appropriate manner in case of intensified, extreme weather events (e. g., facilitating evacuation activities.)

Aiming at a city that is "prepared to absorb and recover from any shock or stress while maintaining its essential functions, structures, and identity as well as adapting and thriving in the face of continual change. Building resilience requires identifying and assessing hazard risks, reducing vulnerability and exposure, and lastly, increasing resistance, adaptive capacity, and emergency preparedness" (ICLEI 2015), the transformation to the resilient city is less clear in terms of the overall targets. The resilient city calls for efficient adaptation and avoiding mal-adaptation while strengthening co-benefits between different interventions such as building dams and dykes in flood-prone areas, additional protection for infrastructures, or the overall flood risk management, including early warning systems. What is more, maintaining or creating new retention areas, improving information, capacity building, etc., need to go hand in hand. Given the strong linkage of resilience to crisis or hazardous events, the resilient city, unlike the post-fossil city, does not pursue clear periods or dates to fulfil distinct targets or measures. Thus, accumulating various adaptation measures and determining factors for the evaluation of success might be a step forward towards the resilient city.

In the preceding discussion about the post-fossil and the resilient city, two topics of transformation to sustainability were addressed and conceptualized. Two major common issues were identified: (1) Measuring, monitoring, and assessing the progress towards the post-fossil city and the resilient city is not an easy tasks, because appropriate indicators are still missing (Sharifil and Yamagata 2014); and (2) The governance of transformation; this will be discussed in more detail in the next section.

4 Goverance of Urban Transformations

Governance and planning are crucial for sustainable urban transformations because this transformation is far more a social, organizational, economic, cultural, and political challenge than a technological one (McCormick et al. 2013). Thus, to pursue transformation, actor constellations, power structures and interests, as well as modes of implementation of transformative actions are at the focus of research and political action. In the following, two contrasting governance approaches will be introduced: Firstly, transformative urban governance as a fundamental new type of governance, illustrated by the transformation concept of the WBGU (see Sect. 2); secondly, growth machines and urban regimes that represent continuity and persistence.

4.1 Transformative Urban Governance – Illustrated by the WBGU Approach

The concept of "transformative urban governance" has been introduced with the central task of initiating a paradigm shift in cities "away from incremental approaches which are driven by short-time requirements to transformative changes with a strategic long-term perspective to the natural livelihood of mankind" (WBGU 2016, p. 9). How this governance should be structured and arranged, and which actors are or will be involved is not outlined in detail and thus needs to be decided on and implemented in the respective city. Explicitly, a number of actor categories are addressed: niche actors, agenda setters, opinion leaders, change agents, and the proactive state (WBGU 2011, p. 7). In general, this transformative governance involves, initially, a new distribution of responsibilities, principles, rules, and material criteria to force and design an overarching and successful transformation of cities towards sustainability in terms of dynamics of fundamental change. "Transformative urban governance must initiate dynamics of fundamental change to meet the stunner and the speed of global urbanization processes" (WBGU 2016, p. 24). Following this, some proposals are formulated that aim to strengthen, stabilize, and maintain the role and the ability of cities or of the "public sector" to act within the national and international arena. Remarkably high requirements for transformative urban governance are outlined, and, in addition, it is diagnosed that the necessary fundamental change is confronted with "mechanisms of blockade that are produced not only by technical pathways, but are caused by fixed constellations of actors and lacking financial and institutional capacities "(WBGU 2016, p. 13). Municipal policy and administration are assigned a positive role per se. Thus, in the light of fixed governance arrangements or structures, they need to be discussed as part of the problem (see the following section on "growth coalitions"). Besides this, non-governmental actors such as social movements, NGO's and relevant actors from civil society are also assigned a decisive role for the transformation. These actors can become active in different fields and levels, can form a "world citizen movement and develop pressure on politicians or enterprises to force climate protection" (WBGU 2014).

Reflecting this ambitious undertaking, the transformative urban governance concept deployed in the WBGU approach is a normative concept that is full of preconditions and nebulous implementations. How the necessary regime change at the urban level could be executed remains unanswered. An urgent appeal is directed at cities to create new and appropriate governance structures, but lacks concrete examples or best practices. Furthermore, requirements dedicated to the national states and the international community of nations concerning the necessary upgrading of cities, as actors in transformation, are not outlined. Moreover, transformative urban governance has to compete with conventional types of governance that are following different targets, in particular, growth coalitions.

4.2 Growth Machines and Urban Regimes – Illustrated by the Smart City Concept

Although, today, far-reaching sustainable development goals and climate protection targets exist, urban governance has not yet fully adopted this approach. Instead of pursuing a sustainable development, a governance-for-growth approach has been, for a long period of time and in many parts of the world, the "undisputed policy goal for a very large number of cities" (Pierre 2011, p. 68). This type of governance, also called "pro-growth" or "growth coalitions", has been diagnosed as relatively stable, with overarching coalitions between elected counselors and private companies whose aim is to combine resources and competences and initiate successful growth policy (Stoker 1995). The idea of urban growth machines (Logan and Molotch 1987) is based on the assumption that growth creates (new) jobs, generates investments, and increases welfare through trickle-down effects. The simple idea of going-for-growth is that the investments of today will be the benefits of tomorrow. Compared with other governance modes, "growth or pro-growth governance is probably the easiest and least challenging to understand" (Pierre 2011, p. 67). It is not surprising that this kind of politics, which focuses on economic growth, requires a close cooperation with local enterprises. The facilitation of economic activity and entrepreneurship is a crucial part of pro-growth politics. In this context, so-called *urban regimes* have emerged, defined as "informal, yet relatively stable group with access to institutional resources that enable it to have a sustained role in making governing decisions" (Stone 1989, p. 4). The example of Atlanta, USA, shows that local enterprises and local politicians can successfully form a coalition that determines local policies for 40 years. The key requisite for their persistency was the combination of several forms of power (political power and financial power), which assures the stability of the regime (Stone 1989). Until recently, such growth coalitions have rarely addressed sustainability issues such as climate mitigation and adaptation or environmental justice. Thus, some examples exist with a clear focus on climate activities that underpin the opportunity of urban politics for fostering sustainability, for example, the cities of Freiburg, Graz (Späth and Rohracher 2011), Mexico City (Delgado 2012; Chelleri et al. 2015) or New York City (Rosenthal and Brechwald 2013). The same is true for a few so-called "Zero Emission Cities" (www.zeroemissioncities.at/) that call for fundamental system changes and new forms of urban governance. These vary between informal arrangements that function as an add-on to the growth coalitions mentioned above and top-down approaches that represent a break in the growth paradigm.

Other, contrasting, examples demonstrate how powerful urban coalitions of interest ("urban regimes") have "captured" the transformation idea by reproducing the economic and political status quo under the disguise of sustainability, rather than pursuing radical changes (see, for Manchester, Hodson and Marvin 2012, or, for Masdar City, Cugurollo 2015).

How growth coalitions emerge and work can be exemplarily demonstrated by the Smart City concept. Notwithstanding the popularity of this concept in urban development practice and theory (de Jong et al. 2015), a clear definition of what actually is a "smart city" is still lacking. Based on a review on smart city concepts, Caragliu et al. (2011) define a city as smart when investment in human and social capital, coupled with investment in traditional and modern information and telecommunication infrastructure, generates sustainable economic development and a high quality of life, while promoting prudent management of natural resources. Besides this holistic understanding of the smart city, many urban researchers interpret the smart city concept more narrowly, addressing only private investments in information technology. Driving forces in a considerable number of smart cities are multinational, powerful, mainly ICT enterprises that hide behind a sustainability concept, aggressive marketing strategies, and huge profits to be made by private companies (Hollands 2015, p. 66). Based on partnerships large projects, as a collaborative effort of public governments and corporate companies, are implemented in top-down approaches (Anttiroiko 2013; Hollands 2015), often neglecting the fact that the challenge of implementing urban sustainability cannot be fully solved through ICT investments. Although a broader smart city approach, as we understand it has, indeed, the potential to contribute to urban sustainability transformations, many existing examples rather represent growth coalitions or urban regimes of powerful actors within an "entrepreneurial city", in the understanding of Harvey (1989), than a contribution to transforming cities towards more sustainability.

To sum up, with respect to the governance of urban transformations, theoretical concepts and top-down approaches predominate to date. In fact, governance is seen as decisive to the implementation of transformation, but it still remains unclear who the key actors are, what the structures or modes of this new governance might be, and which policy paradigms are appropriate considering the specific city context. In addition, up to now, discussions about how the power of growth regimes or coalitions could be broken down or reversed in the direction of urban sustainability are rare. The same holds true for the overarching growth paradigm of the capitalist society. It is clear that, besides very specific examples of so-called transformative governance, a new governance paradigm is not yet visible that could steer transformation towards more sustainable cities in this era of climate change.

5 Conclusion

As shown by the three aspects of urban transformations – extent, topics, and governance – the need to meet the goals of sustainable development constitutes additional challenges for cities. The concepts of Transition Management and the Great Transformation address and conceptualize these challenges; however, in terms of extent, substantial differences exist between small-scale or incremental transitions and fundamental transformations. The same holds true for the time line within which a transformation can take place or become real. As far as the transformation to the post-fossil city is concerned, there is the prospect that it could be carried out within the next 30–35 years. The transformation to the resilient city should be implemented rapidly and in parallel, but without a fixed time frame. With regard to the topics, it is clear that the transformation to the post-fossil city is based on the replacement of fossil with renewable and low-carbon energy sources. In contrast, the transformation to the resilient city is vaguer. The various topics demonstrate that the related transformations are not necessarily compatible but need to be combined or integrated with each other. This is one of the main questions of the New Transformation (Loorbach 2014) or the more recent Great Transformation (WBGU 2016): how can different transitions be linked to each other. In effect, a strategic and tactical approach is needed to force different urban transformations at the same time. In terms of governance, pro-growth coalitions are the dominant type, to date, as diagnosed by the urban regime theory. New types of governance directed at the initiation and regulation of urban transitions or transformations towards sustainable development still only constitute subordinate fields or niches. In many cities, sustainability is still not a central or overarching goal of municipal policy although it has often been propagated as such for more than 20 years. Only in so-called "frontrunner cities", which have decided on ambitious goals related to the transformation to the post-fossil city, have new governance approaches emerged. Here, one has to concede that only cities in the Western hemisphere are such frontrunners and the solutions that they have developed match their conditions.

What is more, urban transformations cannot be "managed" easily or simply because they impinge on the economic and political interests of numerous and powerful actors and, probably, of capitalist society itself (Klein 2014). The change towards a new mode or type of governance is a matter of power relations and constitutes the Achilles heel of the transformation itself: To force urban transformation to sustainability, the ruling traditional or classical growth paradigm has to overcome. The same is true of the power of traditional pro-growth coalitions; they have to overcome by new coalitions and networks that follow the paradigm of sustainability. How this can be "managed" is still an unresolved, practical task and – at the same time – an unanswered theoretical question that needs further research. Examples of functioning new governance modes are still scarce and limited to niches.

Overall, urban transformations towards a more sustainable society are still only just beginning; they are conceptualized in declarations, charters, and masterplans but their implementation into urban practice has started only recently. One actual attempt is the New Urban Agenda adopted at the HABITAT III meeting in Quito, in October 2016 (UN HABITAT 2016). In terms of the phase model, global society is, at the moment, in the phase of "take off" and it is not clear whether it will force the next phase "acceleration". The self-contained and original role cities can play with regard to sustainable development cannot yet be proved. Thus, until now, urban transformations have remained behind the ambitious goals, and transformation, as a concept for radical change, has not yet achieved its far-reaching promises. One of the main questions for further research is how to overcome economic, social, institutional, and political resistance against (urban) transformations towards sustainability.

References

Anttiroiko AV (2013) U-cities reshaping our future: reflections on ubiquitous infrastructure as an enabler of smart urban development. AI Soc 28(4):491–507

Brand U (2016) "Transformation" as a New Critical Orthodoxy. The strategic use of the term "Transformation" does not prevent multiple crisis. GAIA 25(1):23–27

Bulkeley H, Castán Broto V, Hodson M, Marvin S (eds) (2011) Cities and low carbon transitions. Routledge, London

Caragliu A, Del Bo C, Nijkamp P (2011) Smart cities in Europe. J Urban Technol 18(2):65–82

Carter JG, Cavan G, Connelly A, Guy S, Handley J, Kazmierczak A (2015) Climate change and the city: building capacity for urban adaptation. Prog Plan, 95:1–66. doi:dx.doi.org/10.1016/j.progress.2013.08.001

Chan EHW, Choy LHT, Yung EHK (2013) Current research on low-carbon cities and institutional responses. Habitat Int 37:1–3. doi:10.1016/j.habitatint.2011.12.007

Chelleri L, Schuetze T, Salvati L (2015) Integrating resilience with urban sustainability in neglected neighborhoods: Challenges and opportunities of transitioning to decentralized water management in Mexico City. Habitat Int 48:122–130. doi:10.1016/j.habitatint.2015.03.016

Childers DL, Cadenasso ML, Grove JM, Marshall V, McGrath B, Pickett ST (2015) An ecology for cities: a transformational nexus of design and ecology to advance climate change resilience and urban sustainability. Sustainability 7(4):3774–3791

Cugurullo F (2015) Urban eco-modernisation and the policy context of new eco-city projects: where Masdar City fails and why. Urban Stud 1–17. doi:10.1177/0042098015588727

de Jong M, Joss S, Schraven D, Zhan C, Weijnen M (2015) Sustainable – smart – resilient – low carbon – eco – knowledge cities: making sense of a multitude of concepts promoting sustainable urbanization. J Clean Prod 109:25–38. doi:10.1016/j.jclepro.2015.02.004

Delgado M (2012) Water, Energy and Food Security in Mexico City. In: Otto-Zimmermann K (ed) Resilient Cities 2. Dordrecht, Springer, pp 105–111

Geels FW (2002) Technological transitions as evolutionary reconfiguration processes: a multi-level perspective and a case-study. Res Policy 31(8/9):1257–1274

Geels FW (2011) The role of cities in technological transitions. In: Bulkeley H, Castan Broto V, Hodson M, Marvin S (eds) Cities and low carbon transitions. Routledge, London, pp 3–28

Geels FW (2012) A socio-technical analysis of low-carbon transitions: introducing the multi-level perspective into transport studies. J Transp Geogr 24:471–482. doi:10.1016/j.jtrangeo.2012.01.021

Gehl J (2010) Cities for people. Island Press, Washington, DC

Grimm N, Faethl SH, Golubiewski NE, Redmann CL, Wu J, Bai X, Briggs JM (2008) Global change and the ecology of cities. Science 319(5864):756–760. doi:10.1126/science.1150195

Grin J, Rotmans J, Schot J (2010) Transitions to sustainable development. New directions in the study of long term transformative change. Taylor & Francis, London

Harvey D (1989) From Managerialism to Entrepreneurialism: The Transformation in Urban Governance in Late Capitalism, Geografiska Annaler. Series B, Hum Geogr 71(1):3–17

Hodson M, Marvin S (2011) Can cities shape socio-technical transitions and how would we know if they were? In: Bulkeley H, Castán Broto V, Hodson M, Marvin S (eds) Cities and low carbon transition. Routledge, London/New York, pp 54–70

Hodson M, Marvin S (2012) Mediating low-carbon urban transitions? forms of organization, knowledge and action. Eur Plan Stud 20(3):421–439

Hollands RG (2015) Critical interventions into the corporate smart city. Camb J Reg Econ Soc 8(1):61–77

ICLEI (2015). Resilient City Agenda. http://www.iclei.org/agendas.html. Accessed 22 Sept 2016

IPCC – Intergovernmental Panel on Climate Change (2014) Climate Change 2014: Impacts, Adaptation, and Vulnerability. Part A: Global and Sectoral Aspects. Contribution of Working Group II to the Fifth Assessment Report of the Intergovernmental Panel on Climate Change. Cambridge University Press, Cambridge/New York

Kabisch S, Kuhlicke C (2014) Urban transformations and the idea of resource efficiency, quality of life and resilience. Built Environ 40(4):475–485
Kemp R, Schot J, Hoogma R (1998) Regime shifts to sustainability through processes of niche formation: the approach of strategic niche management. Tech Anal Strat Manag 10(2):175–198
Klein N (2014) This changes everything. Capitalism vs. Climate. Klein Lewis Productions, New York
Krellenberg K, Welz J, Link F, Barth K (2016) Urban vulnerability and the contribution of socio-environmental fragmentation: Theoretical and methodological pathways. Prog Hum Geogr:1–24. doi:10.1177/0309132516645959
Kuhlicke C (2010) The dynamics of vulnerability: some preliminary thoughts about the occurrence of 'radical surprises' and a case study on the 2002 flood (Germany). Nat Haz 55(3):671–688
Libbe J (2013) Energiewende: Labor städtischer Transformation. In: DIFU (ed) Urbane Räume in Bewegung. Geschichte, situation und perspektive von Stadt. DIFU, Berlin, pp 211–219
Logan JR, Molotch H (1987) Urban fortunes: the political economy of place. University of California Press, Los Angeles
Loorbach D (2010) Transition management for sustainable development: a prescriptive, complexity-based governance framework. Governance 23(1):161–183
Loorbach D (2014) To transition! Governance Panarchy in the new transformation. DRIFT, Rotterdam
McCormick K, Anderberg S, Coenen L, Neij L (2013) Advancing sustainable urban transformation. J Clean Prod 50:1–11
Müller DB, Liu G, Lovik AN, Modaresi R, Pauliuk S, Steinhoff FS, Brattebo H (2013) Carbon emissions of infrastructure development. Environ Sci Technol 47(20):11739–11746
NEF – New Economics Foundation (2010) The great transition. NEF, London
Nevens F, Frantzeskaki N, Gorissen L, Loorbach D (2013) Urban transition labs: co-creating transformative action for sustainable cities. J Clean Prod 50(1):111–122
O'Brien K (2012) Global environmental change II From adaptation to deliberate transformation. Prog in Hum Geogr *36*(5):667–676
O'Brien K (2015) Political agency: the key to tackling climate change individuals play a central role in the transformations required to avoid dangerous climate change. Science 350:6265
Oekom e. V. (ed) (2011) Post-Oil City. Die Stadt von morgen. OEKOM, München
Parnell S (2016) Defining a global urban development agenda. World Dev, 78:529–540. doi:org/10.1016/j.worlddev.2015.10.028
Patterson J, Schulz K, Vervoort J, van der Hel S, Widerberg O, Adler C et al (2016) Exploring the governance and politics of transformations towards sustainability. Environ Innov Soc Trans. doi:10.1016/j.eist.2016.09.001
Pelling M, O'Brien K, Matyas D (2015) Adaptation and transformation. Clim Chang 133(1):113–127
Pierre J (2011) The Politics of urban governance. Palgrave, Basingstoke
Polanyi, K. (1971). The Great Transformation (1944 first edition). Beacon Press, New York
Raskin P, Banuri T, Gallopín G, Gutman P, Hammond A, Kates R, Swart R (2002) Great transition. The promise and lure of the times ahead. Stockholm Environment Institute, Boston
Renn O (ed) (2015) Aspekte der Energiewende aus sozialwissenschaftlicher Perspektive. Analyse aus der Schriftenreihe Energiesysteme der Zukunft. Acatech, München
Rink D, Banzhaf E, Kabisch S, Krellenberg K (2015) Von der Großen Transformation zu urbanen Transformationen. Zum WBGU-Hauptgutachten Welt im Wandel. GAIA – Ecol Perspect Sci Soc 24(1):21–25
Rosenthal JK, Brechwald D (2013) Climate adaptive planning for preventing heat-related health impacts in New York City. In: Knieling J, Leal Filho W (eds) Climate change governance. Springer, Berlin Heidelberg, pp 205–225
Sharifil A, YamagataY (2014) Major principles and criteria for development of an urban resilience assessment index, international conference and utility exhibition 2014 on green energy

for sustainable development ICUE Pattaya 2014. http://ieeexplore.ieee.org/stamp/stamp.jsp?arnumber=6828942. Accessed 22 Sept 2016

Späth P, Rohracher H (2011) The 'eco-cities' Freiburg and Graz: the social dynamics of pioneering urban energy and climate governance. In: Bulkeley H, Castán Broto V, Hodson M, Marvin S (eds) Cities and low Carbon transition. Routledge, London\New York, pp 88–106

Stoker G (1995) Regime theory and urban politics. In: Judge D, Stoker G, Wolman H (eds) Theories of urban politics. Thousand Oaks, London, pp 54–71

Stone CN (1989) Regime politics: governing Atlanta 1946–1988. University Press of Kansas, Lawrence

UBA – Umweltbundesamt (2015a) Wie Transformationen und gesellschaftliche Innovationen gelingen können. http://www.umweltbundesamt.de/publikationen/wie-transformationen-gesellschaftliche-innovationen. Accessed 22 Sept 2016

UBA – Umweltbundesamt (2015b) Was sind Transformationen? Begriffliche und theoretische Grundlagen zur Analyse von gesellschaftlichen Transformationen. Dessau-Roßlau, UBA

UN – United Nations (2015) Resolution adopted by the General Assembly on 25 September 2015: Transforming our world: the 2030 Agenda for Sustainable Development. http://www.un.org/ga/search/view_doc.asp?symbol=A/RES/70/1&Lang=E. Accessed 22 Sept 2016

UN Habitat. (2016). HABITAT III. New urban agenda. https://www2.habitat3.org/bitcache/97ced11dcecef85d41f74043195e5472836f6291?vid=588897&disposition=inline&op=view. Accessed 22 Sept 2016

UNECE – United Nations Economic Comission for Europe, UNDP - United Nations Development Programme (2012) From transition to transformation. Sustainable and inclusive development in Europe and Central Asia. United Nations, New York

Vandevyvere H, Nevens F (2015) Lost in transition or geared for the S-Curve? An analysis of Flemish transition trajectories with a focus on energy use and buildings. Sustainability 7(3):2415–2436

WBCSD – World Business Council for Sustainable Development (2010) Vision 2050. A new agenda for business. WBCSD, Geneva

WBGU – German Advisory Council on Global Change (2011) World in transition – a social contract for sustainability. WBGU, Berlin

WBGU – German Advisory Council on Global Change (2016) Humanity on the move – unlocking the transformative power of cities. WBGU, Berlin

WBGU– German Advisory Council on Global Change (2014) Climate protection as a world citizen movement. WBGU, Berlin

Wolfram M (2014) Stadt, Wandel, Nachhaltigkeit: Zur Konvergenz von Urbanistik und Transitionsforschung. pnd/online 2:1–11

Wolfram M, Frantzeskaki N (2016) Cities and systemic change for sustainability: prevailing epistemologies and an emerging research agenda. Sustainability 8(2):1–18

Linking Transition Theories with Theories of Institutions – Implications for Sustainable Urban Infrastructures Between Flexibility and Stability

Norman Bedtke and Erik Gawel

1 Introduction

The water, energy, telecommunication, and transport services that are provided through infrastructure systems impact considerably on the quality of life, resilience, and resource efficiency and, thereby, also on the overall sustainability of urban areas.

Urban infrastructures determine the *quality of urban life* significantly by provisioning means for meeting essential needs (related, in particular, to health and hygiene), but also to fulfil the associated residents' desires (e.g., guaranteed mobility). The *resilience* of a city is understood as, in particular, how it can deal with the consequences of shocks or crises such as natural hazards and climate- related extreme weather events. Urban infrastructures are of significant importance for improving the resilience of a city through diverse, flexible, adaptive, and redundant systems but, at the same time, they can also exacerbate the problems when they fail to cope with changing conditions (Schramm and Felmeden 2012).

At the same time, a city's use of resources (land, energy, water) is crucially influenced by the state and the underlying technologies of its infrastructure systems as well as by resource-efficient use patterns of the inhabitants.

Thus, technical infrastructure systems are of major importance for urban development, which is why, in the developed world, significant resources have been invested into their expansion in the past few centuries. As a consequence, in most industrialised countries, an overall high quality of urban infrastructural services is guaranteed today, albeit at varying levels.

N. Bedtke (✉) • E. Gawel
Department of Economics, Helmholtz Centre for Environmental Research-UFZ, Leipzig, Germany
e-mail: norman.bedtke@ufz.de; erik.gawel@ufz.de

S. Kabisch et al. (eds.), *Urban Transformations*, Future City 10,
DOI 10.1007/978-3-319-59324-1_2

The current "modern infrastructural ideal" pursued in this context is characterised by the application of large system solutions, a high degree of centralisation and standardisation, as well as long technological and economic lifespans (Graham and Marvin 2001). Furthermore, the traditional planning approach to long-lived technical infrastructure systems, in particular related to urban water and wastewater services, was long shaped by paradigms of growth and stationarity. This approach was based on the assumption of increasing utilization in the future, with the relevant determinants as largely constant and foreseeable factors (Hug et al. 2010). But this traditional approach is reaching its limits as it becomes more difficult to make reliable predictions about the key influencing factors when framework conditions change rapidly and become more regionally diverse (Milly et al. 2008; Hug et al. 2010). Networked infrastructures are predominantly inflexible and face increasingly challenges from as yet unforeseen changes in the framework conditions such as climate change, demographic change, changing demand structures and altering societal demands, particularly with regard to sustainability targets such as resource efficiency (Koziol 2004; Schramm and Felmeden 2012; Koop and Leeuwen 2016). For instance, the achievement of climate objectives requires a more resource-efficient provision of infrastructure services and cross-sectoral approaches because, e.g., wastewater treatment is increasingly linked to questions of energy efficiency and energy production in the future (Bates et al. 2008).

Against this background of having to cope with dynamic change and new requirements, a transformation of the steady-state infrastructure systems is often regarded as necessary in order to better meet these challenges (Newman 2001; Loorbach et al. 2010; Markard 2011). In the context of urban infrastructures, "transformation" means the use of far-reaching, radical technical innovations and also significant organizational and institutional changes in the provision of services, which may include turning away from existing paradigms of the established "modern infrastructural ideal". For instance, in the course of the transformation to sustainable water management, novel (more decentralised) water supply and wastewater disposal solutions are being discussed that take better account of the aims of ecological sustainability (e.g., decentralised rainwater management) and also increased flexibility (Bedtke and Gawel 2015). In the electricity sector, the transformation is characterised by a shift from a small number of centralised production sites to a large number of local installations, some of which are located in urban areas, also with renewable energy carriers (for the case of the German energy transition, see Gawel et al. 2014). These technical changes also imply far-reaching organisational and institutional changes because, for example, in each case, an adaptation of the tariff systems and investment incentives, of the hierarchy of competencies for state responsibilities, or of the legal framework will be necessary to accompany the technical changes. In addition, new players, such as private investors, are becoming more active in the area of public infrastructure, which is, in turn, changing the requirements for regulation.

In practice, however, the strong pressure to transform is often met by only gradual adaptation measures. In both sectors (water and energy), the modern infrastructural ideal continues to drive policymaking in many industrialised countries. In the energy sector, centralised provision predominates, with fossil fuel-based and nuclear energy production technologies prevailing. Renewable energies continue to play a merely supporting role, although Germany, because of its energy transition, can be considered a pioneer in this area. The sectors of water supply and wastewater disposal are, in particular, characterised across all countries by a marked sluggishness to adapt and low willingness to innovate (Thomas and Ford 2005; Kiparsky et al. 2013). Numerous interdisciplinary studies on the issue of socio-technical transformations emphasize the importance of institutions (Brown 2005; Brown and Farrelly 2009; Kiparsky et al. 2013; Fuenfschilling and Truffer 2014) but neglect the specific mechanisms of institutional change.

This is the starting point for the present analysis. Infrastructure transformations will be an essential part of an overall urban transformation towards more sustainability (NRC 2013). A better understanding of adaptation deficits in urban infrastructure sectors, gained by focusing on institutions, seems to be a promising approach for identifying starting points for infrastructure-related urban transformation analysis. In our chapter, we focus particularly on the urban water infrastructure sector of developed countries, due to both its outstanding importance for urban development and the remarkable institutional inertia within the sector, which raises the question of initiating appropriate urban transformations towards sustainability.

The chapter is organized as follows: With the aim of shedding more light on adaptation deficits in the urban infrastructural sectors, the theories for mapping transformation processes are outlined, to illustrate the underlying mechanisms of socio-technical transformations (Sect. 2). In doing so, the role of institutions is highlighted (Sect. 3) and, drawing on the economic theory of institutional change, drivers of and obstacles to the transformation towards infrastructure sustainability are identified, using the example of water supply and wastewater disposal (Sect. 4). This broadened understanding of transformation as a phenomenon of institutional change makes it possible to identify the opportunities and limits of institutional steering of infrastructure transformations and to derive recommendations for action for steering towards sustainability pathways (Sect. 5).

2 Theoretical Concepts for the Transformation of Urban Infrastructures

To better understand how urban (water) infrastructure transformation can be achieved, we first introduce theoretical approaches to describe socio-technical systems in general as well as insights from the literature on how to steer their transformation.

2.1 Transforming Socio-Technical Systems

Urban infrastructures are complex socio-technical systems. They include technical components such as roads, supply networks, sewer systems, parking lots, power plants, or wastewater treatment plants, as well as non-technical elements such as actors, institutions, and organisations. These elements work together to fulfil a specific long-term purpose; for example, the supply of electricity. The interdependencies between heterogeneous technical, but also social, institutional, cultural, economic, legal and, scientific elements – because they also characterize urban (water) infrastructures – in system transformations were first systematically pointed out in the theory of Large Technical Systems, which sought to understand the long-term development of energy systems (Hughes 1983, 1987; Mayntz and Hughes 1988). The early evolutionary economics studies of Nelson and Winter (1977, 1982) also indicated that technological developments are largely shaped by the cognitive routines and practices of relevant actors such as engineers, technicians, and scientists, which set the boundaries of innovative activities. Their concept of "technological regimes" was later expanded to include the consideration of additional factors, especially rules. Rip and Kemp (1998, p. 338) define a technological regime as a "*rule-set or grammar embedded in a complex of engineering practices, production process technologies, product characteristics, skills and procedures, ways of handling relevant artifacts and persons, ways of defining problems – all of them embedded in institutions and infrastructures*". Geels (2002) goes a step further and summarises, under the term "socio-technical regime", all elements of the production, distribution, and utilisation of technologies that are functionally related to one another. For him, technological development is also influenced by users, policy makers, societal groups, suppliers, scientists, and investors who act within various sub-regimes; in addition, cultural aspects and the regulatory framework exert a considerable influence on technological developments, thereby shaping socio-technical regimes (Geels 2002).

Coming back to the hypothesis that fundamental changes in the urban infrastructure sectors are necessary, it becomes clear that reducing the transformation of infrastructure systems to technological change therefore does not go far enough. Instead, this process also calls for far-reaching changes in institutional factors, which explains the frequent references to "socio-technical transformations" (Markard 2011; Bolton and Foxon 2015). Thus, in the context of far-reaching socio-technical change, the term "transition" rather than transformation is used regularly, denoting the evolution from one socio-technical regime to another (Geels 2004).

In contrast, fundamental social changes such as those that accompanied, for example, the industrial revolution or those that occur during the change from one social order to another (e.g., post-socialist transformation) are called transformations (Polanyi 1944; Smyth 1998; WBGU 2011). In transformation research in the political science context, a distinction is sometimes drawn between a politically influenced change in a social sub-system (transition) and a largely uncontrollable social change on multiple levels (transformation) (Brand 2016). Terminological fuzziness emerges from the fact that the term transition is also used to refer to uncontrolled change ("evolutionary transitions"; see Loorbach and Rotmans 2006) and, at the same time, often in connection with changes in society as a whole ("societal transitions"; see e.g., Loorbach 2007). Due to the widely varying interpretations of the concepts (see also Rink et al., "Exploring the Extent, Selected Topics and Governance Modes of Urban Sustainability Transformations", in this volume), the term transformation is used, for simplification, in the following, whereby our understanding is closer to the term transition, which means that we discuss a goal-oriented, influenceable (finally radical) change in a societal subsystem.

2.2 On the Steering of Socio-Technical Transformations

The question whether a society's technology determines its cultural values and social structure or, vice versa, whether human action shapes technology, has been part of scientific research for many decades. The approach of technological determinism, which dominated the debate for some time, stressed the importance of technologies as significant initiators of socio-economic and institutional change and regarded technological changes as exogenous phenomena that cannot be influenced (Smith and Marx 1994). "Social constructivist" concepts took leave of this paradigm and referred to the importance of social processes as well as the structural, institutional, and cultural framework conditions for technological developments (Bijker et al. 1987; Bijker and Law 1992; MacKenzie and Wajcman 1999). Subsequently, intermediary concepts of a co-evolution of technology and society have been followed, in which a mutual influence of technology and society is recognised although, at the same time, technological or social determinism is not emphasised (e.g., Unruh 2000; Frantzeskaki and Loorbach 2010). Building on this, and with the aim of deriving specific steering proposals for transformation processes, work in the field known as *transition research* (also *transition management*) has attracted greater attention in recent years. Transition management deals with questions relating to the goal-oriented governance of system innovations against a backdrop of complex social problems. It is used to derive transformation strategies for areas such as energy, health care, mobility, or urban water management (Loorbach 2007). The approach follows an integrated, complex systems perspective and essentially understands system change as the outcome of diverse, networked processes that take place on multiple hierarchical levels. This also complies with the approach of complex "urban transformations", which are defined as the necessary alterations to physical, material, and social processes that urban areas need to undergo in order to achieve aims based on the normative idea of sustainability (Kabisch and Kuhlicke 2014; Kabisch et al., "Introduction: Urban Transformations – Sustainable Urban Development Through Resource Efficiency, Quality of Life, and Resilience", in this volume). The "multi-level perspective", as a popular concept within transition theory, identifies processes that can be regarded as ideal for socio-technical transitions (Geels 2002; Geels and Schot 2007):

- in "niches", (radical) innovations gain increasing internal momentum (*Eigendynamik*) from learning processes, their constant improvement, and growing support from influential groups;
- changes on the "landscape scale" (e.g., climate, demography) give rise to increasing pressure for change in the existing socio-technical regime;
- the associated destabilisation of the socio-technical regime opens windows of opportunity for niche innovations.

Transition management sets the stage for purposefully *designing* these processes and can also be a guide for urban (water) infrastructure transformation. It recognises the fact that transformations are beyond the scope of classical steering (top-down), because they represent the outcome of an unmanageable number of processes that, in addition, are often beyond the reach of design (e.g., cultural change). In general, socio-technical transformations result, finally, in radical changes, whereas changes in the socio-technical regime usually occur gradually (Dolata 2011; Geels and Schot 2007). However, several possibilities exist for directly and indirectly influencing the direction and speed of the developments.

The steering philosophy underlying transition management consists of pursuing a policy directed at an overarching theme (such as sustainable urban transport, emission-free energy production) while exploiting transformation dynamics, whereby, in particular, innovative activities that take place in niches (micro-level) are accelerated, thus simultaneously influencing existing socio-technical regimes (meso-level) (Loorbach and Rotmans 2006; Loorbach 2010). According to economic theory, this can be facilitated by the exploitation of market mechanisms when an individual's choice of products and services is influenced by conveying price signals that are aligned with transformation aims. In this way, (sustainable) innovations can be promoted by subsidies, and the existing regime (inter alia, through taxes and charges) can be placed under pressure to change. Furthermore, planning approaches that reflect the aims of transformation represent a centralised steering approach to economic activities. The establishment of "transition arenas" is regarded as another possible option. This is understood as the building of networks aimed at promoting interaction and the exchange of knowledge between key actors in the innovation process, so as to create the foundations necessary for transformation (inter alia transition agendas, transition experiments) (Loorbach and Rotmans 2006). Transition theories identify starting points for steering transformation processes. However, they cannot explain inertia and situations with insufficient activities. This is why we focus on institutions and institutional change in the next section.

3 Socio-Technical Transformations as a Phenomenon of Institutional Change

Although transition theories such as Transition Management do, to some extent, address the role of institutions in transformation processes, they cannot examine their role in detail. Therefore, the section seeks to bridge the gap between them and the theories of New Institutional Economics.

3.1 The Importance of Institutions in the Urban Transformation Process

In general, in New Institutional Economics, institutions are understood as "*a set of formal and informal rules, including their enforcement arrangement*" (Furubotn and Richter 2005, p. 7). According to North, institutions are "*the rules of the game in a society or, more formally, are the humanly devised constraints that shape human interaction*" (North 1990, p. 3). The main function of institutions is to reduce uncertainty and introduce order into everyday life. For this, institutions function as guidelines for human interaction by defining and limiting actors' decision space (North 1990). In the theoretical works on *transition management*, institutions are assigned a significant role in the steering of transformation processes. For instance, on the one hand, institutional starting points for steering (inter alia, market-based support instruments, planning instruments) and, on the other hand, institutional obstacles of a formal (stabilising rules) and informal (routines, ideologies) nature are deemed relevant. In the process, different types of institutions and regulatory systems that characterise a socio-technical regime are identified (Geels 2004, with reference to Scott 1995; see also Table 1).

Table 1 Examples of institutions in the urban water sector

Type of institution	Examples (for the urban water sector)
Regulative institutions	Formal rules, laws (e.g., EU Water Framework Directive, Urban Waste Water Treatment Directive, Drinking Water Directive)
	Technical standards, product specifications (e.g., DWA-work sheets; DIN standards)
	Planning instruments (e.g., urban development plan)
	Incentive structures (e.g., water pricing)
Normative institutions	Values and norms, but also role expectations or obligations (e.g., sustainable corporate culture of water service providers)
	Policy goals (e.g., sustainability strategy)
Cognitive institutions:	Search heuristics, routines, guiding principles, problem solving strategies, paradigms (e.g., modern infrastructure ideal)
	User practices (e.g., saving water)

Source: Authors' compilation, with reference to Geels (2004, p. 906)

Each of these institutions is frequently found in the sub-regimes (technological regime, scientific regime, political regime, socio-cultural regime, user and market regime) that make up a socio-technical regime, as in the case of urban water services, and determine the behaviour of the respective actors and groups of actors (Geels 2004). In this way, regulative institutions, such as technical standards or research subsidies, but also the prevailing cognitive institutions (e.g., routines and problem-solving strategies of engineers), very clearly determine developments in the technological regime. At the same time, widely varying institutional aspects of the scientific regime (e.g., research programmes for innovative sanitation systems), the political regime (amongst other things, water purification standards, subsidy policy) and the socio-cultural regime (e.g., social attitudes to the technology and its consequences), as well as dominant user practices and preferences, influence technological developments.

Irrespective of the diverse connecting points, institutions are nevertheless frequently neglected or only brought up as an afterthought ("*left-over category*") in analyses of innovation processes and socio-technical transformations (Geels 2004). Thus, the mechanisms underlying institutional change remain largely unclear. Although it is recognised that institutions are created by actors, why actors change institutions and in what way this change occurs has not yet been considered. To better explain this situation, the following section seeks to bridge the gap to the theories of New Institutional Economics. This adds an economic perspective to the multiplicity of existing social science concepts of system change (Gawel and Bedtke 2016).

3.2 *The Theory of Institutional Change*

3.2.1 Mechanisms of Institutional Change

First, institutional change can occur as an evolutionary process when new institutions enter into competition with the existing rules and cooperation mechanisms and are adopted due to their advantageousness (e.g., money). Here, institutional change does not occur through conscious centralised steering, but as the "spontaneous" result of the uncoordinated decentralised actions of many individual actors, analogous to the mechanisms of competition. Second, institutions can also be the outcome of conscious (rational) design by single individuals or the result of a collective process of negotiation (e.g., laws) (Kingston and Caballero 2009). Therefore, especially in the second case, the institutional outcome also reflects the actors' interests and their respective bargaining power. Elements of both approaches are found in the institutional economics-based theory of institutional change developed by Douglass C. North (1990, 2005). However, the central hypothesis of his studies is that institutional change occurs if and insofar as it promises the actors a "benefit". Starting with the observation that, in a world of scarcity, organisations have to face omnipresent competition, North sees the key to institutional change in the interdependent interplay between organisations (players) and the existing institutional framework (rules of the game). Here, institutional change represents a cost-benefit consideration of the actors. Established actors or new actors will work towards achieving a change in the rules if a re-configuration of the institutional framework, which is frequently associated with information costs and negotiation costs, improves their position (North 1990, 1995a). The given institutional framework conditions define the limits in this regard and determine the organisations' alternative scope for action. Above all, institutions determine the incentives for further development, because, in addition to exogenous influences (e.g., climate), it is the learning processes that can influence not only the relative prices and the associated incentive structures, but also the perception of chances and possibilities. This perspective is especially relevant for discussing urban water infrastructure transformation, because it could explain why change is insufficient.

Moreover, North explains that these processes are strongly influenced by the intensity of competition. North (1995a, p. 16) states in this regard: "*When competition is 'muted' (for whatever reasons) organizations will have less incentive to invest in new knowledge and in consequence will not induce rapid institutional change. Stable institutional structures will be the result. Vigorous organizational competition will accelerate the process of institutional change.*" The institutional framework of every society sets incentives for distributive (rent-seeking) and also for activities that increase productivity, so that the relative share of these elements determines whether these structures will be preserved or whether they will be changed, in order to realise new possibilities (North 1990, 1995a). Thus, the organisations within that society only learn the competences that promise the highest pay-out under the existing rules. This can also mean acquiring knowledge and skills in order to work towards maintaining the status quo. According to North, the elements that can cause institutional change – insofar as the corresponding incentives are given – are confronted with other elements that, per se, have a stabilising effect. Referring to works on technological path dependencies (Arthur 1994), North (1990) identifies self-reinforcing mechanisms in institutions and imperfect markets that are responsible for the stabilisation of a path already embarked upon. This means that even inefficient paths can be maintained (lock-in) if changing a path involves prohibitively high costs. Furthermore, this allows actors in political and economic markets to create institutions that favour their interests (North 1990). Institutional stability is also promoted through mental models. These are individually learned rules governing the perception and assessment of complex issues in a world of uncertainty, which are based on cultural background or earlier learning processes and frequently give rise to similar thought patterns (shared mental models) (Denzau and North 1994). Mental models change only slowly and stand in the way of change (e.g., by masking new possibilities) (cf., e.g., Hagemann 2016).

North's theorems can be differentiated into flexibilising and stabilising elements (Kretschmer 2006; see Table 2). Building on this basis, it is possible to assess the relationship between the stability and flexibility of institutional orders and derive starting points for forcing institutional change that are also applicable for transforming urban infrastructure.

Table 2 Flexibilising and stabilising elements of institutional change

Flexibilising elements	Stabilising elements
The continuous interaction between institutions and organisations	Perceptions are derived from the mental constructs of the players.
Competition forces organisations to continually invest in skills and knowledge to survive. The kinds of skills and knowledge individuals and their organisations acquire will shape evolving perceptions about opportunities and, hence, choices that will incrementally alter institutions.	The economies of scope, complementarities, and network externalities of an institutional matrix make institutional change overwhelmingly incremental and path- dependent.
The institutional framework provides the incentives that dictate the kinds of skills and knowledge perceived to have the maximum pay-off.	

Source: Compiled by the authors according to North (1995a, p. 15), following Kretschmer (2006, p. 220)

3.2.2 The Concept of Adaptive Efficiency

Institutional change alone would not guarantee a development that favours transformation (towards sustainability) in urban areas. History teaches us that even institutions that are specifically targeted at preserving existing structures are modified over time. At the same time, there is a need for a normative definition along which institutional development should be aligned, in order to pursue desirable change. North proposes the concept of adaptive efficiency, a measure of efficiency that should define rules for the development of a society over time, and determine the actions of policy makers (North 1994, 1995b). North justifies this departure from the classical efficiency criterion of (static) allocative efficiency by arguing that undesired and unintended outcomes will inevitably result if static theoretical concepts form the basis for decision-making in dynamic environments under conditions of uncertainty (North 1995b), because, as he argues, "*in a world of uncertainty, no one knows the correct answer to the problems that we confront*" (North 1990, p. 81). Although North does not give a conclusive definition of adaptive efficiency, he sees it as most likely given when "*the willingness of a society to acquire knowledge and learning, to induce innovation, to undertake risk and creative activity of all sorts, as well as to resolve problems and bottlenecks of the society through time*" (North 1990, p. 80) is particularly pronounced. The idea behind this is that, in the course of an extensive search and discovery process (in the sense of Hayek 1968), and the associated learning processes, new solutions to (social) problems should be found. According to North, societies that allow the largest number of opportunities to experiment are most likely able to resolve problems over time (North 1990, 1994). Even though the factors responsible for this are numerous, for North, the institutional framework conditions play a key role. Therefore, institutions should be designed in a way that fosters as much research as possible into alternative problem solutions. For this, the institutional framework must meet two requirements:

1. Suitable institutions must provide the economic and political flexibility required to seize new opportunities; i.e., as much freedom as possible must be provided for actors to try out new solutions.
2. At the same time, corresponding incentives must be given in order to trigger these processes and steer them in the right direction.

From the statements of the theory of institutional change, it is evident that the relationship between the stabilising and flexibilising elements decisively influences the degree of adaptive efficiency.

In the following, the question of whether the institutional framework of the urban water sector meets the demands for adaptive efficiency will be analysed.

4 Transformation of Urban Infrastructure Systems: The Example of Urban Water Management

In this section, we explain the situation in the sector of urban water infrastructures in more detail. The absence of adaptation in the sector, which is very distant from a far-reaching transformation, will be explained by using the analytical framework based on the theory of institutional change. On the basis of these findings, we will identify starting points for a transformation towards more sustainable urban water infrastructures.

4.1 The Need for a Transformation of Urban Water Infrastructure Systems Towards Greater Sustainability

A series of processes (in the sense of the multi-level perspective on the "landscape level") creates pressure for change in the existing socio-technical system of networked water infrastructures (see also Reese and Gawel, "Sustainable Urban Water Governance – Main Aims, Challenges and Institutional Approaches in Germany and Beyond", in this volume): to date, altered demand structures and migration processes have already led to water infrastructure usage of only 30% in some parts of Eastern Germany (Koziol 2004). The consequences involve technical and functional issues (e.g., recontamination of drinking water, deposits in the pipelines, premature wear and tear) and the threat of increasing water prices (Moss 2008; Hummel and Lux 2007). The massive decline in population, forecast for Germany and due to demographic change (UN 2015), will exacerbate this problem in many rural areas but also in cities. In addition, the effects of climate change will alter water resources qualitatively and quantitatively, causing higher peak demand, and they are expected to cause an increase in heavy rain events, which means a risk of urban flash floods and uncontrolled sewage plant overflows (Wilby 2007; Bates et al. 2008; EEA 2012).

The climatic and demographic developments differ widely from region to region. Reliable predictions about the longer-term development of local overall demand can hardly be guaranteed through dynamic and small-scale changes. In addition, in the area of water resource management, the end of static framework conditions, due to climate change, was recently proclaimed, which means that a central planning factor is now missing (Milly et al. 2008). Furthermore, it is already clear that water infrastructure systems will have to satisfy ever stricter requirements in a sustainability-oriented society: In addition to calls for greater purification capacity with respect to anthropogenic trace substances (Kümmerer 2009), there is a particularly strong demand for resource efficiency of conventional systemic solutions – not only in relation to the recovery of substances such as phosphorus (Cordell et al. 2011) but also in relation to the economic efficiency of infrastructure operations (Gawel and Bedtke 2015).

The key challenge for the future planning and design of the water sector is, therefore, to develop *flexible* supply concepts that can be adapted not only to small-scale models and rapidly changing demographic and climatic changes, but also to new societal requirements. To increase the flexibility of water infrastructure systems, numerous conceivable institutional starting points exist. These include demand management, changed planning processes, or regulations for compulsory connection and usage (Bedtke and Gawel 2015). However, on the whole, it is clear that, due to the technical inflexibility of conventional urban water infrastructure systems, a comprehensive systemic flexibilisation can only be achieved with the help of completely novel system solutions. These include, above all, (semi-) decentralised solutions, which are particularly flexible because of their modularity and because they dispense with long-life networks and thus have a shorter useful life and cause lower costs. These solutions also offer advantages over conventional technologies for meeting the strict requirements in the area of ecological sustainability, through improved energy and resource efficiency or superior treatment performance (Larsen and Gujer 1997; Shannon et al. 2008; Larsen et al. 2009). Turning away from the technological paradigm of networked urban water infrastructure systems also requires an extensive adaptation of the legal and institutional framework. Concepts associated with the transformation of the urban water sector have been widely discussed for some time, primarily in the academic sphere (Larsen and Gujer 1997; Brown et al. 2009). However, in practice, the need for transformation is met with a distinct reluctance to adapt (Thomas and Ford 2005; Kiparsky et al. 2013; Blanchet 2015). The underlying causes of and options for overcoming such obstacles will be examined in the following.

4.2 The Institutional Inertia in the Urban Water Sector as a Challenge

Besides the long lifespans of centralised technical infrastructure systems in urban regions, it is primarily the enormous costs of operating such systems that is responsible for the continuation of the current path. Here, key interests in "system preservation" are reflected in numerous institutions whose aim is to steer the development of the technology: Many institutional regulations, and the established administrative apparatus that has been built up, are directed at supporting this system through a refinancing model based on the principle of solidarity of as many connected users as possible, thus keeping the unit costs low. At the same time, in Germany, municipal provisions for "compulsory connection and usage" ensure that the usage of centralised systems is high and that the user group that finances the system remains as large as possible. In order to enhance the distribution of decentralised elements, an opening up of the municipal provisions on compulsory connection and usage and tariff-based incentives for novel solutions (e.g., reductions in wastewater charges for rainwater with decentralised seepage) would be required. But the local government agencies that have been responsible up to now have little interest in this, because it would be synonymous with a (further) erosion of the funding base for their central systems, which will, of course, still persist. Consequently, institutional innovations are unlikely here.

This is also because path dependency mechanisms take effect here: they can be found, inter alia, in centrally-oriented administrative structures and the corresponding design of incentives for system users, in the building up of specific knowledge about dealing with the administrative requirements on the bureaucratic side, but also in coordination effects (e.g., model statutes for all municipalities). All investments in these structures would become sunk costs in the case of a fundamental path change. Investments in the existing system architecture, on the other hand, generate "increasing marginal returns". In the water sector, these also act through a comprehensive and highly specific set of technical rules governing the operation of water infrastructure, which was developed in lengthy processes to formulate the protection targets of the respective legislative body.

In addition, among the relevant actors in urban water management, the prevailing mental models and ideologies support inefficient institutions and/or lead to them not being examined to a sufficient extent. For instance, the "theory of uniqueness", widespread in the water sector, argues that, due to the special characteristics of the resource water (number one foodstuff), the sector is awarded special status, whereby a purely public provision with the exclusion of market forces would be preferable (Hirshleifer et al. 1960; Schönefuß 2005). This emotionally charged concept (ideology) substantially complicates the implementation of institutional reforms in the economic regulatory framework. Furthermore, among engineers and urban planners, a "monument syndrome" (Thomas and Ford 2005) or the "big pipes in, big pipes out-paradigm" persists (Brown and Keath 2008): a historically evolved and consistently handed down engineers' understanding that water infrastructure systems have to be conceived as large technical systems. This is supported by the conservative and frequently risk-averse self-conception of municipal politicians who, when it comes to urban water management tasks (implementation, organisation and funding), fall back on tried-and-tested solutions (Farrelly and Brown 2011; Brown et al. 2011.) Moreover, there are also the consumers, who have internalised the comfort of "all-round full supply" as a mental model, which means that the possibility of making one's own contribution to urban water management services is seldom recognised (e.g., contribution to decentralised rainwater management by unsealing of land or use of rainwater).

At the same time, due to imperfect monopoly markets, the corrective effect of competitive selection mechanisms is lacking. The natural monopolies of networked central water supply and wastewater disposal exhibit hardly any competitive structures. The pressure for reform caused by the extensive opening-up of other networked sectors (electricity, gas, communication) in the EU in 1990s did not produce resonance in the German urban water sector. Instead, the political focus was placed on a more symbolic "modernisation strategy", in which the fundamental structures of the traditional regulatory framework were deliberately maintained (Gawel and Bedtke 2015). Since then, the water sector has successfully evaded further competition-oriented reform proposals (inter alia, introduction of stricter regulations) (Finger et al. 2007). This development can also be attributed to powerful lobbying groups in which municipal companies are organised and where they speak out against proposals for regulatory reform, as well as to the public's strong opposition to measures aimed at liberalising and privatising water management. The competitive vacuum gives rise to a situation where, for municipal water supply and wastewater disposal companies, the necessity to invest in new skills and knowledge and to apply new technologies in order to remain viable in direct competition hardly exists (Tauchmann et al. 2009).

As a consequence, companies also have less incentives to shape the institutional framework; for example, encouraging the consideration of new technical solutions in sets of rules or through municipal regulatory provisions. The root cause of institutional change is found mainly in other drivers (such as climate change, demographic change, or new water law requirements). However, if municipal disposal companies had an interest in benefitting more from the (also economic) advantages of novel technologies, a much greater intensity of institutional change could be expected.

In addition, in "political markets", there is very little competition for new solutions: For many local government politicians, water infrastructure systems are not a key field of action, because the consequences of a change in direction in the infrastructure sector today will only become visible in the distant future, whereas the costs of restructuring occur in the present (Klinkenberg 2007). For most users, only the quality and price of the service is important, whilst the underlying infrastructure remains a "black box". The problems that result from not transforming water infrastructure systems (e.g., underuse and overuse of systems, slow loss of substance and insufficient resource efficiency) lie far ahead in the future and are usually outside the public's perception. Transformation failures are, therefore, not likely to generate significant and, at the same time, field-specific policy pressure on local government politicians to act in the sense of the exit-voice theory (Hirschman 1970) of local public goods through extensive migration of citizens (exit) or "at the polls" (voice). Furthermore, many of the system transformation measures discussed (e.g., resource recovery) are initially accompanied by investments and represent, in turn, significant additional costs to the individual (burden of charges) and problems of public acceptance (Russell and Lux 2009). This suggests the presence of strong inhibitory effects.

Furthermore, the (political) competition between local governments addressed in the public finance literature (Tiebout 1956) has only marginal effects on the water sector, because of its current high standard nationwide. As a result, only insufficient political attention is paid to the implementation and testing of new technical solutions.

On the whole, therefore, important impulses for change are not expected to come from the demand side (infrastructure users). And, because decentralised solutions are associated with greater planning efforts and lower planning fees, such impulses are not expected to come from the side of planners and engineers either (Sieker and Sieker 2009).

What is clear is that the stabilising mechanisms contributing to the continued existence of inefficient institutions and incentives in the water sector are still very strong (see Table 3). It is thus hard to distinguish a dynamic in which the relevant actors could work towards changing the institutional framework conditions. Impulses for transformation from the scientific field are too weak and fail in the face of ideologies, low incentives for change, and practical constraints (e.g., the budgetary situation of municipalities).

The examples illustrate that the urban water sector must be attested "institutional equilibrium", in North's terms – that is, a situation where numerous (relevant) actors do not find it advantageous to expend resources on the formulation of new institutional arrangements (North 1990). The benefit that emerges from the existing institutional framework for most actors is higher than that expected to result from a change, especially if the switching costs are added. The requirements for an adaptive efficient institutional framework have not been met, because of limited incentives and freedoms for change. Accordingly, non-sustainable and inefficient structures are preserved.

4.3 Starting Points for Steering Transformation in the Urban Water Sector

The question thus arises: How can processes be initiated that contribute to overcoming the described inefficient equilibrium and promote a dynamic of change within the urban water sector? Since a coherent transformation theory does not exist, recommendations for action should be derived on the basis of general theoretical paradigms. According to the theory of institutional change, flexibilising elements should be promoted and the effects of stabilising elements should – if possible – be muted. In this context, the institutional framework should take into account the criterion of adaptive efficiency; this means, above all, promoting the exploration of alternative problem solutions to the greatest extent possible. To achieve this, actors must have as much freedom as possible to try out new solutions and be given the right incentives for initiating processes of transformation towards urban sustainability and steering them in the right direction. Here, the incentive structures should promise

Table 3 Flexibilising and stabilising elements of institutional change with examples from urban water management

Flexibilising elements	Stabilising elements
Continuous interaction between actors and institutions • Lack of competitive pressure due to the monopoly situation for water supply and wastewater disposal companies in the urban water sector. • In the context of political competition, urban infrastructure systems are uninteresting for politicians, because costs for future gains are incurred in the present. • Urban water infrastructure systems do not play an important role in the competition for citizens in the municipalities. ➢ *Overall low competitive intensity in economic and political markets results in a significantly reduced dynamic, whereby only a low level of influence is exerted on the institutional framework.*	**Mental models and ideologies** • The prevailing model of a "modern infrastructure ideal" characterises the perception of the relevant actors: • Urban planners and engineers develop routines, problem-solving strategies, search heuristics etc., which offer little scope for innovation. • (Risk-averse) municipal decision makers favour the use of tried-and-tested (practical) solutions. • Concept of "municipal full supply" among consumers. ➢ *Mental models that encourage actors to view inefficient institutions as right and not question them sufficiently.*
Learning processes and knowledge building • Because of the monopolistic nature of the urban water sector, the necessity to invest in skills and knowledge in order to be able to survive in direct competition hardly exists for water supply and wastewater disposal companies. • Innovative learning takes place primarily in the academic sphere; the signals for change are too weak. ➢ *Limited incentives for innovative learning due to the lack of competition in urban water sectors.*	**Institutional path dependencies** • A complex set of technical rules generates increasing marginal yields (high installation costs, specific knowledge, transaction cost-reducing network effects and expectations about adaptation), thus increasing the path transition costs. • High capital intensity and long lifespans of systems give rise to municipal regulatory institutions (inter alia, compulsory connection and usage, fee systems) that ensure the refinancing and usage of the systems while, at the same time, producing increasing marginal returns ("administrative practice"). • Imperfect markets (inter alia, monopoly structures, limited rationality) prevent efficiency-promoting feedback processes. ➢ *Diverse institutional path dependencies that stabilise the existing (inefficient) path of urban water governance.*

(continued)

Table 3 (continued)

Flexibilising and stabilising elements
Direction for building knowledge and learning (relative weight of structure-preserving and change-inducing incentives) • Lack of full internalisation of the relatively high (environmental) costs due to numerous institutional deficits (such as, in the area of water utilisation charges, non-transparent water pricing) • The long-term nature of the decisions and future benefits of converting the system means that the costs and benefits of transformation efforts are temporally separated • Insufficient reward for the implementation and testing of new technical solutions among property owners • Planning fees are based on investment costs, which is why engineers prefer large technical systems • High customer satisfaction generates low pressure to act on municipal enterprises and political decision makers ➢ *Mainly incentives for distributive activities and related learning processes, whilst productive activities are inadequately rewarded.*

Source: Compiled by the authors

the highest payouts, not for distributive activities, but for productive activities. Building on these considerations, a number of fields of action are described below:

- *Strengthening competitive structures*: Lively competition is seen by North (1995a) as the most important driver for enhancing learning processes and knowledge building, and for increasing the dynamics of institutional change. From this perspective, increasing competitive pressure, as an "impulse from the outside", should be an important concern. Competition leads to search-and-discovery processes and, possibly, the application of economic solutions (Markard and Truffer 2006). Institutional barriers to new solutions (e.g., financing systems, local government statutes) are far more likely to be successively adapted to fit the new requirements. Without effective competition, cost orientation and efficiency-oriented thinking are far less developed in companies, because effective sanctioning mechanisms are lacking and, in addition, costs can frequently be passed on to the consumers. This does not automatically mean that complex sustainability transformation can be entrusted only to competitive vehicles that ensure efficiency. However, even in the water sector, any potential for competition that may exist should be exploited to the full (see Reese and Gawel, "Sustainable Urban Water Governance – Main Aims, Challenges and Institutional Approaches in Germany and Beyond", in this volume).
- *Price signals*: The continued pursuit of a path of infrastructure development that is problematic, especially also from an ecological point of view, is due, not least, to the insufficient internalisation of environmental externalities. Taking all (environmental) costs into account, the relative prices of service provision would shift in favour of novel system solutions that have obvious advantages over conventional systems in relation to ecological aspects (resource efficiency/recovery). Owing to the relative price increase, the relevant actors (in this case, mainly

those demanding water services) would have stronger incentives to work towards a change of the situation and the investments associated with it. This pressure on the demand side is currently almost entirely lacking, owing to countless institutional deficits with full cost recovery and allocation of costs according to the "polluter pays" principle (in the sense of cost recovery in Article 9 of the EU Water Framework Directive) – see Gawel 2016b). Water utilisation charges (water abstraction fees, wastewater charges – see OECD 2010) also play an important role here: they can allocate "environmental and resource costs" and initiate structural change through their incentive effects (Gawel 2016b). It is also important to remove institutionally induced barriers (e.g., non-transparent water billing) that prevent full price perception at the household level. On the local government level, the application of new solutions (e.g., phosphate recovery, energy recovery) should be better rewarded through tariffs. On the other hand, the more consumption-independent tariff models that are currently being discussed (Hoque and Wichelns 2013) are, instead, purely economically motivated and geared towards the current level of system usage; in terms of sustainability, they lead down the wrong path.

- *Information and clarification/enlightenment*: The institutional balance in the water sector exists because decision makers consider the costs of change to be higher than the benefits. In many places, this estimation is possibly based on an assessment of the situation that is too short-sighted, in which the current high quality of service provision is weighed against the generally "diffuse" problematic situations that are mainly expected to arise in the future. The true costs of a "business-as-usual" approach are, therefore, often either unknown or systematically underestimated. A better understanding and consideration of future developments and costs tips the cost-benefit ratio of transformation in favour of change. To strengthen this effect, in addition to identifying the "costs of inertia", the "costs of change" should be simultaneously reduced by removing the existing information deficits related to the transformation itself. For this purpose, municipalities must, for example, be provided with the (frequently lacking) knowledge needed for successful transformation management (*capacity building*). The supply of information that is as comprehensive as possible to decision makers therefore represents an approach for promoting change (Brown and Farrelly 2009) – not least, with support from science.
- *Culture of trying out*: In niches, new technologies can be protected and brought to market maturity, and practical knowledge can be generated. During recent years, several pilot projects, in which novel system solutions were installed for testing, have been initiated in Germany and Europe (Tauchmann et al. 2006). However, in view of the objective of achieving a far-reaching transformation of water infrastructure systems, this is still not enough. What is needed is concrete experience with the practical feasibility of the solutions under a wide variety of framework conditions and influencing factors (climate, settlement structure, socio-economic factors, existing infrastructure, etc.,) ("because the road is unclear, experimentation is essential in order to learn" Van der Brugge and Rotmans 2007, p. 259; see also Farrelly and Brown 2011). However, the risk-

averse approach of municipal service providers is hardly compatible with the possibility of failure arising from a trial-and-error method. In order to establish a culture that is fully in line with North's adaptive efficiency, *real world laboratories* (Schneidewind and Singer-Brodowski 2014) and the financial support of the state (Farrelly and Brown 2011) are needed. In spatially limited, exceptional institutional situations, approaches and technologies that would otherwise not function in the existing administrative network can be temporarily tested and improved (Londong and Hartmann 2014).

- *Changing mental models*: Mental models change when the underlying learning processes undergo a change (Denzau and North 1994). Consequently, planners and engineers should already have greater exposure to new technologies during their education and professional training. Positive testing of new solutions in inter- and transdisciplinary demonstration projects (see previous point) can be expected to lead to (gradual) changes in the mental models of engineers and, in turn, of decision makers. However, this gives rise to the problem that broader application of new solutions must be made possible in the first place. The mental models of consumers can be overcome if public debate is initiated (Gawel 2016a). The aim is to convey the message that water management is facing considerable challenges, and this requires adaptations that might involve a turning away from "all-round full service" and the forfeiting of certain comforts. The sustained successes of earlier campaigns to conserve water show that it is fundamentally possible to bring about a change in awareness and an alteration of attitudes.

5 Conclusions

Against the backdrop of altered framework conditions and increased societal requirements, there is broad consensus that many urban infrastructure sectors are currently, and for the foreseeable future, on a path that, overall, non-sustainable and insufficiently flexible. Therefore, to ensure urban quality of life and resilience in the long term, while at the same time increasing the resource efficiency of urban infrastructures, a cross-sectoral transformation of urban infrastructures towards sustainability is required.

Urban infrastructure systems are complex socio-technical systems. Although they can only be steered to a limited degree (short and medium term), it is possible to influence the intensity and direction of the future development of networked sectors. Nevertheless, in the water sector of developed countries, as investigated in detail here, because it is so crucial to urban sustainability, there is little evidence that the course required for flexibility has been set. If we view the current situation in these countries from a theoretical perspective, an institutional equilibrium causing inertia and lack of adaptive efficiency can be identified. This indicates the presence of a situation in which the necessary institutional change has not occurred because relevant stakeholders have established themselves within the existing structures in such a way that the cost-benefit analysis goes against change, in favour of preserving the current structure. As a result, the necessary measures for change are either not taken or are even obstructed, since change will result in losses for the established stakeholders.

To nonetheless induce "efficient" change, from an institutional economics perspective, various approaches that aim at breaking down this balance of inertia exist. External impulses can, for example, change incentive structures. Reforming the economic regulatory framework of the water sector to implement competitive structures (taking sectoral idiosyncrasies into account) is a promising approach in terms of both static and dynamic efficiency. Furthermore, a more consistent internalisation of the environmental costs should result in the current path becoming more costly than the sustainability transformation path, thus providing greater incentives for a restructuring of urban infrastructure systems. A shift in the relative weight of structure-preserving and change-inducing incentives can also be achieved through information and education, if the real costs of a "business-as-usual policy" are revealed and the transformation costs are simultaneously reduced. At the end of the day, what is needed is more willingness and courage to test new solutions. Mental models and other path dependencies will only be overcome if the flexibilising elements force a change and the stabilising structures become increasingly transparent.

However, none of these measures should be viewed in isolation, since they are based on a series of complex interdependencies. This also applies to the integration of a restructuring of the water sector into an overall set of complex urban transformations, like general urban development or the linking of other infrastructure issues (energy, transport etc.).

References

Arthur WB (1994) Increasing returns and path dependence in the economy. University of Michigan, Ann Arbor

Bates BC, Kundzewicz ZW, Wu S, Palutikof JP (eds) (2008) Climate change and water. Technical paper of the intergovernmental panel on climate change. IPCC Secretariat, Geneva

Bedtke N, Gawel E (2015) Flexibilität von Wasserinfrastruktursystemen – Konzepte und institutionelle Ansatzpunkte. In: Gawel E (ed) Die governance der Wasserinfrastruktur – Vol. 2. Duncker & Humblot, Berlin, pp 77–123

Bijker WE, Law J (eds) (1992) Shaping technology/building society – studies in sociotechnical change. The MIT Press, Cambridge, MA

Bijker WE, Hughes TH, Pinch TJ (eds) (1987) The social construction of technological systems. The MIT Press, Cambridge, MA

Blanchet T (2015) Path dependence and change in the governance of organized systems: the case of water services in three German municipalities (1990–2010). Free University Berlin, Berlin

Bolton R, Foxon TJ (2015) Infrastructure transformation as a socio-technical process – Implications for the governance of energy distribution networks in the UK. Technol Forecast Soc Chang 90(Part B):538–550

Brand U (2016) "Transformation" as a new critical orthodoxy – the strategic use of the term "Transformation" does not prevent multiple crises. Gaia 25(1):23–27

Brown RR (2005) Impediments to integrated urban stormwater management: the need for institutional reform. Environ Manag 36(3):455–468

Brown R, Farrelly MA (2009) Delivering sustainable urban water management: a review of the hurdles we face. Water Sci Technol 59(5):839–846

Brown R, Keath N (2008) Drawing on social theory for transitioning to sustainable urban water management: turning the institutional supertanker. Aust J Water Resour 12(2):73–83

Brown RR, Keath N, Wong THF (2009) Urban water management in cities: historical, current and future regimes. Water Sci Technol 59(5):847–855

Brown R, Ashley R, Farrelly M (2011) Political and professional agency entrapment: an agenda for urban water research. Water Resour Manag 25(15):4037–4050

Cordell D, Rosemarin A, Schröder JJ, Smit AL (2011) Towards global phosphorus security: a systems framework for phosphorus recovery and reuse options. Chemosphere 84(6):747–758

Denzau AT, North DC (1994) Shared mental models: ideologies and institutions. Kyklos 47(1):3–31

Dolata U (2011) Socio-technical change as gradual transformation. Berl J Soziol 21(2):265–294

EEA - European Environment Agency (2012) Urban adaptation to climate change in Europe. EEA Report 2/2012. EEA, Copenhagen

Farrelly M, Brown R (2011) Rethinking urban water management: experimentation as a way forward? Glob Environ Chang 21(2):721–732

Finger M, Allouche J, Luís-Manso P (eds) (2007) Water and liberalisation – European water scenarios. IWA Publishing, London

Frantzeskaki N, Loorbach D (2010) Towards governing infrasystem transitions: reinforcing lock-in or facilitating change? Technol Forecast Soc Chang 77(8):1292–1301

Fuenfschilling L, Truffer B (2014) The structuration of socio-technical regimes – conceptual foundations from institutional theory. Res Policy 43(4):772–791

Furubotn EG, Richter R (2005) Institutions and economic theory: the contribution of the new institutional economics. The University of Michigan Press, Ann Arbor

Gawel E (2016a) “Great transformation” towards sustainability and behavioral economics. In: Beckenbach F, Kahlenborn W (eds) New perspectives for environmental policies through behavioral economics. Springer, Berlin/New York, pp 127–145

Gawel E (2016b) Environmental and resource costs under article 9 of the water framework directive. challenges for the implementation of the principle of cost recovery for water services. Duncker & Humblot, Berlin

Gawel E, Bedtke N (2015) Efficiency and competition in the German water sector between “Modernization” and “Regulation”. J Public Nonprofit Serv 38(2/3):97–132

Gawel E, Bedtke N (2016) Große Transformationen aus Sicht der Institutionenökonomik und der Neuen Politischen Ökonomik. In: Held M, Kubon-Gilke G, Sturn R (eds) Politische Ökonomik großer Transformationen. Metropolis, Marburg, pp 287–322

Gawel E, Lehmann P, Korte K, Strunz S, Bovet J, Köck W (2014) The future of the energy transition in Germany. Energy Sustain Soc 4(1):1–9

Geels FW (2002) Technological transitions as evolutionary reconfiguration processes: a multi-level perspective and a case-study. Res Policy 31(8/9):1257–1274

Geels FW (2004) Socio-technical systems: insights about dynamics and change from sociology and institutional theory. Res Policy 33(6/7):897–920

Geels FW, Schot J (2007) Typology of sociotechnical transition pathways. Res Policy 36(3):399–417

Graham S, Marvin S (2001) Splintering Urbanism: networked infrastructures, technological mobilites and the urban condition. Routledge, London

Hagemann N (2016) Understanding institutional persistence in Ukrainian water service provision sector. Shaker, Aachen

Hayek FA (1968) Competition as a discovery procedure (trans: Snow MS, 2002). Q J Aust Econ 5(3):9–23

Hirschman AO (1970) Exit, voice and loyalty. Responses to decline in firms, organizations and states. Harvard University Press, Cambridge MA

Hirshleifer, J, de Haven JC, Milliman JW (1960) Water supply: economics, technology, and policy. University of Chicago Press, Chicago

Hoque SF, Wichelns D (2013) State-of-the-art review: designing urban water tariffs to recover costs and promote wise use. Int J Water Resour Dev 29(3):472–491

Hug T, Dominguez D, Maurer M (2010) The cost of uncertainty and the value of flexibility in water and wastewater infrastructure planning. Proc Water Environ Fed 2:487–500. doi:10.2175/193864710798284832

Hughes TP (1983) Networks of power: electrification in western society, 1880–1930. The Johns Hopkins University Press, Baltimore

Hughes TP (1987) The evolution of large technological systems. In: Bijker WE, Hughes TP, Pinch T (eds) The social construction of technological systems: new directions in the sociology and history of technology. MIT Press, Cambridge, MA, pp 51–82

Hummel D, Lux A (2007) Population decline and infrastructure: The case of the German water supply system. Vienna Yearb Popul Res 5:167–191

Kabisch S, Kuhlicke C (2014) Urban transformations and the idea of resource efficiency, quality of life and resilience. Built Environ 40(4):497–507

Kingston C, Caballero G (2009) Comparing theories of institutional change. J Inst Econ 5(2):151–180

Kiparsky M, Sedlak DE, Thompson BH, Truffer B (2013) The innovation deficit in urban w: the need for an integrated perspective on institutions, organizations, and technology. Environ Eng Sci 30(8):395–408

Klinkenberg A (2007) Organisationsveränderungen in der kommunalen Wasserversorgung und Abwasserentsorgung. PhD dissertation, University of Duisburg-Essen

Koop SHA, van Leeuwen CJ (2016) The challenges of water, waste and climate change in cities. Environ Dev Sustain 1–34. doi:10.1007/s10668-016-9760-4.

Koziol M (2004) The consequences of demographic change for municipal infrastructure. German J Urban Stud 44(1)

Kretschmer S (2006) Der institutionelle Wandel der EDEKA-Gruppe. Shaker, Münster

Kümmerer K (2009) Antibiotics in the aquatic environment – a review – part I. Chemosphere 75(4):417–434

Larsen TA, Gujer W (1997) The concept of sustainable urban water management. Water Sci Technol 35(9):3–10

Larsen TA, Alder AC, Eggen RIL, Maurer M, Lienert J (2009) Source separation: will we see a paradigm shift in wastewater handling? Environ Sci Technol 43(16):6121–6125

Londong J, Hartmann M (2014) Angepasste Einzelfalllösungen unter wegweisenden politischen Rahmenbedingungen statt detaillierter administrativer Vorgaben – ein Plädoyer für die Ingenieur- und Rechtskunst. In: Pinnekamp J (ed) 47. Essener Tagung für Wasser- und Abfallwirtschaft "Ist unsere Wasserwirtschaft zukunftsfähig". ISA, Aachen, pp 29/1–29/14

Loorbach D (2007) Transition management: new mode of governance for sustainable development. International Books, Utrecht

Loorbach D (2010) Transition management for sustainable development: a prescriptive, complexity-based governance framework. Governance 23(1):61–183

Loorbach D, Frantzeskaki N, Thissen W (2010) Introduction to the special section: Infrastructures and transitions. Technol Forecast Soc Chang 77(8):1195–1202

Loorbach D, Rotmans J (2006) Managing transitions for sustainable development. In: Olsthoorn X, Wieczorek A (eds) Understanding industrial transformation. Springer, Dordrecht, pp 187–206

MacKenzie D, Wajcman J (eds) (1999) The social shaping of technology, 2nd edn. Open University Press, Buckingham

Markard J (2011) Transformation of Infrastructures: sector characteristics and implications for fundamental change. J Infrastruct Syst 17(3):107–117

Markard J, Truffer B (2006) Innovation processes in large technical systems: market liberalization as a driver for radical change? Res Policy 35(5):609–625

Mayntz R, Hughes TP (eds) (1988) The development of large technical systems. Frankfurt am Main, Campus Verlag

Milly PCD, Betancourt J, Falkenmark M, Hirsch RM, Kundzewicz ZW, Lettenmaier DW, Stouffer RJ (2008) Stationarity is dead: whither water management? Science 319:573–574

Moss T (2008) 'Cold spots' of urban infrastructure: 'Shrinking' processes in Eastern Germany and the modern infrastructural ideal. Int J Urban Reg Res 32:436–451. doi:10.1111/j.1468-2427.2008.00790.x
Nelson RR, Winter SG (1977) In search of a useful theory of innovation. Res Policy 6(1):36–76
Nelson RR, Winter SG (1982) An evolutionary theory of economic change. The Belknap Press of Harvard University Press, Cambridge, MA
Newman P (2001) Sustainable urban water systems in rich and poor countries: steps towards a new approach. Water Sci Technol 43(4):93–100
North DC (1990) Institutions, institutional change and economic performance. Cambridge University Press, Cambridge
North DC (1994) Economic performance through time. Am Econ Rev 84(3):359–368
North DC (1995a) Five propositions about institutional change. In: Knight J, Sened I (eds) Explaining social institutions. University of Michigan Press, Ann Arbor, pp 15–26
North DC (1995b) The Adam Smith address: economic theory in a dynamic economic world. Bus Econ 30(1):7–12
North DC (2005) Understanding the process of economic change. Princeton University Press, Oxford
NRC – National Research Council (2013) Underground engineering for sustainable development. The National Academies Press, Washington, DC
OECD – Organisation for Economic Co-operation and Development (2010) Pricing water resources and water and sanitation services. OECD, Paris
Polanyi K (1944) The great transformation. Farrar & Rinehart, New York
Rip A, Kemp R (1998) Technological change. In: Rayner S, Malone EL (eds) Human choice and climate change, 2nd edn. Battelle Press, Columbus, pp 327–399
Russell S, Lux C (2009) Getting over yuck: moving from psychological to cultural and sociotechnical analyses of responses to water recycling. Water Policy 11(1):21–35
Schneidewind U, Singer-Brodowski M (2014) Transformative Wissenschaft. Metropolis, Marburg
Schönefuß S (2005) Privatisierung, regulierung und Wettbewerbselemente in einem natürlichen Infrastrukturmonopol. Duncker & Humblot, Berlin
Schramm E, Felmeden J (2012) Towards more resilient water infrastructures. In: Otto-Zimmermann K (ed) Resilient cities 2. Springer, Dordrecht, pp 177–186
Scott WR (1995) Institutions and organizations. Sage Publications, London/New Delhi
Shannon MA, Bohn PW, Elimelech P, Georgiadis JG, Marinas BJ, Mayes AM (2008) Science and technology for water purification in the coming decades. Nature 452(7185):301–310
Sieker F, Sieker H (2009) Reformschritte zu einem Paradigmen- und Systemwechsel bei der Regenwasserbewirtschaftung – Teil 2: Reformschritte bei Begriffen, Anschluss- und Benutzungszwang, Technischen Regeln und Honorarordnung. Gwf-Wasser/Abwasser 150(11):919–926
Smith MR, Marx L (eds) (1994) Does technology drive history? The dilemma of technological determinism. MIT Press, Cambridge, MA/London
Smyth R (1998) New institutional economics in the post-socialist transformation debate. J Econ Surv 12(4):361–398
Tauchmann H, Hafkesbrink J, Nisipeanu P, Thomzik M, Bäumer A, Brauer A et al (2006) Innovationen für eine nachhaltige Wasserwirtschaft: Einflussfaktoren und Handlungsbedarf. Physica, Heidelberg
Tauchmann H, Clausen H, Oelmann M (2009) Do organizational forms matter? Innovation and liberalization in the German wastewater sector. J Policy Model 31(6):863–876
Thomas DA, Ford R (2005) The crisis of innovation in water and wastewater. Edward Elgar, Cheltenham
Tiebout C (1956) A pure theory of local expenditures. J Polit Econ 64(5):416–424
UN – United Nations (2015) World population prospects, vol 2. Demographic profiles. http://esa.un.org/unpd/wpp/Publications/Files/WPP2015_Volume-II-Demographic-Profiles.pdf. Accessed 1 Juli 2016

Unruh GC (2000) Understanding carbon lock-in. Energy Policy 28(12):817–830
Van der Brugge R, Rotmans J (2007) Towards transition management of European water resources. Water Resour Manag 21(1):249–267
WBGU – German Advisory Council on Global Change (2011) World in transition – a social contract for sustainability. Report. German Advisory Council on Global Change, Berlin
Wilby RL (2007) A review of climate change impacts on the built environment. Built Environ 33(1):31–45

Efficiency-Equity-Trade-Off as a Challenge for Shaping Urban Transformations

Erik Gawel and Christian Kuhlicke

1 Introduction

Over the past few decades, sustainability has emerged as one of the central guiding principles of policy formulation, especially – but not exclusively – in the field of environmental, climate, and development policy. According to the three-pillar model, sustainable development should be considered in terms of ecological, social, and economic aspects. The United Nations recently reaffirmed this triad of targets in its Sustainable Development Goals (UN SDG), defining the direction to be taken in the formulation of global policies up to 2030. They were adopted at the UN Summit at the end of September 2015 (UN 2015). As explicitly emphasised in that context, the three dimensions of sustainability (economic, social, and environmental) are to be pursued "in a balanced and integrated manner" (UN 2015, p. 3). The simultaneous pursuit of all three dimensions of sustainability inevitably involves trade-offs that decision makers have to take into account and address when selecting policy instruments.

Accordingly, sustainable urban development requires a consideration of a complex set of goals, among them efficiency and equity. However, quite often both aims are in conflict with each other (the so called "equity-efficiency trade-off", see e.g., Stiglitz and Walsh 2006, p. 380) when taking into account that resources are not infinite and that redistribution of resources might affect their overall availability. Additionally, various academic disciplines have quite often developed antagonizing

E. Gawel (✉)
Department of Economics, Helmholtz Centre for Environmental Research-UFZ, Leipzig, Germany
e-mail: erik.gawel@ufz.de

C. Kuhlicke (✉)
Department of Urban and Environmental Sociology, Helmholtz Centre for Environmental Research-UFZ, Leipzig, Germany
e-mail: christian.kuhlicke@ufz.de

S. Kabisch et al. (eds.), *Urban Transformations*, Future City 10,
DOI 10.1007/978-3-319-59324-1_3

perspectives on efficiency and equity that challenge interdisciplinary cooperation and exchange.

Economists, on the one hand, usually focus on efficient allocation of resources, to maximize overall welfare for society that can be derived from a given endowment of resources. They usually try to separate equity issues or highlight trade-offs, framing equity, to an extent, as problematic for efficiency considerations. Many critical social scientists, on the other hand, discuss issues of justice and power and thus refer mainly to equity and distributional concerns and neglect efficiency considerations.

However, there are strong linkages between both concepts. The efficient use of sparse or endangered environmental resources may come at the cost of unequal access to, distribution, or use of these resources and, hence, a reduction of quality of life for some groups or areas. Conversely, the pursuit of equity concerns may reduce resource efficiency, i.e., the overall redistribution mass and can, thus, even lead to a situation in which socially vulnerable groups are worse-off than before.

This contribution provides a brief overview on various schools of thought, originating particularly from economics, sociology, and human geography, with a focus on different, quite often opposing, aspects of the efficiency-equity-equation. It aims at emphasizing the as yet greatly underestimated role of efficiency-equity trade-offs in shaping complex transformation processes, not least within the realm of sustainable urban development. After providing a selective overview of theoretical approaches that address equity, efficiency, and their trade-off (Sect. 2), the paper presents two empirical examples, in order to emphasize the practical relevance of the issue for the cases of flood risk management and urban water infrastructure (Sect. 3). Section 4 concludes this article with an outlook for the future.

2 The Efficiency-Equity Trade-Offs – An Overview

2.1 Economics and the Idea of Efficiency and Trade-Offs

Economists have developed various concepts and instruments for dealing with efficiency. In fact, economics can be seen as the "science of efficiency "par excellence, because it analyzes the optimal allocation of scarce resources. The aim is to make "the most" out of a given endowment of resources. This perspective typically prescinds from individual shares and directs attention to the overall outcome for all. In doing so, distributional aspects are not (completely) ignored (see e.g., Atkinson 1975; Feldman 1987); instead, (mainstream) economists rightly abstain from normative issues of what a fair distribution of resources might ever be, because distributional issues require explicit normative input that goes beyond the expertise of economists.

Neoclassical welfare economics provides a stunning and very simple idea of how to deal with efficiency and equity when it comes to societal resource decision-making: The so-called Second Fundamental Theorem of Welfare Economics states that, out of all possible Pareto optimal outcomes (efficient states), one can achieve any particular one by enacting a lump-sum wealth redistribution and then letting the market take over (e.g., Feldman and Serrano 2006; Chipman and Moore 1978). This means that any desired societal distribution can be reached in an efficient way through market allocation, for efficiency purposes, and combined social policy transfers, for equity reasons (not disturbing the price mechanism). Thus, it turns out to be a "separation theorem": social policy and economic policy may conduct their own programs and there is no need to modify efficient allocation by markets (or policies to correct market failure) in order to take equity concerns into account. In concrete terms, this means that, e.g., prices for energy or water services should tell "the economic and ecological truth" thoroughly, whereas social concerns related to affordability issues have to be addressed by additional social policy (Gawel et al. 2015).

However, for practical purposes, we have not gained very much when referring to this separation theorem for public policy. It assumes strong institutions and good governance providing, simultaneously, a social policy that corrects, in a perfect way, any distributional distortions caused by efficiency-oriented policy. Moreover, it is in fact possible, according to the Second Theorem of Welfare Economics, to obtain a certain Pareto optimum (out of many) by starting with any (fair) initial endowments of resources, but the overall outcome will be different, *nota bene*. This means that economic analyses also may depict trade-offs between efficiency and equity. These trade-off analyses are very common; many studies have investigated possible extra societal costs if resources are distributed in a particular way. If taxation, e.g., refers to particular equity objectives, it may decrease the overall GDP, because it could imply disincentives for adding value.

In recent decades, empirically oriented behavioural economics provides new insights into the interaction of efficiency and equity, challenging the previously outlined ideas. In particular, as these studies revealed, explicit fairness preferences of individuals have to be taken into account. In turn, newly emerging theoretical approaches try to integrate equity issues explicitly into economic reasoning (see, i.a., the so-called ERC theory of Bolton and Ockenfels 2000, which theoretically combines "equity, reciprocity, and competition" in a uniform theoretical framework).

Thus, the findings with respect to economic thinking about efficiency and equity are somewhat complex. Welfare economics theoretically provide a relatively smart solution for taking into account both efficiency and equity (by conducting separated but coordinated policies). Therefore, economic efficiency-oriented work, especially policy recommendations, e.g., in the field of environmental policy, often neglects distributive impacts that actually must be integrated, to obtain sound policy advice. In practice, the separation theorem, due to state failure, does not work sufficiently well and thus cannot safeguard public acceptance of efficiency-oriented public policy. A very prominent challenge here is the "right" design of pricing for environmentally relevant urban services (energy, water – see Sect. 3.2). New theories, inspired by behavioural approaches, seek to bridge this gap and to take into account equity concerns from the very beginning. This leads to modified approaches for, e.g., behavioural pricing (Maxwell and Estelami 2009), an economic price theory based on empirically proven consumer behaviour.

However, efficiency considerations provided by economists are highly valuable, by all means. Because no other scientific discipline refers explicitly to the problem of efficient resource allocation equally seriously, this contribution from economics ("talking about efficiency") is particularly needed. Because the idea of resource efficiency is merely an intellectual construct for the entire economy and can neither be observed easily in an empirical way nor attributed to individuals and groups, social sciences other than economics tend to ignore or disesteem the relevance of efficient allocation.

This is all the more true because, in political arenas, efficiency typically lacks political support: efficiency gains (and losses) are invisible (you must imagine the alternative state in order to "see" the added value of efficient resource use) and they are usually distributed across the entire society ("welfare maximum of all") and cannot be captured by groups or individuals exclusively, unlike the case for redistribution policies. This is why "efficiency" turns out to be a public good no pressure group will ever be really interested in. Conversely, rent-seeking for redistributions, for the sake of equity, does make sense and therefore widely dictates political discourses all over the world, marginalizing efficiency impacts (e.g., with respect to the energy transition, see Strunz et al. 2016).

However, economic thinking is just starting to incorporate equity issues into both theory and policy recommendations. Further interdisciplinary work has to be done to fully understand the interaction of efficiency and equity for economic argumentation.

2.2 *Social Sciences' Concepts of Environmental Justice*

The origins of scholarly concerns with equity in environmental social science research are rooted in the civil rights movement in the US (Walker 2012; Elvers 2005). The initial focus, quite often driven by grassroots groups, was predominantly on the relationship between ethnicity, poverty, and the spatial distribution of waste dumps and industrial sites, and how they impact negatively on their surrounding neighbourhoods. There was, hence, a strong orientation towards collecting empirical evidence that would support the claim that environmental resources and risks are distributed unequally and that certain groups are more often exposed to environmental risks than others.

Since then, both the thematic and the geographic scope have been significantly expanded (Walker 2012, p. 2 ff.) as research focuses, in the meantime, on different social (e.g., class, gender, age, disability, future generations) and environmental dimensions (e.g., air pollution, urban dereliction, flooding, noise, greenspace) and is also increasingly established as a research domain outside the US (for Europe, see Elvers et al. 2008; Wolch et al. 2014). It has also become an effective mobilizing concept that attracts attention beyond academic circles (Taylor 2000). As an example, two of the seventeen UN SDGs are directly related to equity: the reduction of "inequality within and among countries" and the promotion of "just, peaceful and inclusive societies" are defined as central goals that should shape national policies until 2030.

However, along with an empirically observable distributional inequality comes the more troubling normative question of whether the unequal distribution of "goods" and "bads" is also unjust? Whilst the link between inequality and injustice seems, at first sight, apparent ("any pattern of inequality is also unjust"), the link between both concepts is usually made without further consideration. Walker (2012) therefore pleads for being analytically more precise and for keeping both concepts separate: Whereas "inequality is a descriptive term, describing a condition of difference or unevenness of something (such as income, health, pollution exposure/creation, opportunity, influence, access to resources, consumption of resources), between different groups of people (old/young, black/white, rich/poor, north/south, this generation/future generation etc.)" (p. 12), injustice is about the normative (some say philosophical) question about the grounds on which an actor (individual/collective) should reason/should decide that a difference is just or unjust.

Similar to the idea of efficiency, the step from an empirically describable unequal distribution of "goods" and "bads" to the normative question, whether this inequality is also unjust, eludes direct observation and demands a relatively presumptive intellectual construction process grounded in philosophical debates. John Rawls, for instance, in his *Theory of Justice* (1971), starts with a thought experiment in order to explicate his specifications of basic principles of justice. He situates different actors in a hypothetical position defined by a "veil of ignorance", implying that they know neither their current nor their future position within a group. Only by being ignorant of their respective social position can all members of a group decide freely about the group's principles of justice, which is then the basis of the actual collective contract the group gives itself. The social contract, however, should consider two basic principles: The first one is liberty (each person is to have an equal right to the most extensive basic liberty compatible with a similar liberty for others) and the second one is difference (inequality should be arranged in a such a way that the least-advantaged members of a group have the greatest opportunities; Rawls 1971). Rawls' reflection became quite influential both in urban (Marcuse et al. 2009; Fainstein 2009) and environmental discourses (Harvey 1996; Walker 2012).

Fainstein, in her essay on "Planning and the Just City", undertakes the endeavour of relating Rawls general reflection to the more specific contexts of cities and engages with the "practical reality of régime formation, social exclusion, and the bases of conflict" (Fainstein 2009, p. 28). Harvey highlights the need to go beyond pure distributional analysis to understand patterns of environmental injustice and puts the socio-political-economic conditions that produce and reproduce socio-spatial disparities at the forefront of his analysis; this argument also taken up and expanded by Walker. He introduces not only the idea of "claim-making" and "framing" in his contestation with the concept of environmental justice (Walker 2012) and by doing so aims at bridging the gap between normative reflections as well as empirical observations, but also outlines three dimension of environmental justice including, next to distributional aspects, procedural aspects ("the ways in which decisions are made, who is involved and has influence") also aspects of recognition ("who is given respect and who is and isn't valued") (ibid., p. 10), a point we return to in Sect. 3.1.

In contrast to economic discussions, the debate on environmental equity is seldom linked with the concept of efficiency, although the interrelation is highly relevant (see introduction) and apparent: Whilst the discussion on justice focuses on distributional and procedural questions, it tends to ignore the availability of (scarce) resources and, hence, seems to imply that resources are available infinitely. At the same time, the idea of "efficiency" is a major driver of how resources (and hence also risks) are distributed within a society and thus shapes not only the distribution of risks but is also a constitutive criterion for procedural aspects, because allocation decisions are often based on the criteria of efficiency, as the two subsequent empirical sections will highlight.

3 Efficiency and Equity in Urban Transformations: Two Empirical Examples

Based on these rather conceptual reflections that aim at showcasing the interlinkages and, at the same time, unravelling the blind-spots of different disciplinary approaches, we now present selected empirical examples that illustrate how efficiency-equity trade-offs are shaping environmental concerns in an urban setting by demonstrating that trade-offs not only relate to social processes (e.g., inner-urban difference or differences between urban and rural areas) but, more fundamentally, also to the future use and exploitation of limited environmental resources. This will be shown using examples from flood risk management (3.1) as well as from network-based urban infrastructures (e.g., water services) (3.2).

3.1 Flood Risk Management in England and the Urban-Rural Divide

The concepts of equity and efficiency are both relevant topics in research and policy-making on flood risks, although they are usually considered separately. On the one hand, concerns for equity prevail in research on social vulnerability. In the course of attempts to better understand which groups or regions are more exposed (than others) to the risk of flooding, or in a worse position to cope with its negative consequences, the "root causes" for the production of unequal distribution of vulnerability patterns have become a concern in many empirical studies conducted in developing and developed countries. Walker for instance, by linking existing flood risk maps with statistical population data, found, for the UK, a strong correlation between social class und flood risk exposure (Walker 2012). The study revealed "a strong social regressive gradient, such that people who are strongly deprived are more likely to live in flood risk areas than those who are less deprived" (Walker 2012, p. 133). Interestingly, the sharp gradient only relates to areas located close to the sea and, hence, to areas exposed to coastal flood risk but not to riverine flood risks. According to Walker, the reasons for this pattern are threefold (ibid., 134 ff.): First, in old industrial ports, working-class housing and population dominated for a long time. Second, lower quality housing was often developed in lower lying and was, therefore, on lower priced land, and, third, although many coastal cities had their "heyday in the first half of the twentieth century, providing holiday destinations for the mass of the British public" (ibid.), they have experienced decline and decay since then and are hence becoming increasingly deprived. As a consequence, more deprived areas are more often exposed to the risk of flooding from the sea.

Efficiency, on the other hand, has evolved as a central concept in flood risk management. As a result of a number of large flood events throughout Europe over the past 20 years, as well as the instalment of the European Floods Directive, many European countries have overhauled their flood management strategies. Specifically, a movement away from solely focusing on state-provided structural protection measures (e.g., dikes and levees), which are funded, planned, and implemented by governmental bodies, has resulted in an increasing emphasis on the importance of developing and implementing alternative measures (Johnson and Priest 2008). In this context, the traditional goal of offering, with public funding, the same protection standard to all flood-exposed areas is challenged because it is considered as "technically, economically, or environmentally [not] feasible" (EA 2010, pp. 16–17). Instead, risk-based management approaches, which consider not only event probabilities and design standards but also try to incorporate the wider economic, ecological, and social consequences, are gaining relevance. They allow the identification of how a risk should ideally be reduced and which level of protection appears to be appropriate, in terms of a reasonable cost-benefit ratio. In England, the Department of Environment, Food and Rural Affairs (DEFRA), for instance, highlights risk-based management as a rational decision-making tool ensuring that "the maximum benefit is achieved with every £1 of taxpayers' money, and that we can be confident that the money could not be better spent elsewhere" (DEFRA 2009, p. 3). Similarly, the Swiss PLANAT underlines the relevance of participation in risk-based approaches, as it would allow to make processes and decisions with regard to funding and the allocation of resources within complex net of actor transparent (Bründl 2009).

This section focuses on the case of England, which is notable for its full-throated commitment to being 'risk-based' (see Kuhlicke and Demeritt 2016, for an overview) and explores in more depth how ideas of efficiency and equity are interlinked.

As a results of England's Making Space for Water (MSFW) (DEFRA 2004) and the Pitt Review (Pitt 2008), a revised policy agenda in line with the UK government's advocacy of risk-based approaches to 'better regulation' (Kuhlicke and Demeritt 2016), flood risk management relies meanwhile on various risk-based technologies and policy instruments (e.g., risk mapping, risk-based protection standards, and risk-based resource allocation) to ensure that flood risk reduction measures are proportionate to their expected costs and benefits (Krieger 2013). In this way, 'risk' is not simply an object to be managed, but a central principle for the organization of flood risk management itself (Rothstein et al. 2006). Rather than trying to eliminate all potential harms, risk-based approaches aim for an optimal balance between acceptable levels of risk and the costs of reducing them further. With references to HM Treasury's (2003, p. v) Green Book, government policy requires cost-benefit analysis of publicly funded flood risk management schemes to ensure that public money is spent proportionally. These strategic-level policymaking requirements are designed to ensure that operational-level decisions "are oriented to gaining the maximum economic benefit for the country as a whole" (DEFRA 2009, p. 12).

Whilst a risk-based approaches for prioritizing flood intervention measures might be efficient with regard to society as a whole, they are not necessarily perceived as fair on the individual level. One can assume, on the one hand, a gap between those who contribute to the funding of flood protection measures and those who benefit from it (Penning-Rowsell and Pardoe 2012) and, on the other hand, the efficient investment of sparse public resources also implies transferring funds to areas in which the cost-benefit ratio is higher than in areas with a lower cost-benefit ratio. This is also reflected in recent scholarly debates about distributional effects of risk-based decision-making processes (Penning-Rowsell and Pardoe 2012) as well as in governmental documents reflecting a wider concern with social justice issues, motivated in part by the concern that "economics-led investment appraisal regimes lead to greater investments where communities are richest" (ibid., 450).

Allocating public resources through risk-based priorisation techniques implies that resources are distributed unequally. As a rule of thumb: the higher the avoided negative consequences of a flood protection measure (i.e., avoided economic impact), the higher its expected benefit. As a consequence, protecting a settlement with 200 buildings by means of a flood protection scheme is more efficient than protecting a settlement with 20 buildings, because the benefit for each € spent is higher. In this sense, risk calculations not only ascribe probability and consequence to future flood events, they also influence, through their reliance on cost-benefit analysis, the frequency and intensity of flood experience events and their negative consequences. Specifically, larger urban areas, usually defined by a higher accumulation of wealth and by a greater presence of critical infrastructure, as well as by a higher share of cultural monuments are, therefore, within a risk-based setting, more likely to be better protected, compared to smaller or more rural areas that are less likely to receive publicly funded protection schemes and experience, as a consequence, the negative effects of flood events more often, as the following interview conducted in the UK reveals:

> … rural areas are going to be the ones that suffer again because there isn't the partners around … in a small community. And the community themselves, being small, are not going to be able to raise the vast thousands upon millions of pounds that are needed towards any flood scheme (interview with a community engagement officer).

The potential increase or creation of inequality was also emphasised by another interviewee:

> well you have got … that difficulty with the small rural communities … they have created a mechanism to try and catch areas of deprivation so that they get a higher score but if it is not scoring high enough they have got very little chance of drawing in the funding … its puts a lot of schemes, you know, out of reach forever" (interview with a flood management officer from a county council).

Therefore, Johnsons et al. argue that the "issue of procedural equality and to addressing the needs of those most vulnerable to flooding" should gain greater weight in policy making (Johnson et al. 2007, p. 376). In this context, Rawls' secondary principle of justice becomes relevant, because it suggests that those most disadvantaged (e.g., areas not protected by a publicly financed protection scheme) should benefit most from alternative management strategies and measures (e.g., private adaptation measures); an idea that is also acknowledged in more recent legislation (Begg et al. 2015).

The empirical case studies show how urban transformation, efficiency, and justice are interlinked. The idea of investing public money efficiently has become a decisive criterion in flood risk management, resulting in an unequal definition of flood protection levels from which particular urban areas will benefit. By introducing the idea of justice (see also Kuhlicke 2014 for the idea of protective justice), the focus is turned towards procedural as well as onto distributional aspects and the question of whether and how unequal patterns of protection can and should be mitigated by relying, for instance, on the second principle of justice, as outlined by Rawls (1971).

3.2 *Sustainable Infrastructure: Pricing Water Services and the Efficiency-Equity Trade-Off*

Sustainability policies based on the economic rationale of providing incentives to get prices right and to internalize external costs inevitably place, at the same time, a significant burden on societal groups and often raise distributional concerns. This holds for the German energy transition (Gawel et al. 2015), as well as for the sustainability-oriented transformation of urban water infrastructure in developed countries (see Bedtke and Gawel, "Linking Transition Theories with Theories of Institutions – Implications for Sustainable Urban Infrastructures Between Flexibility and Stability", in this volume). Internalizing external costs or just enforcing the principle of full cost recovery typically results in rising commodity prices for consumers. When this affects basic commodities that are consumed to a similar extent by consumers in different income groups, such as water or electricity, the burden is typically felt relatively more strongly in the lower-income groups than in the higher income groups. Against this background, sustainability policies that lead to rising consumer prices are often classified as socially unjust because they would impose a disproportionately large burden of the costs of sustainability policy on low-income households (see, in general, Heindl and Löschel 2014; for water OECD 2009; for energy Bardt and Niehues 2013).

However, resource efficiency – with respect to both the human-made water infrastructure and the natural water resources – requires full cost recovery for water services including "environmental and resource cost", as stated in Article 9 of the EU Water Framework Directive (WFD) (Gawel 2016). On the other hand, equity considerations, as laid down in target 6.1 of the UN SDGs, call for "universal and equitable access to safe and affordable drinking water for all". Thus, the pricing water services is challenged because it simultaneously addresses economic, ecological, and social aspects of water use. Full cost recovery and incentive-oriented pricing may support resource efficiency and adequate refinancing of services but can jeopardize equitable and affordable access for all water users, particularly under conditions of water shortage.

Remarkably, in Goal 6, the UN SDGs emphasize the right to water but, at the same time, in an explicit sustainability context that also includes resource efficiency considerations. Unlike the Millennium Development Goals, which focused entirely on social objectives, the UN SDGs explicitly attach special importance to other sustainability aspects of water supply, e.g., integrated water resource management (target 6.5) and water use efficiency (6.4), as well as "sustainable management of water and sanitation for all" (overall title of Goal 6). Obviously, this requires a certain kind of integration of these different targets that have to be fulfilled simultaneously. This is well in line with theoretical considerations of interacting dimensions of sustainability for water infrastructure (Gawel and Bretschneider 2016a): only if water resources are sustainably protected and a water infrastructure is adequately financed and operating without waste and for least-cost, is it possible to ensure, in the long term, those full-coverage supply services that a right to water represents. The fact that water management issues are related to all interacting sustainability dimensions thus becomes a prerequisite for a sustainable understanding of the right to water (see, for the concept of "sustainable access to water": Gawel and Bretschneider 2016b).

Against this background, prices for urban water services are key factors for sustainable urban development, because prices (must) fulfill many important (economic) functions (cf. Table 1):

As a part of the revenues, they refinance the efforts of a service provider to the benefit of the users because

- they express the degree of efficiency of production,
- they set incentives or scarcity signals for providers regarding the application of scarce production factors (e.g., raw water) and also for users, who must estimate whether the services provided justify the resource consumption (efficiency demand), and
- they represent a hurdle to the access of goods and must justify that they are reasonable under social aspects (affordability, right to water).

Obviously, prices in general – including charges in urban water management – fulfill very different functions that have to be expressed in one and the same value (e.g., rate or tariff, if applicable): economic efficiency, ecological sustainability, social reasonableness, and adequate refinancing. At first sight, there are harmonious

Table 1 Water pricing policy objectives between efficiency and equity

Environmental sustainability	**Financial sustainability (Refinancing)**
Discourage depletion of critical natural capital	*Guarantee long-term reproduction of physical assets*
Guarantee the preservation of ecological functions of water as natural capital.	Compensate for the resources that are used as inputs in water-related activities.
Minimise the use of "supply side" solutions to water scarcity.	Cash flow should guarantee the conservation of value of physical assets.
Use efficiency	Cost recovery should be chosen only for efficient costs.
encourage water saving	
discourage wasteful water use	
Minimise the alteration of natural flow patterns.	
Economic efficiency	**Social concerns**
Water is allocated to the most beneficial uses (user efficiency) and economic resources are not wasted (production efficiency)	*Adequate access to affordable water at fair and equitable conditions*
Allocate water with priority to uses with highest value to society as a whole.	Identify "water needs" and allocate water in a way that is not skewed by concentration of power.
Compare costs of water management and water-related services with their value, i.e., do not misallocate economic resources.	Structure tariffs so that lower-income users can have access to and afford to use water services.
Cost efficiency: minimise lifecycle costs of services, i.e., the creation of physical capital and operation and maintenance costs	Achieve an equitable way the sharing of the cost of managing water resources.
Regulation should ensure optimal risk allocation among stakeholders (including users and taxpayers).	

Source: Modified after OECD (2009, p. 25)

as well as adversarial relations between the four basic principles of a water pricing policy (Table 1): Thus, an "efficient" charge that is free from monopoly profits contributes to social reasonableness (low prices) at the same time. An economic full cost recovery, including so-called environmental and resource costs according to Article 9 WFD (Gawel 2016) however, demands – just as much as the justified request for adequate refinancing of the investments made – appropriately "high" charges.

Consequently, "sustainable charges" can be understood as prices for water services that consider all facets of the previously mentioned price impacts in an appropriate way and lead to a reasonable balance of possible trade-offs. In contrast, approaches that argue with only a single aspect (e.g., "only" right to water or "only" cost efficiency) and, thereby, derive legal demands for the regulatory framework of water management, are not convincing.

Interestingly, against this background, it is possible that charges that are either too high or too low emerge at the same time – depending on the price function according to which the charge is measured. Indeed, water prices in Germany are currently "too high" (containing monopoly profits) as well as "too low" (subsidized and calculated without complete environmental and resource costs or without necessary preservation measures, e.g., leakages in the pipe system). Consequently, there are significant doubts whether current charges for water services in Germany can be regarded, altogether, as economically "cost-covering" in terms of the requirements of Article 9 WFD (Gawel 2012). The obvious backlog in redevelopment (lack of reinvestments in run-down systems) shows that current operations are "below cost" – mainly due to political reasons and in deference to reasonableness of charges. This is valid independently of the question whether charges seem "too high", because monopoly profits emerge out of "unnecessary" depreciation caused by a lack of competition or equivalent cost pressure. Both phenomena can appear simultaneously, which makes a simple external diagnosis about the reasonableness of charges difficult. In economic theory, the problem of "too high" or "too low" price demands has long been discussed by taking the example of a monopoly provider – a typical example in water management –who manufactures a product with external environmental and resource costs (Buchanan 1969; Barnett 1980). In this case, a purely competition-oriented monopoly policy leads to welfare loss, as long as environmental impacts are disregarded. Only an integrated consideration and instrumentally modified monopoly control can secure a socially optimal monopoly policy.

Similar conflicts can emerge – as current intensive discussion in Germany shows – when new tariff models that are strongly independent of quantity are intended to stabilize the revenues for service providers if water demand declines (Hoque and Wichelns 2013). These models serve the sub-goal of adequate refinancing and also refer – with regard to high fixed costs of pipelines for water services –to economic principles of pricing.

However, it must be clear that, in the long term, it is in fact the infrastructure that must follow the demand. Moreover, the use of the natural water reserves has to be kept as small as possible – and therefore it cannot be the (sole) task of prices to provide for comfortable utilization of a historically developed infrastructure that is probably no longer appropriate, and to protect it permanently from economic pressure to change. Furthermore, Art. 9 para. 1 sub-para. 2 WFD does not call for a water pricing policy of EU member states that provides adequate incentives for permanent security of revenues for providers; instead, incentives should be provided to "use water resources efficiently", which refers to the resources of natural water circulation. For sustainability reasons, these measures should prevent charges being determined predominantly by consumption-independent basic prices (or even "flat rates"). In fact, a reasonable balance between goals and interests must be found and secured.

In conclusion, it is clear that charges play a key role in economic "institutions" regarding governance in urban water management, because many targets can and must be addressed within a single variable. On the one hand, requirements of ecological sustainability must be fulfilled by providing incentives to use resources efficiently (production efficiency) and to behave environmentally consciously (user efficiency). At the same time, prices reflect scarcities and should, ideally, contribute to always use water in the most beneficial way. Moreover, charges serve the goals of refinancing and "preservation of substance", or financing of system rebuilding for demographic or climatic reasons. Social matters such as the affordability of water services, acceptance or equity concerns regarding allocation of burden are directly connected to charges. It is obvious that the broad set of price policy goals must evoke contradiction and, thus, balanced integrative solutions for charges ("sustainable charges") are a significant challenge.

Approaches that derive attempts at far-reaching reforms only from a sub-goal and without considering the impacts on other goals are not constructive. This applies especially for the basically justified attempt to introduce more cost efficiency in "sustainable water management" or to implement the "right to water" comprehensively. This is also true for unnecessarily unilateral tariff models that strive only to secure revenue. In fact, "sustainable charges" must try to address all basic challenges for water prices (Gawel 2016), each in an adequate manner and, thus, also address the pricing-related efficiency-equity trade-off .

4 Conclusion

Sustainable urban transformation, following the widely accepted three-pillar model of sustainability, requires pursuing its three dimensions (economic, social, and environmental) "in a balanced and integrated manner" (UN 2015, p. 3). However, the simultaneous pursuit of all three dimensions of sustainability inevitably involves manifold trade-offs, especially between equity and efficiency. This contribution used an interdisciplinary social science perspective to highlight the relevance of this topic for sustainability-oriented urban transformations. To date, resource efficiency and equity issues are often discussed in fragmented discourses and in an isolated way, although the flood risk management case study has demonstrated that there are also examples that consider both the idea of efficiency and, to a certain extent, the idea of equity. Moreover, trade-off issues, if covered at all, are widely framed as mere conflicts that have to be solved by "prioritization". Thus, economists and other social scientists often find themselves in unnecessary opposition when dealing with sustainability transformation.

Instead, efficiency and equity have to be taken into consideration simultaneously and with full acknowledgement of their possible trade-off relationships. Instead of tackling this challenge by mere prioritization, more attention should be paid to emerging integrative theoretical frameworks, e.g., in behavioural social sciences, as well as to smart solutions in practice. This implies overcoming strongly antagonistic perspectives on efficiency and equity offered, till now, by different academic disciplines. Improving interdisciplinary cooperation and exchange is needed in the field of the efficiency-equity trade-off.

Moreover, this analysis pinpointed promising further avenues for research in the field of urban transformations. For instance, more general hypotheses on the interaction of efficiency and equity could be investigated. For example, whilst urban environmental resource-efficiency might result in an increasing rural-urban inequity (e.g., the quality of life in urban areas depends, to a large extent, on the exploitation of resource and ecosystems located in more rural areas), the inequity between environment and society (i.e., human society acts at the cost of the natural environment) could be reduced through urbanization in general, because scarce and limited environmental resources can be used more efficiently in urban areas. The authors feel that intensified interdisciplinary research on the impact of efficiency-equity trade-offs in the field of urban transformations could yield improved insights into transformation processes and, thus, considerable added value for both theory and practice – as has been demonstrated here with two examples of urban development.

References

Atkinson AB (1975) The economics of inequality. Oxford University Press, London

Bardt H, Niehues J (2013) Distribution effects of the Renewable Energies Act. Zeitschrift für Energiewirtschaft 37(3):211–218

Barnett AH (1980) The pigouvian tax rule under monopoly. Am Econ Rev 70(5):1037–1041

Begg C, Walker G, Kuhlicke C (2015) Localism and flood risk management in England: the creation of new inequalities? Environ Plan C Gov Policy 33(4):685–702

Bolton GE, Ockenfels A (2000) ERC: a theory of equity, reciprocity, and competition. Am Econ Rev 90(1):166–193

Bründl M (ed) (2009) Strategie Naturgefahren Schweiz. Umsetzung des Aktionsplans PLANAT 2005–2008. Projekt A 1.1. Risikokonzept für Naturgefahren – Leitfaden. Nationale Plattform für Naturgefahren PLANAT, Bern

Buchanan JM (1969) External diseconomies, corrective taxes, and market structure. Am Econ Rev 59(1):174–177

Chipman JS, Moore JC (1978) The new welfare economics 1939–1974. Int Econ Rev 19(3):547–584

DEFRA – Department for Environment Food & Rural Affairs (2004) Making space for water: developing a new Government strategy for flood and coastal erosion risk management in England. A consultation exercise. Department for Environment, Food, and Rural Affairs. http://www.look-up.org.uk/2013/wp-content/uploads/2014/02/Making-space-for-water.pdf. Accessed15 Feb 2017

DEFRA. – Department for Environment Food & Rural Affairs (2009) Appraisal of Flood and Coastal Erosion Risk Management: A Defra Policy Statement. http://www.defra.gov.uk/publications/files/pb13278-erosionmanage-090619.pdf. Accessed 9 Aug 2016

EA – Environment Agency (2010) Flood and Coastal Erosion Risk Management Appraisal Guidance (FCERM-AG). http://webarchive.nationalarchives.gov.uk/20131108051347/http://a0768b4a8a31e106d8b0-50dc802554eb38a24458b98ff72d550b.r19.cf3.rackcdn.com/geho0310bsdb-e-e.pdf. Accessed 9 Aug 2016

Elvers H-D (2005) Umweltgerechtigkeit (Envirionmental Justice): integratives Paradigma der Gesundheits- und Sozialwissenschaften? UFZ-Diskussionspapiere 2005. UFZ, Leipzig

Elvers H-D, Gross M, Heinrichs H (2008) The diversity of environmental justice. Eur Soc 10(5):835–856

Fainstein SS (2009) Planning and the just city. In: Marcuse P, Conolly J, Novy J, Olivio I, Potter C, Steil J (eds) Searching for the just city: debates in urban theory and practice. Routledge, London/New York, pp 19–39

Feldman AM (1987) Equity. In: Durlauf SN, Blume LE (eds) The New Palgrave: a dictionary of economics, vol 2, 1st edn. Palgrave Macmillan, London, pp 183–184

Feldman AM, Serrano R (2006) Welfare economics and social choice theory, 2nd edn. Springer, New York

Gawel E (2012) Pricing municipal water services in Germany – are they already really cost-covering? J Public Non-Profit Serv 35(3):243–266

Gawel E (2016) Environmental and resource costs under article 9 water framework directive. Challenges for the implementation of the principle of cost recovery for water services. Duncker & Humblot, Berlin

Gawel E, Bretschneider W (2016a) Content and implementation of a right to water. An institutional economics approach. Metropolis, Marburg

Gawel E, Bretschneider W (2016b) Sustainable access to water for all: how to conceptualize and to implement the human right to water. J Eur Environ Plan Law 14(2):190–217

Gawel E, Korte K, Tews K (2015) Distributional challenges of sustainability policies – the case of the German energy transition. Sustainability 7:16599–16615. doi:10.3390/su71215834

Harvey D (1996) Justice, nature and the geography of difference. Wiley-Blackwell, Oxford

Heindl P, Löschel A (2014) Addressing social implications of green growth – energy sector reform and its impact on households. Issue note prepared for session 1 of the green growth and sustainable development forum, 13–14 November 2014, Paris. OECD, Paris

HM Treasury (2003) The Green Book: appraisal and Evaluation in Central Government. The Stationery Office/Tso. https://www.gov.uk/government/uploads/system/uploads/attachment_data/file/220541/green_book_complete.pdf. Accessed 9 Aug 2016

Hoque SF, Wichelns D (2013) State-of-the-art review: designing urban water tariffs to recover costs and promote wise use. Int J Water Resour Dev 29(3):472–491

Johnson CL, Priest SJ (2008) Flood risk management in England: a changing landscape of risk responsibility? Int J Water Resour Dev 24(4):513–525

Johnson CL, Penning-Rowsell E, Parker D (2007) Natural and imposed injustices: the challenges in implementing 'fair' flood risk management policy in England. Geogr J 173(4):374–390

Krieger K (2013) The limits and variety of risk-based governance: the case of flood management in Germany and England. Regul Gov 7(2):236–257

Kuhlicke C (2014) Hochwasservorsorge und Schutzgerechtigkeit: Erste Ergebnisse einer Haushaltsbefragung zur Hochwassersituation in Sachsen. UFZ-Discussions Paper 15/2014. https://www.ufz.de/export/data/global/77689_Hochwasservorsorge%20und%20Schutzgerechtigkeit_UFZ_final.pdf. Accessed 9 Aug 2016

Kuhlicke, C., & Demeritt, D. (2016). Adaptive and risk-based approaches to climate change and the management of uncertainty and institutional risk: the case of future flooding in England. Glob Environ Chang, 37, 56–68 (2016). doi:10.1016/j.gloenvcha.2016.01.007

Marcuse P, Conolly J, Nocy J, Olivo I, Potter C, Steil J (eds) (2009) Searching for the just city. Debates in urban theory and practice. Routledge, London/New York

Maxwell S, Estelami E (eds) (2009) Behavioural pricing. J Product Brand Manag (Special Issue) 18(7)

OECD - Organisation for Economic Co-operation and Development (2009) Managing Water for All. An OECD perspective on pricing and financing. OECD, Paris

Penning-Rowsell EC, Pardoe J (2012) Who benefits and who loses from flood risk reduction? Environ Plann C: Gov Policy 30(3):448 466

Pitt M (2008) Learning lessons from the 2007 floods. The National Archives. http://webarchive.nationalarchives.gov.uk/20100807034701/http://archive.cabinetoffice.gov.uk/pittreview/_/media/assets/www.cabinetoffice.gov.uk/flooding_review/pitt_review_full%20pdf.pdf. Accessed 15 Feb 2017

Rawls J (1971) A theory of justice. Clarendon Press, Oxford

Rothstein H, Huber M, Gaskell G (2006) A theory of risk colonisation: the spiralling regulatory logics of societal and institutional risk. Econ Soc 35(1):91–112

Stiglitz JE, Walsh CE (2006) Economics, 4th edn. W.W. Norton & Co, London/New York

Strunz S, Gawel E, Lehmann P (2016) The Political Economy of Renewable Energy Policies in Germany and the EU. Util Policy 42:33–41. doi:org/10.1016/j.jup.2016.04.005

Taylor DE (2000) The rise of the environmental justice paradigm: injustice framing and the social construction of environmental discourses. Am Behav Sci 43(4):508–580

UN – United Nations (2015) Transforming our world: the 2030 agenda for sustainable development. Resolution adopted by the General Assembly on 25 September 2015, A/70/L.1. http://www.un.org/ga/search/view_doc.asp?symbol=A/RES/70/1&Lang=E. Accessed 22 July 2016
Walker G (2012) Environmental justice: concepts, evidence and politics. Routledge, London
Wolch JR, Byrne J, Newell JP (2014) Urban green space, public health, and environmental justice: the challenge of making cities 'just green enough'. Landsc Urban Plan 125:234–244. doi:10.1016/j.landurbplan.2014.01.017

On the Connection Between Urban Sustainability Transformations and Multiple Societal Crises

Annegret Haase, Norman Bedtke, Chloe Begg, Erik Gawel, Dieter Rink, and Manuel Wolff

1 Introduction

Crises are a frequent companion of global urbanization; within the last decades, a wide range of crises has influenced the fates of cities and their inhabitants. When reviewing the first two decades of this century, we might feel that we are living in a "time of constant crises"; starting with the real estate and financial crisis in 2007/2008, a wide range of subsequent crises has spread all over the world: various economic, real estate, and bank crises, as well as the Euro debt crisis, which have led to political crises with significant social repercussions, also across Europe (Funke et al. 2015). In addition, a number of wars and violent conflicts such as, e.g., those in Syria and Northern Iraq have caused new crises; in particular, the increased refugee migration towards Europe (preferably to large cities of some European countries), an area that still has to cope with the consequences of the above-mentioned crises. Multiple environmental crises are receiving much less attention at the moment (Gawel 2014). These include biodiversity loss or climate change that evokes extreme events, such as floods or droughts that also regularly impact on cities or urban regions. Last but not least, the increasing demographic polarization of the world's population, with population decrease and ageing in the developed countries and further population growth in the Global South, are perceived as a long-term crisis that is leading, e.g., to increasing south-north migration at a global scale. Given a situation of "multiple crises", i.e., crises that differ in their nature but

A. Haase (✉) • C. Begg • D. Rink • M. Wolff
Department of Urban and Environmental Sociology, Helmholtz Centre for Environmental Research-UFZ, Leipzig, Germany
e-mail: annegret.haase@ufz.de; chloe.begg@ufz.de; dieter.rink@ufz.de; manuel.wolff@ufz.de

N. Bedtke • E. Gawel
Department of Economics, Helmholtz Centre for Environmental Research-UFZ, Leipzig, Germany
e-mail: norman.bedtke@ufz.de; erik.gawel@ufz.de

S. Kabisch et al. (eds.), *Urban Transformations*, Future City 10,
DOI 10.1007/978-3-319-59324-1_4

are characterized by many interlinkages concerning their roots and impacts, we can observe emerging crisis-driven policies. These policies vary according to the type of crisis: bank rescues and austerity policies to tackle the financial crisis, diplomacy and border closures to regulate the immigration crisis, massive international climate-protection agendas and meetings to address the climate crisis that are characterized by particular interests and symbolic declarations, and so on.

These policies, however, are in sharp contrast to discussions on the need for sustainability transformations in our societies. What is more, sustainability and climate change policies are based on normative principles and largely ignore the context of multiple crises. A transformation literature has developed that, however, does not reflect the role and impact of multiple crises and their role in urban transformations towards sustainability.

A prominent example of this body of literature, the report of the German Scientific Council on Global Change (WBGU 2011), coined the temporal framework for a transformation to the so-called "low carbon society" within the next decades, until 2050, as a process that is occurring and being shaped by changes in our cities or urban environments. The report, in fact, considers the case of a failure of the transformation, which would lead to a "climate crisis", but it completely ignores existing crises (such as demographic ageing, the financial crisis since 2008, and the Euro debt crisis) and their impacts on ongoing transformations. In fact, the most recent WBGU report on urban transformations (WBGU 2016; Rink et al. 2015) considers crises to be important for the realization and continuation of transformations, but it does not address the relationship in depth.

Assuming that the outlined transformations will create new conflicts and crises, any transformation concept or theory should reflect actual and potential crises systematically: On the one hand, transformations towards sustainability are often caused by conflicts or crises, e.g., crises that operate as a "wild card" and trigger new policies such as the German "*Energiewende*" (exit from energy production based on nuclear and fossil fuel) in the wake of the nuclear accident in Fukushima, Japan. In this case, the crisis worked as a driver of transformation. On the other hand, various crises such as financial, debt, or economic crises and crisis-driven (austerity) policies may lead to a re-shaping of priorities that define the political priority of issues such as climate adaptation and sustainability as second rank, in comparison to economy or labour market policies, or "of lower priority". Here, the crisis represents an obstacle or barrier for a transformation towards sustainability, and originally prioritized goals, such as climate protection or renewable energy generation, might be undermined by the necessity, or the frequently quoted "lack of alternatives", to cope first and foremost with the crisis. In such a case, the development might even lead to a situation that questions the procedure and feasibility of the intended transformation. In effect, the relationship between (multiple) crises and transformation is substantial but variegated and poses a number of questions.

Set against this background, this chapter seeks to shed light on the connections between (multiple) crises and current as well as potential transformations of cities towards sustainability. The urban scale of multiple crises is insofar relevant because cities are densely populated areas, economic hubs, centres of political power and innovation and "labs of societal change"; according the UN, in 2014, 54 percent of the world's population lived in cities, towns, and urban areas.[1] This chapter explains how cities affect and are affected by crises in three ways: Firstly, they drive crises, due to their need for resources, and they often participate actively in the creation of crises (e.g., as speculating actors in the financial crisis 2008 and beyond); secondly, cities are extremely vulnerable to the consequences of crises due to their nature as population, economy, and infrastructure hubs. Thirdly, cities play a decisive role in coping with crises, not least since most of the technical and non-technical solutions for crises are developed in urban contexts.

Given this situation, the chapter will, apart from revealing the mentioned connections between crises and the urban context, investigate the impact of multiple crises on sustainability transformations or the relevant policies. It will also look systematically at the underpinnings between crises and sustainability transformations in order to suggest more robust or resilient policies in terms of buffering or adequately responding to crises.

The chapter is explorative: it supports its conceptual bases and theses with examples of empirical research from different fields related to urban development. It looks at the relationship between crises and transformation from an interdisciplinary perspective that includes sociological, economic, scale-based (geographical) and governance/policy aspects. In doing so, the chapter seeks to demonstrate the complex interlinkages between sustainable transformations and multiple crises; it also seeks to highlight how crises (of different types) may affect transformations towards sustainability (e.g., either support or hinder). With regard to the examples from various fields of urban (sustainability) research, the paper will not provide great detail; rather, it intends to present some initial thoughts about the hitherto under-researched linkage between crises and sustainability transformations; empirical material is used mainly for illustration and support of our theoretical/conceptual arguments.

The key terms of the chapter – crisis and urban (sustainability) transformation – are defined as follows: According to Booth (1993, p. 86), a crisis is "a situation faced by an individual, group or organization which they are unable to cope with by the use of normal routine procedures and in which stress is created by sudden change". Based on this, when we speak of a *crisis*, we mean the culmination or dramatic intensification or aggravation of problems or phenomena that may endanger societal cohesion or a societal system, or bring a society to the brink of destruction (Leusch 2014). Crises can have different scopes (long-term or short-term, creeping or sudden) and can play different roles/have different impacts at different scales (global, regional, local) (for a survey, see, e.g., Keown-McMullan 1997; Shaluf et al. 2003; Birkland 2006). In our chapter, we refer to literature that operates either directly with the term "crisis" or with related terms such as "disaster", "catastrophe" or, even more generally, "potential focusing events" (Birkland 1997). We are in line with Birkland (2006, p. 4), who states: "The distinction between crisis, disaster, and catastrophe is useful, but the line between what constitutes a disaster and what constitutes a crisis is unclear." Thus, instead of using a clear-cut distinction between similar terms (see, for example, Faulkner 2001, p. 137), we

[1] http://www.un.org/en/development/desa/news/population/world-urbanization-prospects-2014.html (accessed 30 August 2016).

decided to define crisis and stick to the defined context. We are looking here at different types of crises, including climate change, demographic change, energy crisis, financial/debt/austerity crisis, ecological crisis, etc.

When we speak of *transformation*, we mean qualitative, (comprehensive/far-reaching/deep) changes that have different dimensions: (1) the act or process of transforming; (2) the state of being transformed; (3) change in form, appearance, nature, or character. Because our focus is on urban transformations towards more sustainability, we are in line with Kabisch and Kuhlicke (2014, p. 478) who argue that "our understanding of urban transformation is based on the normative idea of sustainability and the necessary alterations to physical, material as well as social processes that urban areas need to undergo in order to achieve this aim. More specifically, we understand sustainable urban transformation as 'structural transformation processes – multidimensional and radical change – that can effectively direct urban development towards ambitious sustainability goals'" (McCormick et al. 2013, p. 1). Being aware of the fact that both terms, transition and transformation, are frequently used interchangeably in the debate on sustainability futures and are, in single strands of debates, especially in the political sciences literature (Brand 2016), used distinctively, we decided to use both terms for the theoretical perspective, according to how they are used in different debates and by different authors and, considering the terminology of this book, stick to transformation in a more formal way in the example, discussion, and conclusion sections of this article. As mentioned above for the case of crisis, we included literature based on how transformation and transition are defined and not according to which terms are actually used.

2 Theoretical Perspectives: Connections Between Transformation and Crises

In this section, we expand on the transformation literature and its link to crises. Even though a systematic treatise on the connection between crises and (urban) (sustainability) transformations is lacking, much theoretical and empirical literature is devoted to underlying relevant aspects. The following review demonstrates that the effects of crises on transformation processes (and vice versa) are complex and diverse. It seems appropriate, for a first approximation, to distinguish between crises as either drivers of or obstacles to transformations and to also consider ambiguous impacts (see Table 1 for an overview).

The most obvious link between crises and transformation processes is that a crisis can be a powerful driving force for change (political, institutional, social) because it is essential for sustainability transformations. An example in the transition theory literature is the widely accepted multi-level perspective (MLP) framework, which distinguishes between niches (form the micro-level, where radical novelties emerge), socio-technical regimes (the locus of established practices and associated rules), and a socio-technical landscape (the exogenous environment that usually changes slowly and influences niches and regime dynamics, e.g., climate, demo-

Table 1 Relationships between crises and transformation (selection)

Category	Author(s) and main statement(s)	Context
Crisis as driver	Alesina and Drazen (1991): "crises begets reform"-hypothesis	Economic crisis and economic reforms
	Drazen and Easterly (2001): Clarifications on hypothesis • Magnitude of the deterioration is important • Perception of crisis required • Acceptance of uncertain policy results increases with severity of crisis	Crises in general
	Empirical support for hypothesis: • Fidrmuc and Tichit (2013): Post-socialist reforms; crises serve as a catalyst for reforms • Rochet et al. (2008): Reorganization of strongly change-resistant public organizations as result of crises • Gawel (2014); Gawel et al. (2016): Environmental crises drive political action ("issue attention cycle", public choice approach) • Campos et al. (2010): Political crises are more important determinants of structural reforms than economic crises	Multiple crises/ environmental crisis
	Geels (2013): • Particularly the simultaneous occurrence of several crises (e.g., financial, socio-economic, environmental crises) can foster sustainability transitions • The financial–economic crisis may lead to socio-political protest for institutional reform and may trigger a shift to a new (green) growth path	(Financial) crisis and sustainability transition
	Geels (2002): Crises may destabilize socio-technical regimes	Sustainability transition
	North (1990, 1995): Institutional change occurs wherever it promises to be profitable; link to theories of public choice, e.g., "issue-attention cycle" (Downs 1972)	Institutional and policy change in general
	Birkland (2006): About "dynamics of policy change after sudden events known as focusing events"	Policy change in general
Crisis as an obstacle	Geels (2013): Fiscal and economic problems as central political issues may hinder or delay sustainability transitions	Financial crisis and sustainability transition
	Fidrmuc and Tichit (2013): Economic tensions undermine support for pro-reform parties (voting out)	Economic crisis and economic reforms
Ambiguous results	Prittwitz (1990, 1993): Policy action is less oriented to the extent of a crisis, and more to the available resources/political capacity	Environmental crises
	Fuller (2010): "crisis talk" used as a political tool to pursue individual/personal targets	Crises in general

Source: compiled by the authors

graphic trends) (Rip and Kemp 1998; Geels 2002). The MLP regards transitions as an integrated process at these levels that is characterized by: (a) stabilization of the niche innovation; (b) external environmental changes at landscape level, which create pressure; and (c) weakening of the existing regime (Geels 2002). Crises that occur at the landscape level (e.g., climate crisis) may put pressure on existing socio-technical regimes and, therefore, lead to windows of opportunity for change and novelties. As Geels (2013) emphasizes, particularly the simultaneous occurrence of several crises (e.g., financial, socio-economic, environmental) can foster sustainability transitions driven by the recognition of fundamental problems and the need for necessary, far-reaching political reforms. Such a "crisis begets reform"-hypothesis is also supported by the economic literature. Alesina and Drazen (1991) argue that (economic) reforms are often postponed because of a "war of attrition" over the distribution of reform costs. Reform costs often affect only a part of society and unaffected groups will oppose reforms until the costs of postponing a reform exceed the cost of implementing it. This is in line with theories of institutional change. According to North (1990, 1995), institutional change occurs wherever it promises to be profitable. In an institutional equilibrium, where none of the social and political actors find it advantageous to devote resources to restructuring the prevailing institutions (North 1990), crises may have the potential to change the underlying incentive structures for institutional change and reforms. A similar view is offered by public choice approaches: Here, the "issue-attention cycle" (Downs 1972) is driven by publicly perceived crises and crisis debates that change, sometimes abruptly, politicians' supply of "regulation markets" and environmental policy (Gawel 2014; Gawel et al. 2016). However, for Drazen and Easterly (2001), the "reform follows crisis"-argument seems to be tautological, because a reform policy generally requires a state where change is necessary. In their opinion, the magnitude of deterioration in the status quo is crucial. How the need for reform is perceived is vital for change, and a severe crisis serves as a potential trigger. Therefore, Drazen and Easterly argue that problems have to accumulate and become severe in order to induce reforms. In such a situation, people realize that there is a permanent problem that requires major policy changes. As a result of crisis, they are more likely to accept these policy changes and reforms with uncertain outcomes. Furthermore, the crisis may be able to weaken the power of interest groups that block reforms. Empirical support for the driver hypothesis is found in numerous studies on reforms and institutional change in different contexts. Within the scope of post-socialist reforms, crises serve often as catalysts of reform and institutional change, whereby the severity of the crisis has a positive impact on the subsequent pace of the change (Fidrmuc and Tichit 2013). For example, the fundamental change in Germany's energy policy is a direct consequence of the Fukushima nuclear crisis (Strunz 2014). Another example is that political changes in flood risk management in Germany result from several extreme flood events in rapid succession during recent decades (Gawel et al. 2016). Rochet and colleagues (Rochet et al. 2008) offer various examples of radical change, as result of a crisis, in strongly change-resistant public organizations (e.g., NYPD, French Forestry Commission).

Less obvious, but equally important: A crisis can also be an important barrier to change and therefore to transition. Geels (2013) offers a compelling example: Global or local financial crises and environmental problems are competing for both political and societal attention. Due to different time frames, causes of and solutions to a financial crisis may hinder (ecologically oriented) sustainability transitions by changing social and political priorities. Voters will usually reward a policy that, first of all, supports their social security, which is why political priorities are often tied to fiscal and economic rather than on environmental issues. By contrast, financial cuts and utility losses, which regularly occur as result of reforms (at least in their early stages), will lead to opposition. Earlier post-socialist transformations have shown that economic tensions undermined the support for pro-reform parties (Fidrmuc and Tichit 2013).

Apart from these obvious effects, the influence of crises on transformation processes is often ambiguous. For instance, it is plausible that significant challenges (severe crises or catastrophes) can sometimes be neither tackled nor communicated by policymakers, whereas relatively "small" problems trigger specific political action. One example is air pollution, which is addressed by environmental policy at a time when significant improvements have already been achieved. Another example is the dioxin scandal: high dioxin levels in eggs received attention whilst values in meat that were up to 10 times higher were ignored (Prittwitz 1990). Prittwitz (1993) explains this "catastrophe paradox" as the political capacity to react: political action, in his opinion, does not depend primarily on the acuteness of a problem, but rather on the resources available for reacting. Furthermore, the occurrence of crises has to be distinguished from "crisis talk": difficult situations are framed by interest groups as a crisis (even though a crisis does not actually exist) in order to ensure that institutional change is beneficial to them (Fuller 2010).

Based on the previously mentioned considerations, Fig. 1 summarizes our understanding of the (main) relationships between crises and political transformation processes. The greater the perception of problems, the higher is the likelihood of political reforms due to the political pressure to act. Therefore, especially urgent crises might act as drivers for change. However, this does not necessarily mean that (radical) reforms foster a transition towards sustainability. Inappropriate, symbolic political actions may also result. Likewise, resulting policy changes might turn out to be inadequate with respect to the sustainability challenge. At the same time, urgent crises may be an obstacle, due to changed political priorities or the lack of political capacities to react. However, creeping crises (crises which have not reached their peak) do not usually trigger political action. If these kinds of crises are perceived at all, it is much more likely that they will result in symbolic political actions or will become the basis for crisis talk. In any case, the triggering of political action might also be merely "crisis talk", initiated by interest groups that are able to set the agenda and to frame the public perception of an issue as a "crisis", in order to meet their special interests.

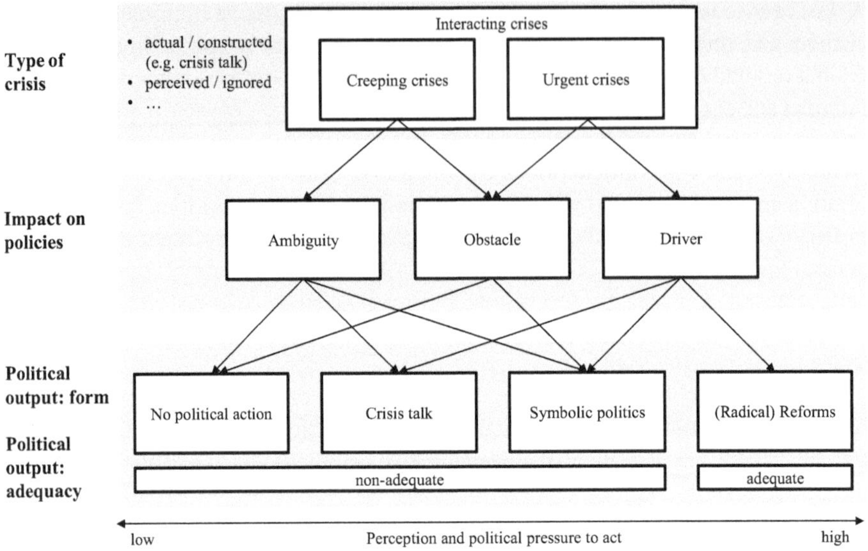

Fig. 1 Relationships between crises and political output (Source: authors' compilation)

3 Connection Between Urban Sustainability Transformation and Multiple Crises, Based on Empirical Evidence from Various Arenas of Urban Development

3.1 Introduction to the Urban Context and Examples

In this section, we will operationalize our theoretical reflections with the help of examples from our research and referring to the urban context.

As already noted in the introduction, cities are cases in point that demonstrate the complex relationship between transformation and crises. Cities may be drivers of crises (heat islands, hot spots of energy and water use) or places specifically affected by or suffering from crises and their consequences (e.g., financial or debt crises but also heat waves, water and energy scarcity, cities as hubs of vulnerable populations, etc.). In some cases, cities are both source (or driver) and victim of a crisis at once. This applies for the case of massive resource use by cities and the subsequent scarcity of these resources (e.g., land, sometimes drinking water) but also for the active role of city governments as actors in financial deals and speculations that contributed to the crash in 2008, which caused those cities to suffer or go bankrupt. As for the future, the major decisions on the reachability of global sustainability will be made in cities, also for areas outside the urban realm; sustainability is meant here in both a narrower, environment-based sense and in a wider understanding including political stability, peace, and democracy. Cities are "labs of innovation" and the places from where most of the strategies, technologies, and tools that help to make our lives more sustainable are coming. In recent years, a plethora of labels describing sustainable urban futures has developed (green cities, eco-cities, smart cities etc.; for an overview, see de Jong et al. 2015).

In keeping with this context, in the following, we first briefly present examples of crises as drivers or obstacles for sustainable urban development and then describe two examples in-depth in which urban contexts are contested by crises: the urban water infrastructure crisis and the financial crisis of 2008 and its consequences. The examples include different types of crises, impacts on policies and forms of political output, as summarized in Fig. 1.

3.2 *Examples of Crises' Impacts on Urban Sustainable Development*

A broad spectrum of examples for the function of crises as drivers of or as obstacle to (sustainable urban) transformation exists. While some environmental crises, such as flood events or the nuclear catastrophe in Fukushima, have functioned as drivers for transformations (of disaster/risk management and national energy policy in Germany), others, such as political crises and warfare in various regions of the world, have endangered transformations towards more wealth and a more just distribution of economic and environmental goods and burdens on a global scale. Economic crises regularly question pathways towards sustainability through altering the focus and tying up finances and resources that were originally planned for reaching sustainability goals.

In most cases, however, the relationship between crises and transformation, or, more precisely, the impact of crises on transformations, is anything but linear or one-dimensional – it is complex. Its complexity is shaped by and dependent on numerous framing and context factors as well as the actions of the different actors involved. The two examples, which we will expand on now, demonstrate this.

3.2.1 Urban Water Infrastructure Crisis

Water services provided through (technical) infrastructure systems have a considerable impact on the quality of life, resilience, and resource efficiency and, as such, on the level of sustainability in urban areas. Currently, the inflexible pipe-dependent central water supply and disposal systems in developed countries are increasingly exposed to uncertain, changing conditions such as substantial climate and demographic changes as well as more demanding requirements related to economic and technical resource efficiency (Grafton et al. 2015 for an overview). In particular, the increasing underutilization of capacities has the potential to generate severe functional (e.g., recontamination of drinking water as a result of long retention times, deposits in the sewers, premature wear and tear) and financing problems (significant loss of revenues or acceptance of increasing water prices) that already provide, in some places, an impression of the anticipated "infrastructural crisis". For instance, altered demand structures and population decline have already led to a water infrastructure usage of only 30 percent in some parts of Eastern Germany (Moss 2008; Koziol 2004). A number of new (often more decentralized) technological options and approaches that are better able to cope with uncertain conditions as well as with new requirements for flexibility and resource efficiency has been discussed or some time (Larsen and Gujer 1997; Brown et al. 2009; Howe et al. 2011). Such a technological paradigm shift is a major challenge for urban planning and infrastructure regulation because it questions many established practices and routines as well as institutionalized decision- and policy-making processes. However, in practice, we observe an institutional inertia and adherence to old structures of a "modern infrastructure ideal" (Graham and Marvin 2001) that are characterized by centralization, large technical systems, and predominantly public provision. As a consequence, water supply and wastewater disposal is characterized across many countries by a strong inertia to adapt and low willingness to innovate (Brown and Keath 2008; Kiparsky et al. 2013). Thomas and Ford (2005) label this situation as a "crisis of innovation in water and wastewater". Although we are now heading towards a crisis of water infrastructures in many cities, it is a creeping crisis and currently invisible to the general public, due to the (still) high quality of water services. Therefore, the crisis is not sufficiently severe to induce sustainability reforms, especially because the system is characterized by path dependency (self-reinforcing mechanisms, which are likely to generate a lock-in; see Arthur 1989) and by individual interests of established, powerful players (see Bedtke and Gawel, "Linking Transition Theories with Theories of Institutions – Implications for Sustainable Urban Infrastructures Between Flexibility and Stability", in this volume). This may change in the future, when the crisis abruptly becomes visible, e.g., in the form of urban flash floods (as a consequence of insufficient capacities for heavy precipitation events and overload of sewage systems) or requirements that households boil their drinking water (to cope with poor water quality as a result of long retention times in the pipes). Until then, the political output may be limited to forms of crisis talk, for example, when water suppliers argue that an orientation towards more economic efficiency in the sector (by regulation and elements of competition) will inevitably imply a decrease in quality. This, in turn, may also lead to symbolic political acts such as the "German modernization strategy for the water sector", which is inappropriate for addressing efficiency goals because of design weaknesses and implementation problems and, furthermore, because it neglects other sustainability concerns (Gawel and Bedtke 2015).

3.2.2 Financial Crisis and Urban Austerity: Trade-Offs Between Environmental, Social, and Financial Sustainability

The 2007/2008 global financial crisis and its consequences have become one of the most discussed crises of the last decades. It started with real estate crises, such as the subprime crisis in the US or the real estate bubble in Spain, which is quite typical for financial crises. The crisis began in cities, and cities suffered most from its consequences. It had a prehistory in the piling of public debts in the Euro zone by a number of countries, which culminated in the near bankruptcy of these states and their banks. As a result of the bank rescues, two approaches were followed: on the one hand, countries tried to support their economies with plans to stimulate investments, on the other hand, in many countries of the Euro zone and at all political and administrative levels, austerity policies and measures that included cuts in pensions, social and other welfare spending were also established (IILS 2011). Financial sustainability was thus, first disturbed by the debts and, later on, by the policy response to counteract the debts; both processes, which are clearly distinguished here as written descriptions, have been, in reality, multiply entangled. Numerous cities – or, more precisely – particular cities, notably weak cities or cities in countries hit hard by the crisis, have suffered from the crisis and will probably continue to do so in the medium and long run. This is due to 'scalar jumping': "cities are [...] where austerity bites [...] the projection downward of these pressures establishes a socially regressive form of scalar policies – with cities poisoned at the shark end" (Peck 2012, p. 629). Austerity measures that had been occasionally introduced as a precaution, not only in those countries that were most markedly affected by the crisis, have impacted on the social and cultural infrastructures of cities; cuts and closures – even if of a different scope – have been the main consequences (Vila and Freixes 2015; Matsaganis 2013). As a result, both processes – the increased state indebtedness and the austerity measures to counteract these debts – have led to a decrease in the general welfare level, in the individual quality of life of the affected urban inhabitants, and an increase in poverty. Cities became a symbol of instability and impoverishment, because of, e.g., evictions in Spain or mass out-migration of young and well-educated people from the GIPS countries (Greece, Ireland, Portugal, Spain), and there, mainly from large cities. Many cities fell into a debt trap because they were embedded in the global financial system through credits and ratings and, thus, inseparably linked to the national policies that – until the late 2000s – supported or tolerated increasing debts. The bankruptcy of cities as a result of indebtedness and impoverishment is a global phenomenon that can also be found outside the Euro zone: the bankruptcy of the city of Detroit in 2013 became a symbolic case of the impacts of financial crises on cities (Chung 2015). This case also makes it clear that poor, economically weak cities, such as, e.g., shrinking cities, are more affected by crises than growing or economically successful cities. Based on this situation, a two-fold "crisis talk" has emerged: On the one hand, austerity measures and cuts in welfare are promoted as TINA (**t**here **i**s **n**o **a**lternative) policies and as being inevitable to keep "the system" working. On the other hand, the opponents of austerity measures usually exclude the process of indebtedness and the lack of responsibility for financial sustainability shown by the affected governments from their mode of talking.

As a consequence of decade-lasting indebtedness and subsequent austerity policies, sustainability goals or the welfare needs of marginalized or weak social groups often become of secondary importance, because the priorities are located in the consolidation of markets, the economy, and budgets. The goals are either completely cancelled or they are postponed and subordinated to the "crisis-related priorities". Even if market consolidation and a stabilization of national economies doubtlessly contribute (in the long run) to a more solid base for an overall welfare system and general sustainability, in the short term, many of the introduced measures had bad consequences for many population groups that were most affected by immediate results of the crisis, including unemployment, cuts in social spending, etc. Even if "financial responsibility", as a background of austerity and as an important part of the crisis talk, seemingly supports the sustainability concept, the measures currently employed to achieve financial stability are, in fact, a continuation and acceleration of privatization and market-friendly reconfiguration of national assets and economies. It is debatable whether these measures will really lead to economic, financial and – as a consequence – social sustainability.

The example of financial crisis and the associated austerity policy supports the view of crises as a hindrance for sustainability transformation (Geels 2013). It addresses the question of how seemingly contradictory goals such as austerity and (comprehensive) sustainability can be brought into accordance at all, and it challenges the feasibility and sense of current coping strategies, such as cuts in expenses and welfare, without discussing fundamental changes of the economic and financial system that created the crises. The example of austerity policy and the opposing rhetoric about it (either totally in favour or against, see above) in the public and the media might be also an example of "crisis talk" (Fuller 2010) that should help to legitimize long-term cuts and make them the new 'normality'.

Last, but not least, the financial crisis and austerity conditions also have an impact on political shifts and stability: A recent study confirmed that financial crises, in particular, result in an upswing of right wing populism and have a strong potential to destabilize political systems (Funke et al. 2015). Thus, financial crises and austerity regimes responding to them might be a case of a crisis that brings about other crises in its wake.

3.3 Cross-Case Assessment

Cross-referencing our cases, following the heuristics introduced in Fig. 1 above, we can identify *several issues related to the types or character of crises, responses to crises and policy formulation, as well as contextual factors impacting on the perception and response to crises.*

It is obvious, that *different types of crises* exist. In the case of the urban water infrastructure sector, the crisis is creeping and, to date, predominantly invisible to the public. But it is already foreseeable that the situation has the potential to become a severe crisis if appropriate actions are not taken. However, so far, the crisis is not trigger for a transformation of the sector, which is characterized by high stability and inertia. By contrast, urban austerity, which had been a creeping development for a long time, suddenly became a severe financial crisis from 2008 onwards. The austerity measures, introduced by many countries when the situation worsened (another confirmation of the argument by Drazen and Easterly 2001) may be a hindering factor for (sustainable) transformations (Geels 2013). Austerity, as the new, predominant "crisis talk", often changed certain political priorities (e.g., save money and spend it first for jobs and investment while cutting down on other spending), which could have negative consequences for other policy fields, e.g., climate protection, resource protection, social inclusion and justice (see argument of Fuller 2010).

Consequently, we also have *different impacts of the described crises on (transformative) policies*. A fundamental change of the technological approaches in the urban water infrastructure sector would also produce a need for numerous, far-reaching reforms of institutions and policies. However, since the crisis is creeping, the consequences are ambiguous and have not lead to a change in policies so far. The lack of financial sustainability (debt and austerity crises) has to be seen, in general, as an obstacle to an overall sustainability transformation. Particular in the case of financial sustainability, the phenomenon of "crisis talk" plays an important role in framing and guiding public action in the field.

With respect to the *policy output*, several results are also observable. In the case of urban water infrastructures, necessary policies that are adequate for dealing with the sustainability challenges have not been implemented or even designed. On the other hand, symbolical political acts and "crisis talk" are performed to address the increasing pressure to provide water services efficiently. In the case of the financial crisis, many reforms were, in fact, introduced but there are many doubts about whether austerity policy leads to more sustainable outcomes or can prevent future crises in a system that, as such, remained untouched by coping policies.

The *adequacy of policy output* is also a complicated issue. The need for a technological change and far-reaching reforms of the urban water sector is obvious. Therefore, no political action, "crisis talk", or symbolical political acts are apparently adequate forms of political output for dealing with the challenges. In the case of financial crisis and associated urban austerity, the situation is as follows: while political decision-makers by and large see policy interventions as helpful and responsible, TINA policies and consequences are perceived as negative and disproportionally hard by many affected people; the way in which the issue is dealt with is also criticized, because it did not question the system as such.

Generally, the *severity of a crisis does not necessarily lead to transformative policies*. As the examples of water infrastructure and austerity policy show, systems or traditional settings are often not touched, even in the case of an acute crisis or when a creeping crisis threatens to become acute in the near future.

4 Conclusion

Crises play an important role in both the debate on and the framework conditions of urban transformations towards sustainability. This observation is supported by the theoretical reflections and empirical examples provided in this chapter. It does not apply only to the urban realm, although cities are meaningful examples that show the variegated relationships between crises and transformation. As we have demonstrated, there are variegated relationships between crises, their emergence, character, and policies, as a form of response, and the way they impact on sustainability transformations (in cities). These relationships range from crises being a push factor for transformation up to being an obstacle for (further or successful) transformation. In most cases, a variety of factors and context determine what the impact of a crisis on an on-going or insipient transformation could look like, exactly. With the help of our examples, we could show what is missing in the transformation debate, when crises are not considered as a factor or context. The debate on sustainable transformations should not be held in a kind of "laboratory mode", i.e., without being set into real world contexts. Here, crises have to be considered as repeating context factors, too. Discourses and institutions bringing forward the debate on sustainable transformation at a global scale, such as the WBGU, or the German Advisory Council on the Environment (SRU), or the UN, the Club of Rome, or the Global Agenda Council on Sustainable Development, should actively integrate the crisis perspective into their further dealings with potentials, limits, and conditions of future transformations. The relationship between crisis and transformation, which was highlighted in this article only in an explorative, mosaic-like way, should be systematically investigated and systematized.

In conclusion, we stress that the consideration of the factor crisis is important but needs dispassion and balance; it is surely best located somewhere between "taboo" and "alarmism". Crises should be addressed without falling into the trap of "crisis talk" or assuming a permanent crisis mode. Therefore, a clear definition of what a crisis is and how we define it seems to be crucial for a balanced and matter-of-fact discussion. As researchers, we have to critically assess whether we are facing a crisis context and, if so, how it is constituted, but also whether it is losing importance or has even ended.

Last, but not least, the temporality of crises requires some attention. As shown in the examples and theoretical reflections above, crises may last for a long time, be creeping and latent, or they can become acute or even emerge in a sudden or event-like way. In a discussion on their impact on long-term sustainability, acute crises might even be misleading, because they could blur the view for fundamental objectives and long-term challenges. Acute crisis policies have rarely led to sustainable solutions so far; they were driven more by the wish to find a short-cut out of the problem.

References

Alesina A, Drazen A (1991) Why are stabilizations delayed? Am Econ Rev 81(5):1170–1188

Arthur WB (1989) Competing technologies, increasing returns, and lock-in by historical events. Econ J 99(394):116–131

Birkland TA (1997) After disaster: agenda setting, public policy and focusing events. Georgetown University Press, Washington, DC

Birkland TA (2006) Lessons of disaster: policy change after catastrophic events. Georgetown University Press, Washington, DC

Booth S (1993) Crisis management strategy: competition and change in modern enterprises. Routledge, New York
Brand U (2016) "Transformation" as a new critical orthodoxy – the strategic use of the term "transformation" does not prevent multiple crises. Gaia 25(1):23–27
Brown R, Keath N (2008) Drawing on social theory for transitioning to sustainable urban water management: turning the institutional super-tanker. Aust J Water Res 12(2):73–83
Brown RR, Keath N, Wong THF (2009) Urban water management in cities: historical, current and future regimes. Water Sci Technol 59(5):847–855
Campos NF, Hsiao C, Nugent JB (2010) Crises, what crises? New evidence on the relative roles of political and economic crises in begetting reforms. J Dev Stud 46(10):1670–1691
Chung CS (2015) Zombieland/the Detroit bankruptcy: why debts associated with pensions, benefits, and municipal securities never die and how they are killing cities like Detroit. Fordham Urb Law J 41(3):771–847
de Jong M, Joss S, Schraven D, Zhan C, Weijnen M (2015) Sustainable – smart – resilient – low carbon – eco – knowledge cities; making sense of a multitude of concepts promoting sustainable urbanization. J Clean Prod 109:25–38. doi:10.1016/j.jclepro.2015.02.004
Downs A (1972) Up and down with ecology-the 'Issue-attention cycle'. Public Interest 28:8–50
Drazen A, Easterly W (2001) Do crises induce reform? Simple empirical tests of conventional wisdom. Econ Polit 13(2):129–157
Faulkner B (2001) Towards a framework for tourism disaster management. Tour Manag 22(2):135–147
Fidrmuc J, Tichit A (2013) How I learned to stop worrying and love the crisis. Econ Syst 37(4):542–554
Fuller C (2010) Crisis and institutional change in urban governance. Environ Plan 42(5):1121–1137
Funke M, Schularick M, Trebesch C (2015) Going to extremes: politics after financial crises, 1870–2014. Centre for Economic Policy Research, discussion paper 10884, London, UK. http://cepr.Org/active/publications/discussion_papers/dp.Php?Dpno=10884#. Accessed 20 Apr 2016
Gawel E (2014) Institutionen ökologischer Nachhaltigkeit in Zeiten der Krisen. In: Held M, Kubon-Gilke G, Sturn R (eds) Unsere Institutionen in Zeiten der Krisen (= Normative und institutionelle Grundfragen der Ökonomik, 13). Metropolis, Marburg, pp 67–95
Gawel E, Bedtke N (2015) Efficiency and competition in the German water sector between "modernization" and "regulation". J Pub Nonprofit Serv 38(2/3):97–132
Gawel E, Lehmann P, Strunz S, Heuson C (2016) A public choice framework for climate adaptation – barriers to efficient public adaptation and lessons learned from German flood disasters. J Inst Econ 12:1–27. doi:10.1017/S1744137416000163
Geels FW (2002) Technological transitions as evolutionary reconfiguration processes: a multi-level perspective and a case-study. Res Policy 31(8/9):1257–1274
Geels FW (2013) The impact of the financial-economic crisis on sustainability transitions: financial investment, governance and public discourse. Environ Innov Soc Trans 6:67–95. doi:10.1016/j.eist.2012.11.004
Grafton Q, Daniell KA, Nauges C, Rinaudo JD, Chan NWW (2015) Understanding and managing urban water in transition. Springer, Dordrecht/Heidelberg/New York/London
Graham S, Marvin S (2001) Splintering urbanism: networked infrastructures, technological Mobilites and the urban condition. Routledge, London
Howe CA, Vairavamoorthy K, van der Steen NP (2011) Sustainable water management in the city of the future. UNESCO-IHE, Delft
IILS - International Institute for Labour Studies (2011) A review of global fiscal stimulus, IILS Joint discussion paper series no. 5. IILS, International Labour Office (ILO), Geneva
Kabisch S, Kuhlicke C (2014) Urban transformations and the idea of resource efficiency, quality of life and resilience. Built Environ 40(4):497–507
Keown-McMullan C (1997) Crisis: when does a molehill become a mountain? Disaster Prev Manag: Int J 6(1):4–10

Kiparsky M, Sedlak DE, Thompson BH, Truffer B (2013) The innovation deficit in urban water: the need for an integrated perspective on institutions, organizations, and technology. Environ Eng Sci 30(8):395–408

Koziol M (2004) The consequences of demographic change for municipal infrastructure. German J Urb Stud 44(1)

Larsen TA, Gujer W (1997) The concept of sustainable urban water management. Water Sci Technol 35(9):3–10

Leusch P (2014) Der Begriff Krise ist eigentlich unbrauchbar http://www.deutschlandfunk.de/soziologiekongress-der-begriff-krise-ist-eigentlich.1148.de.html?dram:article_id=299913. Accessed 28 Apr 2016

Matsaganis M (2013) The Greek crisis: social impact and policy responses http://library.fes.de/pdf-files/id/10314.pdf. Accessed 13 Sept 2016

McCormick K, Anderberg S, Coenen L, Neij L (2013) Advancing sustainable urban transformation. J Clean Prod 50:1), 1–1),11

Moss T (2008) 'Cold spots' of urban infrastructure: 'Shrinking' processes in eastern Germany and the modern infrastructural ideal. Int J Urb Reg Res 32(2):436–451

North DC (1990) Institutions, institutional change and economic performance. Cambridge University Press, Cambridge, MA

North DC (1995) Five propositions about institutional change. In: Knight J, Sened I (eds) Explaining social institutions. University of Michigan Press, Ann Arbor, pp 15–26

Peck J (2012) Austerity urbanism. American cities under extreme conditions. Cities 16(6):626–655

Prittwitz Vv (1990) Das Katastrophenparadox. Elemente einer Theorie der Umweltpolitik. VS Verlag, Wiesbaden

Prittwitz Vv (1993) Katastrophenparadox und Handlungskapazität. Theoretische Orientierungen der Politikanalyse. In: Héritier A (ed) Policy-analyse: Kritik und Neuorientierung, vol 24. VS Verlag, Wiesbaden, pp 328–357

Rink D, Banzhaf E, Kabisch S, Krellenberg K (2015) Von der „Großen Transformation" zu urbanen Transformationen. Zum WBGU-Hauptgutachten Welt im Wandel. Gaia, 24(1):21–25

Rip A, Kemp R (1998) Technological change. In: Rayner S, Malone EL (eds) Human choice and climate change, vol 2. Battelle Press, Columbus, pp 327–399

Rochet C, Keramida O, Bout L (2008) Crisis as change strategy in public organizations. Int Rev Adm Sci 74(1):65–77

Shaluf IM, Ahmadun F, Mat Said A (2003) A review of disaster and crisis. Disaster Prev Manag: Int J, 12(1):24–32

Strunz S (2014) The German energy transition as a regime shift. Ecol Econ 100(1):150–158

Thomas DA, Ford R (2005) The crisis of innovation in water and wastewater. Edward Elgar, Cheltenham

Vila JL, Freixes T (2015) The impact of the crisis on fundamental rights across member states of the EU. European Parliament, Directorate-General Internal Policies of the Union. http://www.europarl.europa.eu/RegData/etudes/STUD/2015/510020/IPOL_STU%282015%29510020_EN.pdf. Accessed 13 Sept 2016

WBGU - German Advisory Council on Global Change (2011) World in transition – a social contract for sustainability. Report. German Advisory Council on Global Change, Berlin

WBGU - German Advisory Council on Global Change (2016) Der Umzug der Menschheit. Die transformative Kraft der Städte. Zusammenfassung. German Advisory Council on Global Change, Berlin

How to Measure Progress Towards an Inclusive, Safe, Resilient and Sustainable City? Reflections on Applying the Indicators of Sustainable Development Goal 11 in Germany and India

Florian Koch and Sohail Ahmad

1 Introduction

Cities[1] seem to be keystones within global policies towards sustainability. Urban areas are hot spots that drive environmental change at multiple scales (Grimm et al. 2008) and a large share of CO_2 emissions has urban origins (Sethi and Puppim de Oliveira 2015). At the same time, a global urbanization process that increases the total number, and also the share of urban dwellers worldwide, is taking place. Therefore, ambitious global goals for sustainability that do not consider urban areas seem to be predestined to fail and documents such as the UN's New Urban Agenda

[1] Even though a myriad of urban scholars have struggled to define what a city or an urban area actually is, and "the urban is not a pre-given, self-evident reality, condition or form" and "cannot be plausibly understood as a bounded, enclosed site of social relations" (Brenner and Schmid 2013, p. 19f), we chose, for this article, a political science perspective, defining cities as the political entities and territories limited through administrative borders within which local governments act. Nevertheless, we acknowledge that these entities are nested in multi-level political arrangements (Kübler and Pagano 2015), that public actors are not necessarily most crucial for the pursuit of collective goals (governance perspective) (Pierre and Peters 2015), and that other definitions – for example, urbanism as a way of life (Wirth 1938) and the city as the place where a specific form of living appears (Simmel 1903 and Koch 2011) – exist and reveal the shortcomings of defining cities through political borders. Still, for the sake of this article, which deals with public data and indicators, we stick to this "traditional" form of definition.

F. Koch (✉)
Department of Urban and Environmental Sociology, Helmholtz Centre for Environmental Research-UFZ, Leipzig, Germany
e-mail: florian.koch@ufz.de

S. Ahmad
Mercator Research Institute on Global Commons and Climate Change (MCC), Berlin, Germany

Department Economics of Climate Change, Technische Universität Berlin, Berlin, Germany
e-mail: ahmad@mcc-berlin.net

S. Kabisch et al. (eds.), *Urban Transformations*, Future City 10,
DOI 10.1007/978-3-319-59324-1_5

(UN Habitat 2016) highlight the need for profound changes in current urban developmental paths. These paradigm changes, also labeled *urban transformations towards sustainability*, demonstrate the normative dimension of urban transformations (see Rink et al., "Exploring the Extent, Selected Topics and Governance Modes of Urban Sustainability Transformations", in this volume). *Urban transformations* receive increasing political recognition, as demonstrated by the much-cited statement of UN Secretary General, Ban Ki Moon, that "our struggle for sustainability will be won or lost in cities" (UNESCAP 2014, p. 1) (Rudd 2015 later complemented, rightly, that this struggle "will be won or lost by cities"). That is why one of the 17 Sustainable Development Goals (SDGs, also known as Global Goals, UN 2015), which, together, form the 2030 Agenda, has an explicit urban focus (Goal 11: Make cities and human settlements inclusive, safe, resilient and sustainable). Cities are also important for the implementation of other SDGs. Goals such as "End poverty in all its forms everywhere" (Goal 1) or "Ensure inclusive and equitable quality education and promote lifelong learning opportunities for all" (Goal 4) also need to encompass the urban dimension of poverty and learning opportunities in cities, in order to be realized. Estimates based on the wording of the SDG zero draft indicate that 21% of the 169 targets of all 17 SDGs can only be implemented with urban stakeholders, 24% should be implemented with urban stakeholders, and a further 20% should have a much clearer orientation towards urban stakeholders, although current wording does not suggest this (Misselwitz et al. 2015). Nevertheless, this chapter focuses only on Goal 11, acknowledging the crosscutting character of cities for the other SDGs as well.

The argument that cities are increasingly seen as driving forces for reducing global environmental change and as facilitators of a more sustainable development globally is strengthened by ongoing political discussions: Besides the above-mentioned New Urban Agenda, also the UNFCCC's Paris Agreement, as well as the latest report of the German Advisory Council on Global Change (WBGU), highlights the power of cities for sustainable transformations (WBGU 2016). These perceptions of cities as solution-providers for global sustainability are relatively new: Whereas, for a long time, cities were considered mainly as polluters and threats to the environment, the pendulum has swung back and urban areas are now seen as one option for combining economic, social, and environmental development in a sustainable way. The Sustainable Development Goals play a crucial role in this process because the importance of cities for achieving global sustainability is acknowledged and, at the same time, targets and indicators for inclusive, safe, resilient and sustainable cities are made explicit. The predecessor of the SDGs, the UN Millennium Goals, were adopted in 2000 and focused on the developing world through goals such as halving extreme poverty, halting the spread of HIV/AIDS, and providing universal primary education. In contrast, the Sustainable Development Goals have a global character, including developing and developed countries. This implies that Goal 11 applies to cities of the Global South and the Global North equally, i.e., Goal 11 can be considered as a global normative framework for urban transformations.

As welcome as the intent to create global targets for sustainable cities worldwide is, it is nonetheless of crucial importance to also think about how to implement these goals. In this context, several issues, including governance aspects, funding, the negotiation of emerging trade-offs, as well as the general character of global agreements and responsibility, need to be considered. Implementation needs also to be measured and therefore indicators and the availability of data play a major role (Koch and Patterson 2015). In contrast to the Millennium Goals, a fundamental change about the data provision from the SDGs is that progress measurement – not only for Goal 11, but for all goals – should take place not only at national but also at subnational levels. This is a challenge for the official statistics (Schnorr-Bäcker 2016) because comparable data on the indicators mentioned in the SDGs are often more easily available on the national than on the city or regional level.

The UN's Statistical Commission, as well as the Inter-agency and Expert Group on Sustainable Development Goal Indicators (IAEG-SDGs), are currently discussing indicators and data availability for the SDGs. Regarding Goal 11, several adjustments have already been made (United Nation Statistic Division 2016), and further changes are expected. Nevertheless, agreement on the 10 core principles of the SDG indicators exists (Simon et al. 2016)[2]: Based on these principles, it is stated that, wherever possible, data from the UN institutions should be used. With respect to this point, the WBGU proposed that UN Habitat could play an important role regarding indicators for and monitoring of Goal 11 (WBGU 2016, p. 445). Nevertheless, the question of disaggregation, which is of special importance for Goal 11, is not yet fully resolved. Two key concerns exist:

1. As stated in the latest UN documents, the indicators should contain, as the disaggregation level, only a distinction between urban and rural on the national level (as well as disaggregation by sex and age). It is argued that disaggregation with other characteristics should be realized when relevant and possible, but without further specification.
2. It is not yet clear whether data on the level of urban agglomerations (i.e., the city and its surrounding area/the urban region) or from the city area, defined through administrative boundaries, should be used. Whilst UN statistics frequently use urban agglomerations as a statistical unit, most publicly used data refer to the area of administrative boundaries. This divergence would require a somewhat complex re-aggregation after the disaggregation is done in order to ensure comparability.

Based on the UN's principles of using well-established data sources and making data collection transparent, we investigate which publicly accessible data on the SDGs already exist on the city/urban agglomeration level. This is based on the idea that monitoring the process of the SDGs needs to be performed locally. Aggregated

[2]The 10 principles are: (1) indicators that are limited in number and globally harmonized, (2) simple, single-variable indicators with straightforward policy implications, (3) allow for high-frequency (annual) monitoring, (4) consensus-based, in line with international standards and information already collected by national and environmental/economic information systems, (5) constructed from well-established data sources, (6) disaggregated, (7) universal, (8) mainly outcome-focused, (9) science-based and forward-looking, (10) a proxy for broader issues or conditions

UN data on the proportion of urban populations living in slums, informal settlements or inadequate housing in all urban areas in one country do not indicate where the actual problem is and in which city action is needed. This also holds true for the other SDGs. As Satterthwaite questions, "What does it serve to know the proportion of a nation's territory that is public space? Or the national average in the time or distance to public transport?" (Satterthwaite 2016). In contrast, if it were possible to evaluate, for each city, how the various indicators of Goal 11 are fulfilled or what progress has been made, accountability as well as priorities on the urban level could be more easily addressed.

2 Aims and Approach

Because the SDGs have a global character and should be applied not only in developing but also in developed countries, two test case countries – India as an example of a developing and Germany as an example of a developed country – were chosen in order to analyze the SDG 11 indicators. This chapter looks at German and Indian cities with a focus on how the implementation of SDG 11 can be measured in these two contrasting contexts. The manifold differences between German and Indian cities, for example, those related to population development, social welfare, informal building activities, institutional capacities, or municipal competencies make India and Germany a testbed for whether the indicators are globally applicable. Those countries were chosen because the authors are familiar with the respective national context. First, we briefly review the relevancy and availability of data pertaining to targets referring to Goal 11. Thereafter, the availability of data for the indicators related to Goal 11 is examined. Methodologically, we refer to existing documents and official data for German and Indian cities (for Germany: data of the *Statistisches Bundesamt* as well as other publicly accessible data; for India: data from the Office of the Registrar General and Census Commissioner, and the National Sample Survey Office, among others), to our own experience with urban planning processes in both countries, as well as to secondary literature on research that has been done on the SDGs and their implementation.

The chapter builds on previous work of the authors related to the Issues Papers of the Habitat III process and the ISSC seminar on sustainable urbanization in Taipei, in November 2014 (Ahmad and Koch 2015).

This chapter describes Goal 11, the related targets and indicators, and evaluates the relevancy and existing data for German and Indian cities. Furthermore, we discuss the findings and highlight common problems as well as country- or city-specific obstacles.

3 The Urban Dimension of the Sustainable Development Goals

The targets of Goal 11 demonstrate, on the one hand, the complexity of urban sustainability and, on the other hand, the difficulties concerning the data, which should reveal whether the goal has been fulfilled or not. Table 1 provides a survey of data availability and reliability, as well as the relevance of the respective targets in the German and Indian contexts. We have divided the parameters (relevancy, availability, and reliability) of data referring to the target into low, medium, and high, based on the approach of Simon and Arfidsson (2015) and Simon et al. (2016). The relevance of an indicator depends on whether the related target is of importance for the respective national urban context. For example, in high-income countries with good provision of housing, the measurement of the proportion of the urban population living in slums, informal settlements, or inadequate housing is less relevant than in countries in which the major proportion of the population lives in precarious forms of housing. In the table, we classify the relevancy, availability and reliability with L (=low), M (=medium), and H (=high). The relevance of an indicator for the German and the Indian context was determined by consulting the drafts of the national Reports for Habitat III (SRL 2015; UTC 2015).[3] High availability is given if the data are easily accessible and usable for long-term monitoring. In order to track availability, we analyzed whether publicly accessible databases contain data on the respective indicator. Therefore, we cannot prove whether data on the selected indicators exist at all but we can estimate how easy available the data are. Reliability refers to the fact that gaps between official statistics and the actually occurring developments may exist, especially with regard to classifications in the so-called informal sectors. This may be the case for indicators that explicitly refer to informal (as yet unofficially documented aspects such as, for example, work relationships lacking legal definitions or housing construction outside of the existing formal regulatory framework) forms of development (Simon and Arfvidsson 2015).

4 Discussion

Earlier studies have pointed out priorities for the UN SDGs: devise metrics, establish monitoring mechanisms, evaluate progress, enhance infrastructure, and standardize and verify data (Lu et al. 2015). All of these priorities are still major challenges. More precisely and for urban contexts involving a comparative framework of five fairly representative cities (globally) and broadly relevant, acceptable, and practicable targets, Simon et al. (2016) conclude that each city faced problems

---

[3] The final reports haven't been published by the time we wrote this article (April 2016) and could therefore not be used as sources. Nevertheless, the reports have now been published and are available at www.unhabitat.org

Table 1 Targets and indicators of Goal 11 of the SDGs and data availability for German and Indian cities

			Data related to targets (either L (Low), M (Medium), or H (High)		Comments
Target	Indicator	Country	Relevancy	Availability and reliability	Sources
11.1 By 2030, ensure access for all to adequate, safe, and affordable housing and basic services, and upgrade slums;	11.1.1 Proportion of urban population living in slums, informal settlements, or inadequate housing	India	H	L	Census of India (2011), National Sample Survey (NSS), and Indian Human Development Survey (IHDS)
		Germany	L	L	No official data about slums/informal settlements exist in the German context. The German National Report for Habitat III mentions that no slums exist in Germany (SRL 2015, p. 77) and, therefore, that no need for data on these issues exists because the goal has already been achieved
11.2 By 2030, provide access to safe, affordable, accessible and sustainable transport systems for all, improve road safety, notably by expanding public transport, with special attention to the needs of those in vulnerable situations, women, children, persons with disabilities, and older persons;	11.2.1 Proportion of population that has convenient access to public transport, by sex, age, and persons with disabilities	India	H	L	The national travel survey is missing, and there are limited data sets on transport systems, especially disaggregated by sex, age, and person with disabilities.
		Germany	M	H	The draft for the German National Report to Habitat III indicates that 100% of the urban population has access to public transport, thus the target has already been achieved (SRL 2015, p. 78). Official data on transport issues only exist on the federal state level, partly without disaggregation levels.

11.3 By 2030, enhance inclusive and sustainable urbanization and capacity for participatory, integrated, and sustainable human settlement planning and management in all countries;	11.3.1 Ratio of land consumption rate to population growth rate	India	M(11.3.1)/H(11.3.2)	M(11.3.1)/H(11.3.2)	11.3.1 Good data sets available, as mentioned above in target 11.1. 11.3.2 The devolution of power at the city level is still a major milestone to achieve and, similarly, civil society participation. In this context, data are obviously limited.
	11.3.2 Proportion of cities with direct participatory structures in the civil society that operate regularly and democratically in urban planning and management	Germany	H (11.3.1)/L (11.3.2)	H (11.3.1)/L (11.3.2)	11.3.1 Good data set on land use change and population change available on the city level. 11.3.2 Direct participation is part of the national planning law; therefore, it can be assumed that 100% of German cities already fulfill this goal.
11.4 Strengthen efforts to protect and safeguard the world's cultural and natural heritage	Total expenditure (public and private) per capita spent on the preservation, protection, and conservation of the entire cultural and natural heritage, by type of heritage (cultural, natural, mixed, and World Heritage Center designation), level of government (national, regional, and local/municipal), type of expenditure (operating expenditure/investment), and type of private funding (donations in kind, private non-profit sector, and sponsorship)	India	H	L	Limited data available, and often not in the public domain. Details can be found at Archeological Survey of India (www.asi.nic.in)
		Germany	H	L	Very poor data availability for private spending for cultural and natural heritage, official data exist only for public spending on UNESCO World heritage sites

(continued)

Table 1 (continued)

Target	Indicator	Country	Data related to targets (either L (Low), M (Medium), or H (High)		Comments
			Relevancy	Availability and reliability	Sources
11.5 By 2030, significantly reduce the number of deaths and the number of people affected and substantially decrease the direct economic losses, relative to global gross domestic product, caused by disasters, including water-related disasters, with a focus on protecting the poor and people in vulnerable situations;	11.5.1 Number of deaths, missing persons and persons affected by disaster per 100,000 people	India	H	L	11.5.1 Limited data available from National Disaster Management Authority, Government of India, and also lacking in urban/ rural distinction and completeness.
					11.5.2 Data almost impossible to find, given that these issues have only limited insurance and losses are, therefore, not documented.
	11.5.2 Direct economic loss due to disaster in relation to global GDP, including disaster damage to critical infrastructure and disruption of basic services	Germany	M	H	11.5.1 Data on victims of disaster is available, but the official data do not indicate numbers of missing persons and persons affected by disasters
					11.5.2 Cooperation with insurance companies necessary, no official data exist
11.6 By 2030, reduce the adverse per capita environmental impact of cities, including paying special attention to air quality and municipal and other waste management;	11.6.1 Proportion of urban solid waste regularly collected and with adequate final discharge out of total urban solid waste generated, by cities	India	H	M	11.6.1 Weak data infrastructure for municipal governments; however, the Central Pollution Control Board (CPCB) can provide such data sets but not comprehensively.
					11.6.2 The CPCB provides good data regarding particulate matter.
	11.6.2 Annual mean levels of fine particulate matter (e.g., PM2.5 and PM10) in cities (population-weighted)	Germany	M	H	11.6.1. German law requires that urban solid waste is collected, data exist
					11.6.2 Cities are obliged to measure fine particulate matter; therefore, it can be assumed that data exist

11.7 By 2030, provide universal access to safe, inclusive and accessible, green and public spaces, in particular for women and children, older persons, and persons with disabilities	11.7.1 Average share of the built-up area of cities that is open space for public use for all, by sex, age and persons with disabilities	India	H	L	11.7.1 Master Plans of cities, but often not comparable; however ,RS/ GIS could be used.
					11.7.2 Limited data, including under- reporting.
	11.7.2 Proportion of victims of physical or sexual harassment, by sex, age, disability status, and place of occurrence, in the previous 12 months	Germany	M	M	11.7.1 Data on land use in cities exist. Due to the law, it can be assumed that open space can be used by all
					11.7.2 Data exist on sexual harassment against women (Europäischen Union für Grundrechte (FRA); thus, data are partially available
11.a Support positive economic, social and environmental links between urban, peri-urban and rural areas by strengthening national and regional development planning;	11.a.1 Proportion of population living in cities that implement urban and regional development plans integrating population projections and resource needs, by size of city	India	H	L	Required by Indian planning law, but great variation between provinces.
		Germany	H	L	Required by German planning law and therefore it can be assumed that all German cities implement such plans.

(continued)

Table 1 (continued)

Target	Indicator	Country	Data related to targets (either L (Low), M (Medium), or H (High)		Comments
			Relevancy	Availability and reliability	Sources
11.b By 2020, substantially increase the number of cities and human settlements adopting and implementing integrated policies and plans towards inclusion, resource efficiency, mitigation of and adaptation to climate change, resilience to disasters, as well as develop and implement, in line with the Sendai Framework for Disaster Risk Reduction 2015-2030, holistic disaster risk management at all levels;	11.b.1 Proportion of local governments that adopt and implement local disaster risk reduction strategies in line with the Sendai Framework for Disaster Risk Reduction 2015-2030	India	H	L	11b.1 Not available, but could be compiled. 11b.2 Some data exist; could be strengthened.
	11.b.2 Number of countries with national and local disaster risk reduction strategies	Germany	L	H	11b.1 No official data on the proportion of local governments that adopt and implement disaster risk reduction strategies 11.b.2 Disaster risk strategies already exist in Germany
11.c Support least developed countries, including through financial and technical assistance, to build sustainable and resilient buildings, utilizing local materials	11.c.1 Proportion of financial support to the least developed countries that is allocated to the construction and retrofitting of sustainable, resilient, and resource-efficient buildings utilizing local materials	India	L	L	Not applied directly for India. Nevertheless, India has supported neighboring countries, e.g., Nepal and Afghanistan. Data can be compiled.
		Germany	H	H	Data on financial support to the least developed countries exist on national as well as on European levels.

Source: Authors' assessment based on the UN (2015) and UNSTATS (2016) datasets

in providing all the data required; however, each city also proposed various changes to maximize the local relevance of particular targets and indicators.

The analysis of Indian and German cities has confirmed these results and, again, demonstrated the challenges concerning data for the SDG as well as the difficulties related to having universal, but at the same time, city-specific data. National or local data on all of the indicators for Goal 11 do not exist in either the German or in the Indian context. Even though some indicators (not only for Goal 11 but also for other SDGs) can be already measured on local levels (Schnorr-Bäcker 2015), the provision of data for all the indicators on a subnational and, even more, on an urban level remains challenging.

Of course, the provision of data on the indicators of the targets does not automatically guarantee that Goal 11 will be realized and cities will become "inclusive, safe, resilient and sustainable", but it is necessary to measure the progress towards this goal. If data on the indicators do not exist, Goal 11 becomes an issue of wishful thinking that some cities might, and others might not achieve. Through data monitoring, the necessary actions to avoid the non-achievement of the SDGs can be identified. As our analyses have shown, data on most of the indicators for Goal 11 do not exist, currently, for cities in Germany and India. Adjustments and modification of the official statistics are necessary, not only in order to provide data for Goal 11 but also for the other Goals (Schnorr-Bäcker 2015). From an urban point of view, we propose four major points that need to be considered in further discussions on monitoring and evaluating progress on the SDGs and, especially, on Goal 11.

- *Reach and limits of data:* The selected indicators try to sketch which issues need to be considered in order to make a city inclusive, safe, resilient and sustainable. It was beyond the scope of this article to question whether these indicators are the right ones for defining inclusive, safe, resilient and sustainable cities (for a detailed discussion of the SDGs, see ICSU and ISSC 2015, or Loewe and Rippin 2015). We acknowledge that these indicators can serve only as a first approximation towards what future urban development should look like; other issues, such as social coherence on an urban level, the institutional capacity of city–/region-wide forms of governance, or the strengthening of urban circular economies, also need to be considered.
- *Disaggregation level:* As stated above, a disaggregation of data that only distinguishes between rural and urban areas on the national level is not helpful for monitoring and developing concrete actions in order to achieve Goal 11. We suggest, therefore, that the focus of such datasets should be beyond the urban/rural disaggregation, but also city- specific. Only if city authorities can identify which target's implementation poses a problem, can specific actions be taken. Nevertheless, the indicators should not lead to a situation in which cities that have already achieved the targets should lean back and cease actions towards inclusiveness, safety, resilience and sustainability. Furthermore, it has to be acknowledged that not all targets of Goal 11 can be realized by local authorities; several require national politics (see, for example, targets 11.4 or 11.c). In cases where data on a city level are not accessible, data should be compiled on types of

cities, e.g., size of the city, or based on specific typologies of cities, e.g., coastal cities. Such a data compilation could help to develop urban public policy interventions for specific types of cities.

- *Transparency – open access data*: One of the targets of Goal 17 is to encourage civil society partnerships to strengthen the means of implementation and to revitalize the Global Partnership for sustainable development. In keeping with this line, we would like to emphasize the need for open access data on the SDGs' indicators. This is especially true for the city level, where public interest and civil society involvements may be greater than for issues that are discussed on a more abstract, global level. If civil society involvement should be facilitated in the implementation of the SDGs, free access to data that monitor the SDGs' implementation is needed. Civil society's access to data can be considered as a powerful way of feeding information on the implementation level of the SDGs back into the policy and political arena, to hold responsible stakeholders to account. If we do not create these sorts of 'feedback loops', based on civil society, how will we make sure that the SDGs are actually being implemented? This is in line with current discussions on adjustments and modifications of the official statistics and public access to SDG-relevant data (Schnorr-Bäcker 2015).
- *City and country specifics:* As our analysis has shown, the relevance of the various targets of Goal 11 varies and depends, to a high degree, on the respective national context. Therefore, we suggest that each country should assess the relevance, availability, and reliability of datasets for measuring and monitoring SDGs/targets and, thereafter, efforts should be made to create and compile the respective datasets. Nevertheless, this process should be made transparent and decisions about why some indicators are considered to be of greater importance than others should be explained. Cities should be encouraged to perceive the task of monitoring the implementation of the SDGs as municipal business and include them in their urban development strategy.

5 Outlook

Measurement is acknowledged to be crucial for monitoring and implementation of the SDGs and freely accessible data for each city, preferably divided between the city and the wider urban region, would be useful. However, considering the related costs and the restricted resources of municipal or regional statistical offices, not only but especially in developing countries, it is obvious that trade-offs between an expansive, but hardly realizable set of indicators and the focus on a more limited set of indicators need to be made (Cities Alliance 2015). Discussing these trade-offs should not be left to national and international institutions. Perspectives from urban stakeholders, as well as from cities and urban civil societies, are needed. Therefore, the UN Habitat III conference in Quito, characterized as the first implementation conference on the SDG agenda, seems to be an appropriate context to discuss these issues, and a close linkage between the Habitat III New Urban Agenda and the

SDGs, as envisaged, is fundamental (Cities Alliance 2015). Whether our cities have the adequate data and mechanisms to measure whether they are becoming more inclusive, safe, resilient, and sustainable thus depends on how urban stakeholders, such as local administrations, mayors, civil society and urban enterprises, but also urban scholars, raise their voices and become more visible in the global debates on the Sustainable Development Goals.

From a theoretical point of view, the SDGs and, especially, Goal 11 can be considered as a normative and universal vision of what cities should look like in 2030. Therefore, SDG Number 11 can also be read as global guideline for the minimum standards that a city should fulfill – irrespective of its location in the Global South or North. This makes the SDGs especially valuable for discussions on urban transformations towards sustainability, such as those included in this book: The SDGs attempt to operationalize sustainable development and present indicators on how to measure progress towards sustainability. Even though shortcomings concerning the selection of indicators and the conceptualization and ambition of the SDGs exist (Unmüßig 2016), the SDGs can be a starting point for thinking about indicators of how to measure sustainability transformations and about how the shift from declarations of interest towards implementation of sustainable urban development can be realized.

References

Ahmad S, Koch F (2015) Urban housing and basic services. In: Griffith C, Watkins M (eds) Comments on Habitat III issue papers, The Urbanization and Global Environmental Change (UGEC) project. Arizona State University, Tempe, pp 38–42

Brenner N, Schmid C (2013) The 'urban age' in question. Int J Urb Reg Res 38(3):731–755

Census of India (2011) Primary census abstract for slum http://www.censusindia.gov.in/pca/final_pca.aspx. Accessed 12 June 2016

Cities Alliance (2015) Sustainable development goals and Habitat III: opportunities for a successful new Urba agenda. Cities Alliance discussion paper no. 3. http://www.citiesalliance.org/sites/citiesalliance.org/files/Opportunities%20for%20the%20New%20Urban%20Agenda.pdf. Accessed 26 Sept 2016

Grimm N, Faethl SH, Golubiewski NE, Redmann CL, Wu J, Bai X, Briggs JM (2008) Global change and the ecology of cities. Science 319(5864):756–760. doi:10.1126/science.1150195

ICSU – International Council for Science, ISSC (2015) Review of the sustainable development goals: the science perspective. International Council for Science (ICSU), Paris

Koch F (2011) Georg Simmels „Die Großstädte und das Geistesleben" und die aktuelle Gentrification-Debatte. Eine Annäherung. In: Sundsboe A, Bieniok M, Mieg H (eds) Georg Simmel und die aktuelle Stadtforschung. Interdisziplinäre Betrachtung zu Dichte, Diversität und Dynamik der Großstadt (pp 91–113). VS Verlag, Wiesbaden

Koch F, Patterson J (2015) How can science policy help to deliver the global goals?. The Guardian, Science Policy Blog. https://www.theguardian.com/science/political-science/2015/oct/09/how-can-science-policy-help-to-deliver-the-global-goals. Accessed 26 Sept 2016

Kübler D, Pagano MA (2015) Urban politics as multilevel analyses. In: Mossberger K, Clarke SE, John P (eds) The Oxford handbook of urban politics. Oxford University Press, New York, pp 114–129

Loewe M, Rippin N (eds) (2015) Translating an ambitious vision into global transformation the 2030 agenda for sustainable development. Deutsches Institut für Entwicklungspolitik (DIE), Bonn

Lu Y, Nakicenovic N, Visbeck M, Stevance A-S (2015) Policy: five priorities for the UN sustainable development goals-comment. Nature 520(7548):432–433

Misselwitz P, Salcedo Villanueva J, Meinert G (2015) The urban dimension of the SDGs: implications for the new urban agenda. In: Cities Alliance (eds), Sustainable development goals and Habitat III: opportunities for a successful new urba agenda. Cities Alliance discussion paper no. 3. http://www.citiesalliance.org/sites/citiesalliance.org/files/Opportunities%20for%20the%20New%20Urban%20Agenda.pdf. Accessed 4 Oct 2016

Pierre J, Peters G (2015) Urban governance. In: Mossberger K, Clarke SE, John P (eds) The Oxford handbook of urban politics. Oxford University Press, New York, pp 71–86

Rudd A (2015) An explicitly urban sustainable development goal has been adopted by the UN (#11). Now what? Where could it go wrong? Blog on the Nature of Cities. http://www.thenatureofcities.com/2015/12/08/an-explicitly-urban-sustainable-development-goal-has-been-adopted-by-the-un-11-now-what-how-can-it-be-effective-in-the-ways-it-was-intended-and-in-the-ways-that-we-need-where-could-it-go-wrong/. Accessed 4 Oct 2016

Satterthwaite D (2016) Where are the local indicators for the SDGs? Blog of the IIED. http://www.iied.org/where-are-local-indicators-for-sdgs. Accessed 4 Oct 2016

Schnorr-Bäcker S (2015) Inspire, sustainable development and measuring progress from German official statistics perspective. Presentation at the Geospatila world forum 2015 in Lisbon/Portugal

Schnorr-Bäcker, S. (2016). New data sources for official statistics for monitoring comprehensive sustainability strategies. Abstract for the 34th IARIW general conference, Dresden. http://www.iariw.org/dresden/schnorr-abs.pdf. Accessed 4 Oct 2016

Sethi M, Puppim de Oliveira J (2015) From global 'North-South' to local 'urban-rural': a shifting paradigm in climate governance? Urb Clim 14(4):529–543

Simmel G (1903) Die Großstädte und das Geistesleben. In: Petermann T (ed) Die Großstadt. Vorträge und Aufsätze zur Städteausstellung. Dresden, Gehe–Stiftung, pp 185–206

Simon D, Arfidsson H (2015) Pilot project to test potential targets and indicators for the urban sustainable development goal 11. Final report. http://www.mistraurbanfutures.org/en/pilot-project-test-potential-targets-and-indicators-urban-sustainable-development-goal. Accessed 29 May 2016

Simon D, Arfvidsson H, Anand G, Bazaz A, Fenna G, Foster K et al (2016) Developing and testing the urban sustainable development Goal's targets and indicators–a five-city study. Environ Urban 28(1):49–63

SRL – Vereinigung für Stadt- Regional- und Landesplanung (2015) Nationalbericht Habitat III Stand 20.03.2015 (draft). http://www.srl.de/dateien/dokumente/de/Habitat%20III%20Nationalbericht%20Deutschland%20Stand%20200315.pdf. Accessed 29 May 2016

UN – United Nations (2015) Sustainable development goals: 17 goals to transform our world http://www.un.org/sustainabledevelopment/. Accessed 7 Mar 2016

UN Habitat (2016) Habitat III. Zero draft of the new urban agenda. https://www.habitat3.org/node/529098. Accessed 15 Sept 2016

UNESCAP - United Nations Economic and Social Commission for Asia and the Pacific (2014) Expert group meeting on sustainable urban development in Asia and the Pacific: towards a new urban agenda background paper http://www.unescap.org/sites/default/files/Background%20Document%20EGM_Final.pdf. Accessed 29 May 2016

Unmüßig, B. (2016). Agenda 2030: a new path for all?. https://www.boell.de/en/2016/01/22/agenda-2030-new-path-all. Accessed 15 Aug 2016

UNSTATS – United Nations Statistic Division (2016) Provisional proposed tiers for global SDG indicators Retrieved from: http://unstats.un.org/sdgs/files/meetings/iaeg-sdgs-meeting-03/Provisional-Proposed-Tiers-for-SDG-Indicators-24-03-16.pdf. Accessed 24 Apr 2016

UTC – Urban Thinkers Campus (2015) India national report to Habitat III http://utc.niua.org/utc-india-plenary-session-2/. Accessed 7 Mar 2016

WBGU – German Council on Global Change (2016) Der Umzug der Menschheit – Die transformative Kraft der Städte. WBGU, Berlin

Wirth L (1938) Urbanism as a way of life. Am J Sociol 44(1):1–24

Part II
Urban Resources and Governance

Outline

Erik Gawel and Andreas Zehnsdorf

An analysis of how to organize and provide living conditions in cities that enable a dignified life for all includes the efficient provision and fair distribution of resources, to enable an acceptable quality of life within a resilient urban environment. Cities demand, use, and, in many cases, even overuse natural resources (land, water, energy, etc.). They also, increasingly, grab natural resources far from their territory, produce three-fourth of global carbon emissions, and use 60% of residential water (Grimm et al. 2008, p. 756), as well as 65% of energy consumption (Solecki 2013). Thus, efficient use and fair distribution of resources is crucial for a sustainable development in the context of urbanization.

This part of the book focusses on urban transformations towards a sustainable management of urban resources and on appropriate governance structures to implement them. With respect to resource efficiency, the focus is on the analysis and evaluation of resource use in terms of unprecedented urban land consumption, urban energy services, water quality and water scarcity, deficient sanitation, or air pollution, as well as built and technical infrastructures. Thus, Part II is dedicated to the improved use of urban resources and infrastructures through a variety of governance options.

Klassert et al. discuss, in an empirical case study, the challenges of sustainable urban water infrastructures in Amman, Jordan, under conditions of severe water scarcity and the current refugee influx. They highlight the crucial role of demand-oriented measures beyond mere technical or supply-side approaches for long-term sustainability. **Hansjürgens and Brenck** present "The Natural Capital Germany Approach", an application of the TEEB initiative (The Economics of Ecosystems and Biodiversity) in the context of urban transformations. They show how we can make ecosystem services visible and incorporate them into urban decision-making. **Nivala et al.** outline the role of green infrastructure for increased resource efficiency in urban water management. They show that communities are aware of the benefits of green technologies for minimizing the negative impacts of urbanization and that a green infrastructure is a real alternative to traditional approaches to gray infrastructure planning. **Reese and Gawel** discuss the concept of sustainable urban water governance in Germany and in the EU, emphasizing its main aims, challenges, and institutional approaches. **Lenz et al.** refer to the energy context of sustainable infrastructure and discuss, in a case study, the challenges of the "Wärmewende" in urban areas of Germany. The authors argue that a complete conversion of the heat supply to renewable energies ("Wärmewende") is only possible if heat from solar thermal systems, heat pumps, and geothermal plants, as well as heat from excess electricity from renewable energies (electricity heat) and options for heat recovery are used. **Haase et al.** analyze the nexus between urban dynamics, land use, and ecosystem service provision. They note that shorter shrinkage and regrowth times are likely for the future, and that the effects of nonlinear development need to be taken into account in the discussion of transformation. Finally, **Banzhaf et al.** deal with the existence of urban brownfields – a spatial resource – in terms of integrated urban development regarding ecological, social and economic characteristics. They identify the revitalization of brownfield sites as an important tool for intensifying building activity in such urban districts and, thus, for the regeneration of formerly disadvantaged locations.

The results gained here are also highly relevant when aiming at increased urban resilience and quality of life: improving resilience (e.g., with respect to floods) (see Part IV in this volume) or providing living conditions with high a quality of life (as discussed in Part III of this volume) is costly and requires a large input of scarce resources of all kinds, as well as high-performance infrastructures. Moreover, both are reliant on sound governance structures that ensure efficient use and fair distribution of resources, resource-based services, and infrastructures. For example, the human right to water can only be ensured in the long term if natural water resources are sustainably protected, if a powerful water infrastructure is adequately financed and operated in a thrifty and efficient way, and if water tariffs are fair and affordable. In this respect, the interlinkages of different resources also have to be kept in mind because they represent the manifold nexus challenges for governance, e.g., water and energy, land and energy, water and food, amongst others.

Sustainable Transformation of Urban Water Infrastructure in Amman, Jordan – Meeting Residential Water Demand in the Face of Deficient Public Supply and Alternative Private Water Markets

Christian Klassert, Erik Gawel, Katja Sigel, and Bernd Klauer

1 Introduction: Transforming Urban Water Infrastructure

Rapid urban growth processes pose severe challenges to the existing water infrastructure, particularly in developing countries (see Bedtke and Gawel, "Linking Transition Theories with Theories of Institutions – Implications for Sustainable Urban Infrastructures Between Flexibility and Stability", in this volume). Responding to these challenges might exceed the scope of a gradual change and require a sustainability-oriented system transformation (Kabisch and Kuhlicke 2014). This chapter examines the prospects for such an urban transformation in Amman, the capital of the Hashemite Kingdom of Jordan, where the challenges of supplying water to all residents are particularly demanding. Since the year 2000, the population of the greater Amman municipality is estimated to have grown by more than one third (DOS 2014), and this trend can be expected to persist, due to continually high rates of immigration of Syrian refugees. In addition, Jordan is among the most water-scarce countries in the world (Yorke 2013) and is currently overexploiting its renewable groundwater sources by about 65% above the sustainable extraction rate (IRG 2015). Thus, making progress towards a more sustainable use of its freshwater resources is a matter of urgency. The pressing scarcity of water has led *Miyahuna*, the public water utility of Amman, to introduce a water quantity rationing scheme by which households only receive water for a limited number of hours per week, leading to perceived and actual water quality problems (supply interruptions can, e.g., lead to contaminant infiltration and the development of biofilms; see Hashwa and Tokajian 2004; Yorke 2013; Potter and Darmame 2010). Both the supply intermittency and the quality concerns have forced residents to intensify the use

C. Klassert (✉) • E. Gawel • K. Sigel • B. Klauer
Department of Economics, Helmholtz Centre for Environmental Research – UFZ, Leipzig, Germany
e-mail: christian.klassert@ufz.de; erik.gawel@ufz.de; katja.sigel@ufz.de; bernd.klauer@ufz.de

S. Kabisch et al. (eds.), *Urban Transformations*, Future City 10,
DOI 10.1007/978-3-319-59324-1_6

of various coping strategies (e.g., maintaining private storages, ordering private water tankers, purchasing bottled water), which further complicate any targeted steps to initiate a transformation of the public water supply system and water use patterns towards sustainability.

Sustainable development is generally characterized by a use of resources that does not "[compromise] future needs" (WCED 1987). The corresponding sustainability requirements are commonly directed at the environmental, the social, and the economic spheres of development (UN General Assembly 2005). Specifically, with regards to urban development, Kabisch and Kuhlicke (2014) propose that three guiding concepts should be emphasized in pursuing a transformation towards sustainability: quality of life, resource efficiency, and resilience. These guiding concepts represent important facets of a broader sustainability perspective. They are interlinked and interdependent, which creates synergies, trade-offs, and conflicts between them (ibid.). In the context of an analysis of urban water supply systems, they can be interpreted as follows: Supplying water to the residents of a city has a fundamental effect on their *quality of life*, because water is essential to various basic needs, such as drinking, food preparation, personal hygiene, washing, and cleaning. In addition to the individual utility derived from the use of resources, Kabisch and Kuhlicke (2014) also include aspects of distributional equity in their definition. Moreover, in order to be able to provide water services to its customers in the long run, the water utility or authority in charge of urban supply needs to ensure the sustainability of its freshwater resources, i.e., the surface- and groundwater bodies from which it abstracts water. *Resource efficiency*, defined in a broad sense that comprises both technical and economic efficiency, contributes to this objective by improving the ratio between the water inputs into the supply system and the welfare contribution generated by these inputs, e.g., by influencing residential consumption patterns or changing supply technology. The welfare contribution of water use in this definition encompasses all positive impacts of water on the city's residents and economic entities that are conceptually amenable to economic valuation. Whenever resource efficiency entails a reduction of buffering capacities, it may conflict with the transformation objective of *resilience*, defined as the capacity of the water supply system to recover from exogenous shocks, such as refugee influxes, interruptions in supply sources, or extreme weather events (ibid.). The benefits of any reductions of water inputs into the system should, therefore, be weighed against their effects on resilience. Infrastructure improvements can, however, ameliorate this conflict (see Bedtke and Gawel, "Linking Transition Theories with Theories of Institutions – Implications for Sustainable Urban Infrastructures Between Flexibility and Stability", in this volume). A system that is exposed to shocks due to rapid changes, e.g., in demographic, climatic, or economic conditions, can only be sustainable if it exhibits a certain degree of resilience to shocks. Thus, all three guiding urban transformation concepts are directly related to the overall objective of reaching a sustainable state, ensuring the continued supply of water to all residents, although, individually they do not, of course, provide a comprehensive description of such a state.

This raises the question of how to govern the transformation of Amman's urban water infrastructure system towards a more sustainable state. In view of the debate about the term transformation (see Kabisch et al., "Introduction: Urban Transformations – Sustainable Urban Development Through Resource Efficiency, Quality of Life, and Resilience", in this volume), we adopt the following definition: The transformation under investigation in this chapter (1) concerns the water sector, not the whole urban system; (2) is a controlled process, rather than a spontaneous one; and (3) contains gradual as well as disruptive elements. Whilst the term infrastructure can have various meanings, we use it to refer to technical public infrastructure, including all technical means for providing man-made water supply services. In Amman, this mainly includes the piped water network and a small fleet of public water tankers, both run by *Miyahuna* (Gerlach and Franceys 2009). Against this background, we analyze governance options to promote a sustainability-oriented transformation. In addition to *Miyahuna*, the Jordanian Council of Ministers, which sets water tariffs, and the Water Authority of Jordan (WAJ), which controls water transfers to Amman, fulfill important roles in water sector governance for the city (Humpal et al. 2012). Due to this important role of government entities in governing Amman's water supply, we will focus our analysis on transformation governance through public policy.

Conceptually, public water supply policy can be divided into supply-side and demand-side measures. On the supply side, government entities make decisions about investments in establishing, spatially expanding, or improving the supply infrastructure (e.g., repairing leakages), the abstraction of water from freshwater resources, and the service level with regards to the spatial allocation, temporal availability, and quality of supply (Griffin 2001). Demand-side measures include decisions about the water pricing rate, consisting of a general price level and specific tariff structures (Goldstein 1986), installing metering devices, reducing illegal abstractions from the network or directly from freshwater resources, improving the technical efficiency of household water appliances, and information and awareness campaigns (Green 2003). Efforts to employ these governance instruments to initiate a transformation towards sustainability are, however, impeded by various path-dependencies and mutually reinforcing sources of inertia that characterize urban water infrastructure systems (Bedtke 2015). In order to overcome this inertia and switch to a more sustainable development path, governance initiatives need to apply both supply-side and demand-side measures efficiently. Only such an integrated governance approach, which ensures a sustainable use of freshwater resources, sufficient financing of the supply infrastructure, and a wastage-free and efficient distribution, can guarantee the provision of water services and their affordability in the long run (Gawel and Bretschneider 2016). In practice, however, initiatives often fail much earlier, by focusing only on supply-side measures, aimed at accessing new sources of water, and ignoring the need to address the demand side with measures targeted at the behavior of water users. In Jordan's water sector, this tendency is observable in a clear prioritization of large infrastructure projects, such as the *Disi Water Conveyance Pipeline* and the *Red Sea-Dead Sea Canal* (Bonn 2013).

The remainder of this chapter addresses the significance of this neglect of the demand side in Amman's governance of the water supply system as follows: Sect. 2 describes the institutional setting of water provision and consumption in Amman as a basis for the argumentation. Subsequently, Sect. 3 identifies the sustainability problems that result from the water sector's focus on supply-side measures and the governance challenges arising in solving these. Finally, Sect. 4 derives the implications of our analysis for the governance of a sustainability-oriented water infrastructure transformation in Amman.

2 Institutional Settings in Amman

Despite plans in the National Water Strategy (MWI 2009) to increase the role of water demand management, Jordan's water policy is still mainly shaped by supply-side measures (Bonn 2013). Giving a greater role to pricing policies addressing the demand side could serve to fulfill the functions of both financing water services and allocating water resources (OECD 2010; Dalhuisen et al. 2000). Pricing policies can be used to pursue the four sustainability objectives of economic efficiency, environmental and financial sustainability, and social affordability (OECD 2010). If the price level for piped water is, however, set too low to reflect its scarcity, a shortage occurs. This means that the quantity demanded exceeds the available water supply. In such situations, either a coordinated non-price rationing mechanism is introduced deliberately, or spontaneous rationing mechanisms, such as queuing, will emerge (as is often observed for limited stocks of retail goods). In the case of piped water, a simple restriction of the water quantity input into the network would, however, not lead to queuing, as households are generally supplied through an in-house connection. Households could, therefore, independently withdraw water from the network until the water pressure becomes too low. In the absence of pricing, this can only be avoided by a coordinated quantity rationing mechanism, such as temporarily interrupting water supply to parts of the network.

Up until today, piped water prices in Jordan have reflected the scarcity of freshwater resources and the funding needs of the water sector only to a limited degree (Kubursi et al. 2011; Humpal et al. 2012; FDA/AFD 2011). Instead of rationing piped water through higher prices, in 1987, Amman's water utility reacted to the pressing water shortage and to leakages in the piped network by restricting piped water supply to a limited number of hours per week, varying from district to district (Potter et al. 2010). This deliberate introduction of supply intermittency provided a coordinated quantity rationing mechanism. It also helped to reduce the amount of water lost through leakages, which is necessarily larger when leaking pipes are under pressure for longer periods. Compared to increasing piped water prices to reach market equilibrium, however, this intermittency rationing approach has the same disadvantage as any quantity rationing scheme: it cannot account for differences in the willingness to pay for water between households and fails to clear the market, keeping it in a state of disequilibrium. The water utility simply cannot have the necessary information to determine which households barely make ends meet with the water quantity they receive and which households could more easily reduce their consumption because they use water for less important purposes. As an example, a family whose members regularly eat outside of the home (e.g., at their school or workplace) might have a lower demand for potable water than one that prepares most meals at home. The first family may, however, be using more water for baths, without this difference being readily observable for the water utility. Apart from having limited information about differences in the demands of each household's members, the water utility will also most likely have insufficient technical information about the water distribution and storage facilities within their buildings (e.g., leakages, storage capacity). This lack of distinction can impose a high cost on water users: In a study of water supply in Sydney, Grafton and Ward (2008) estimate that the use of a quantity rationing mechanism instead of price rationing created welfare losses per household that were equivalent to almost half the value of an average water bill. This shortcoming will often be further exacerbated by measures undertaken in the interest of technical and administrative practicality. Differentiating the piped water supply durations in Amman by distribution districts may be much more feasible than trying to target each household individually, but it also means arbitrarily treating different households within one district in the same way and similar households across district borders differently. Furthermore, technical errors in the implementation of supply interruptions can cause unintended deviations from the planned distribution schedule. In summary, a lack of price signals, arbitrariness in the allocation, and distortive technical factors create a mismatch between demands and supply quantities, making intermittency rationing economically as well as technically inefficient.

Intermittency rationing will, therefore, satisfy demands to a much smaller extent than price rationing and keep the market in disequilibrium, thus creating pressure to look for alternative sources of water to meet the excess demand. In fact, the quantity rationing of piped water in Amman by means of supply interruptions has caused residents to rely strongly on a combination of water sources: besides the direct use of water from the piped network, these mainly include the purchase of water from private tanker truck operators, the release of previously stored water from in-house storage tanks, the purchase of bottled water from retail stores, and the delivery of containers of water filtered in local water stores (Rosenberg et al. 2008).

A consistent economic foundation for this type of behavior is provided by so-called new consumer theory, which allows for the assumption that water users have independent demands for various characteristics of water, leading to the simultaneous use of different water sources (Lancaster 1966). In their development of a practical specification for a fundamental *right to water*, Gawel and Bretschneider (2016) identify four characteristics of water that are regularly included in assessments of individuals' access to water: (1) its quality, (2) its temporal availability, (3) its spatial accessibility, and (4) its price. The three non-pecuniary dimensions help to explain the demands for different water sources in Amman, given their respective prices. Since all four dimensions affect the suitability of a water source for various household uses, water sources differing along these dimensions are imperfect substitutes. This can complicate any efforts to initiate a transformation towards sustainability.

The intermittency of piped water supply in Amman means that the temporal availability of this resource is limited. Based on the data in a 2011 report about Amman's piped water network (Abu Amra et al. 2011), households only received piped water for about 50 h per week, on average, depending on their district of residence. As an indirect consequence, intermittency can also lead to negative impacts on water quality, which can potentially create health risks, e.g., by facilitating the infiltration of contaminants into the piped network and creating slow water flow conditions that promote the development of biofilms (Hashwa and Tokajian 2004; Yorke 2013). Residents are especially aware of such risks due to a severe water pollution crisis that occurred in 1998 and reduced their confidence in the safety of the public supply (Theodory 2000). Both the limited temporal availability and the low quality of water have substantial negative impacts on residents' quality of life.

The temporal availability of water is improved by all of the alternative water sources. In-house storage tanks are essential for extending the temporal availability of piped water supply across the whole week. This possibility is only limited by the tanks' capacity, which has been estimated to be, on average, about three cubic meters for low-income and 16 cubic meters for high-income households (Potter and Darmame 2010). Because capacity enhancements are relatively inexpensive, storage size is unlikely to be a constraining factor under normal conditions, but it can determine the resilience of households when faced with unexpected water shortages (Klassert et al. 2015). As a downside, the necessary storage of water in private households leads to a deterioration of the quality of water drawn from the piped network (Humpal et al. 2012): In-house storage tanks provide favourable conditions for the growth of bacterial contaminations caused by piped network infiltrations or by blending lower quality tanker water with the piped water in the tank, thereby reducing the usability of water over time and creating additional health risks (Hashwa and Tokajian 2004). These quality reductions, again, negatively affect residents' quality of life.

Private tanker operators make the usually distant rural groundwater sources spatially accessible to urban residents at any time, although interviews conducted with tanker drivers have revealed that up to 5 h can pass between request and delivery (Sigel et al. 2017), meaning that a substantial temporal access hurdle remains. Water sold on the private tanker water market is partially purchased legally at government licensed private wells in and around Amman, and partially obtained from unlicensed wells or illegal surface water abstractions (Gerlach and Franceys 2009). In 2004, 1267 tanker trucks with capacities between 3 and 20 m^3 were registered in Amman, highlighting the significance of this market (ibid.). The quality of tanker water varies, due to the variety of sources used and the conditions of transportation, and is generally lower than the quality of piped water, whereas its price is usually significantly higher (Rosenberg et al. 2008). Of the truck drivers interviewed, 84%, however, stated that their customers cared about water quality, and of these, 98% reported that the quality was more important to customers than the price (Sigel et al.

2017). The fact that many customers are willing to pay higher prices for lower quality tanker water is a strong indicator of the inefficiency of the intermittency rationing scheme for satisfying demands.

Bottled water and filtering store deliveries provide a type of supply that is both constantly available and qualitatively safe for drinking and cooking purposes (Rosenberg et al. 2008). Both sources are prevalent across all income classes, but because they are vastly more expensive than piped and tanker water, their use clearly increases with rising income (Gerlach and Franceys 2009). Spatial access to bottled water is impeded by the fact that bottles have to be transported from the store to the place of residence. Thus, households have to invest time and effort in order to use this source.

In summary, households' access to water along the three non-pecuniary dimensions differs significantly between sources. The degree to which Amman's residents can choose between different sources, therefore, has a strong impact on their quality of life. Compared to a situation in which sufficient high-quality piped water is always available at an affordable price, however, the money, time, and effort spent on the various coping strategies implies that intermittency rationing shifts part of the public cost burden of maintaining the piped water supply infrastructure to private water users.

3 Governance Challenges for Residential Water Supply

An urban transformation that ensures the long-term provision of affordable and accessible water services requires governance efforts that address current shortcomings on both the demand side and the supply side. The previous section has identified inefficiencies of Amman's intermittency rationing scheme in satisfying water demands. The resulting combination of water sources used by Amman's residents is characterized by temporal interruptions of piped water supply, quality issues with regards to piped, stored, and tanker water, and high prices for bottled and filtered water, all of which negatively impact their quality of life through resource inefficiencies. We now examine the consequences of the intermittency rationing scheme for the sustainability of Amman's water supply system and attempt to identify which obstacles to a sustainable development of the urban water sector need to be tackled by governance efforts.

3.1 Sustainability Problems

Amman's piped water system delivers vast quantities of water to its residents. In 2009, Amman received 129.7 million m^3 of piped water, of which 76.7 million m^3 appeared on water bills, whereas the remainder was either lost through leakages or consumed, but not paid for (Abu Amra et al. 2011). In order to generate this supply quantity, about 50 million m^3 are pumped about 1000 meters uphill from the Jordan valley, while the rest is obtained from various sources in the highlands (Abu Amra et al. 2011). Since 2014, the Disi Water Conveyance project has been delivering an additional 100 million m^3 per year from the far South of the country to the northern governorates, where Amman is located (IRG 2015). Jordan relies heavily on groundwater abstractions and exploits its renewable groundwater resources at a rate that is about 65% beyond the safe yield (ibid.), making an adjustment of consumption patterns inevitable in the long run (Yorke 2013).

Only rough estimates of the exact composition of the supply quantity that does not appear on water bills exist: according to a water infrastructure study, *Miyahuna* was only able to collect payments for 51.6% of the net water supplied to the governorate of Amman in 2014 (IRG 2015). These losses are termed unaccounted-for or non-revenue water (NRW), depending on the calculation formula used. The IRG (2015) study states that about half of them are due to leakages (technical NRW), whereas water meter malfunctions, theft, and other tariff collection problems account for the other half (administrative NRW). Whilst technical NRW lowers resource efficiency directly, administrative NRW further reduces the potential of the already low water prices for making users internalize the full cost of water. This lack of resource efficiency weighs even heavier in the light of the above-mentioned extensive efforts that are necessary to maintain Amman's current water supply. The prices for piped water determined by the government, however, reflect neither the scarcity of freshwater resources, nor the operation and maintenance costs of the supply infrastructure (Kubursi et al. 2011). The combination of low prices, leakage, and organizational inefficiencies aggravate the sustainability problems that are already inherent in the prevailing consumption patterns.

Rather than through prices, the scarcity of freshwater resources is currently reflected in Amman's intermittency rationing scheme. This creates further detriments to resource efficiency: Firstly, the quality deteriorations during the necessary storage of piped water exacerbate the general scarcity of water, because a large part of the available resources ceases to be available for important purposes of water use. Secondly, creating and maintaining this storage capacity requires costly investments thus shifting costs of supply from the public to the private sector. A certain amount of storage capacity in an arid country might be desirable from a resilience perspective, but the dimensions of the available storage in Amman would seem unnecessary without the scheduled supply interruptions. Finally, households have to invest substantial amounts of time and effort in filling tanks and developing and executing other coping strategies, which could be more beneficially employed for other purposes. The substitution of public responsibility for maintaining the piped water supply infrastructure with these private efforts is clearly not the least-cost solution, and it can impose a substantial cost and time burden on low-income households, who spend larger shares of their income on water than high-income households and who, more often, have to organize some of their weekly activities around obtaining and using water (Potter and Darmame 2010).

The most significant direct impact of the intermittency rationing scheme on the sustainability of freshwater resources has probably been the emergence of the partially illegal private tanker water market. Whereas water in the piped network can usually be considered to be a private good, rural groundwater often has the characteristics of a common pool resource, requiring effective regulation to ensure its sustainability (Schomaker 2010). Whilst the WAJ is struggling to maintain the sustainability of its own groundwater pumping, rules about extraction quantities and quality standards are even more difficult to enforce across widely dispersed private wells and surface water bodies. Illegal tanker water extractions, therefore, circumvent most efforts to ensure a sustainable use of groundwater. In addition, the decentralized supply solution provided by the private tanker water market is simply technically less efficient than delivery through the public network could be. Delivering water via tanker trucks is vastly more energy intensive than pumping it through pipes, especially if a network is already established (see, e.g., Sima et al. 2013). A similar conclusion holds for other substitutes for piped water: The use of water bottles and containers is also technically inefficient for most purposes (apart from travel-related or spontaneous purchases) and harmful to the environment, because it creates substantial quantities of plastic waste.

3.2 Sustainable Urban Water Supply – The Economic Perspective

From an economic perspective, resource efficiency in a wider sense constitutes a key prerequisite for sustainability, but trade-offs between resource efficiency and other sustainability requirements, such as quality of life and resilience, also have to be considered. For an urban water supply system to be resource-efficient, it has to be optimized with regard to (1) the abstraction of renewable and non-renewable water resources, (2) investments in establishing or improving the supply infrastructure, (3) the service level in terms of the spatial allocation, temporal availability, and quality of supply, and (4) an adequate price level and structure (Griffin 2001). Decisions about water abstraction rates need to take all opportunity costs into account, including those related to future uses (ibid.). Investment decisions should lead to an optimal capacity and spatial scope of the supply network and take all expected future benefits and costs into account (Dandy et al. 1984). Especially in the context of developing countries, an optimal expansion of the supply network does not necessarily entail piped water connections to all households, but might consist of a combination of centralized and decentralized solutions (WBWDRT 1993; Altaf et al. 1993; Madanat and Humplick 1993). Similarly, although the current rate of supply intermittency in Amman seems to be inefficiently high, avoiding any risk of an interruption might not be worth the cost (Griffin and Mjelde 2000). Finally, pricing should pursue all of the four OECD (2010) sustainability objectives of economic efficiency, environmental and financial sustainability, and social affordability by not only taking into account full cost recovery and economic efficiency, but also aspects of equity and administrative feasibility (Dalhuisen et al. 2000).

Even in theory, therefore, the welfare-maximizing solution would not necessarily be a fully centralized system with comprehensive continuous supply, where all the coping strategies that residents have adopted cease to exist and every user is charged the same full cost price per cubic meter. As a case in point, whilst transporting large shares of the water supply in a city via tanker trucks will usually be economically and technically inefficient, providing water to remote locations and during unplanned network interruptions in this way might increase efficiency. In Amman, this service could continue to be a role for private tankers or it could be ensured through an expansion of the public tanker water supply. A small fleet of public water tankers already exists in Amman but, currently, most households do not seem to have access to this service (Gerlach and Franceys 2009). Furthermore, maintaining some in-house storage capacity could be preferable to investments necessary for completely eliminating supply interruptions. Lastly, whilst ecological sustainability might require an increase in the piped water price level, the necessity to choose a tariff structure that shifts part of the financial burden from low-income to high-income users, to ensure affordability, will most probably still remain.

Despite these qualifications, the size of the private tanker water market, the lack of trust in the quality of piped water, the large share of NRW, and the actual water quality deteriorations occurring during storage times in the network and in private tanks indicate that the current system in Amman does not supply water efficiently. Therefore, some progress towards a more comprehensive role of the piped network in water supply and towards a greater balance between supply- and demand-side policies could contribute to an increased sustainability on both the supply and the demand side. This raises the question of whether this type of progress could be implemented in practice.

3.3 The Transformation Conundrum

3.3.1 Four Obstacles to Sustainably Transforming Amman's Water Supply Sector

In practice, the fact that demand-side governance was largely neglected in Amman has led to a self-sustaining system, which poses several obstacles to a sustainability-oriented transformation of its urban water infrastructure, in addition to pre-existing obstacles inherent to the institutional environment:

1. *Political infeasibility of tariff increases:* Firstly, as pointed out above, water tariffs are not determined by Amman's water utility, but by the Council of Ministers. Raising residential piped water price levels to reflect full costs might, therefore, be politically infeasible, since equity and political stability concerns make it a highly sensitive societal issue. In 2009, for example, public resistance against

price increases led the King of Jordan to repeal a water tariff increase passed by parliament (Bonn 2013).

2. *Intermittency rationing:* Secondly, while the piped water price level is low, the low level of service, especially the supply intermittency and the resulting quality issues, imposes high non-pecuniary barriers to accessing piped water. These access barriers might actually be serving an important function in preventing an excessive use of water resources (Gawel and Bretschneider 2016). If *Miyahuna* attempted to improve its service level without a simultaneous increase in tariffs, the system might simply not be able to meet all demands, unless it increases its water intake.
3. *Non-revenue water (NRW):* Thirdly, organizational inefficiencies (as well as technical obstacles) might make it difficult to reduce leakages and unpaid uses of water, creating additional losses if a more continuous supply was introduced. Despite considerable efforts to reduce NRW in the *Miyahuna* network, little progress has been made (Yorke 2013).
4. *Private tanker water market:* Finally, the tanker water market has emerged as a private solution to the demand problem neglected by the public water sector. Whilst the quantity of tanker water used in Amman is estimated to be much lower than the quantity of piped water, the fact that tanker water is the marginal good for many households makes it very relevant for their consumer surplus (Klassert et al. 2015). The private tanker water market fulfills the function of balancing shortcomings in the public water supply by providing water during scheduled interruptions, in other crisis situations where additional supply is needed, to customers who do not have a sufficient connection, and, generally, during the water-scarce summer time (Humpal et al. 2012). Therefore, while delivering water via tanker trucks may often not be the most efficient solution, in general, their role of filling public supply gaps still creates substantial welfare gains under the status quo (Klassert et al. 2015).

3.3.2 Interdependencies Between the Four Obstacles

A major challenge for initiating any transformation to a more sustainable water supply system in Amman is that the four transformation obstacles – political infeasibility of tariff increases, intermittency rationing, NRW, and the private tanker water market – are highly interdependent:

1. Combatting unsustainable tanker water extractions requires relaxing the intermittency rationing scheme. The reason is that, without a more continuous supply, any attempt to reduce the size of the private tanker water market could have strongly negative effects on water users' welfare, because a key element of many households' strategy for coping with the shortcomings of public supply would be eliminated (Klassert et al. 2015).
2. Relaxing the intermittency rationing scheme, however, requires both reducing NRW and overcoming the political infeasibility of tariff increases. Firstly, short-

ening supply interruptions while the current rate of NRW persists would increase both the technical losses of piped water through leakages and provide more opportunities for obtaining water free of charge (administrative NRW) through theft, meter tampering, etc., (Humpal et al. 2012). Secondly, the need for planned intermittency as a quantity rationing mechanism will only disappear if an alternative rationing mechanism, such as a higher water price level, is in place.

3. Reducing NRW also requires overcoming the political infeasibility of tariff increases. *Miyahuna* will need to finance continued investments in an improved water supply infrastructure and in greater organizational efficiency if it aims to limit leakages and unbilled uses of piped water. Increased tariffs are a necessary precondition for achieving financial sustainability with regard to at least part of these costs (Humpal et al. 2012).
4. In turn, overcoming the political infeasibility of raising water tariffs might require improvements related to all three of the other obstacles. Firstly, residents may not be willing to pay a higher price for piped water unless they receive a better service with regard to the supply intermittency. Secondly, they also might not be motivated to accept a higher price level for the sake of sustainability if they observe wastage and theft due to unresolved NRW problems and know about unsustainable tanker water extractions from illegal sources. Finally, the reasons for raising the piped water tariff might be defied if the higher price drives people to substitute more tanker water for piped water.

The reinforcing tendencies of the four transformation obstacles create a conundrum that makes it difficult to address any one of the four obstacles before the others have been addressed. The subsequent section will develop potential solutions to this transformation conundrum.

4 Governance Implications: Water Demand Policy Is Key

4.1 Overview: Governance Options for Sustainable Water Policy

The challenge in solving this transformation conundrum is to find starting points where some progress is possible, despite the self-sustaining tendencies of the current water supply system in Amman, and which might initiate a transformation to a more sustainable use of Jordan's freshwater resources, especially of its groundwater. This section will, therefore, examine supply- and demand-side policies that could contribute to this kind of progress, such as: implementing an adequate price level and structure, ensuring a sustainable use of freshwater resources, fighting illegal abstractions, improving infrastructure and repairing leakages, avoiding intermittency, and improving water quality.[1] These will be assessed with regard to their

[1] The other demand-side measures mentioned above: installing metering devices, improving the technical efficiency of household water appliances, and information and awareness campaigns, will not be addressed in this chapter.

effectiveness, economic efficiency, environmental and financial sustainability, equity and acceptability, and their political feasibility (OECD 2010; Dalhuisen et al. 2000). The remainder of this section is organized as follows: Firstly, we discuss whether there is an option to use a better designed tariff structure to improve the allocation of piped water and to contribute to financing the public supply network in a fair and equitable way (Sect. 4.2). Secondly, we analyze the potential for fighting illegal water abstractions, especially by private tanker truck operators (Sect. 4.3). Thirdly, we identify funding requirements for a sustainable supply infrastructure (Sect. 4.4). Finally, we outline a possible implementation strategy for such policies to promote a sustainability-oriented transformation (Sect. 4.5).

4.2 Water Pricing

In order to find starting points for transformation, it is important to recognize that the unsustainable state of the water supply system originated from a neglect of demand-side governance. This is especially apparent in the fact that a higher price level for piped water could contribute to a sustainable urban transformation with both of the water price functions mentioned in Sect. 2: raising funds for improvements to the public supply and making water users internalize the scarcity value of water. The latter point is especially important, because the lack of price signals about water scarcity created the need for intermittency rationing in the first place. This rationing mechanism, in turn, led to the emergence of the tanker water market. Apart from being partially illegal, this market most likely exhibits an excessive size (compared to a welfare maximizing optimum). This is due to the fact that water supplied by tanker trucks does not provide obvious comparative advantages relative to the piped water supply system for the entire market volume supplied. Adequate pricing, to clear the market is, therefore, a necessary precondition for reducing supply interruptions and combatting illegal tanker water abstractions. The question is, of course, whether in practice, prices could actually be implemented to fulfill these functions in an effective, efficient, and equitable way.

4.2.1 Effectiveness of Pricing

Among economists, pricing is commonly seen as a powerful means to govern resource use, especially to promote economic efficiency, by ensuring that markets will clear, and to initiate sustainability-oriented path changes (Griffin 2016). However, the question here is whether pricing is actually effective under the real-world conditions in Amman. The price elasticity of demand is a concept that may answer this question, in so far as it indicates to what extent demand would ever react to price changes.

In a review of price elasticity estimates for water demand in developing countries, Nauges and Whittington (2010) find that most studies report values between

−0.3 and −0.6. Even though the economic terminology refers to such values as inelastic, they imply that a doubling of the price would lead to a demand reduction of 30–60%. Nauges and Whittington (2010) point out that these elasticities are similar to findings for industrialized countries, and that they are large enough for prices to have a substantial effect on water demand. The authors conclude that, under such conditions, it is possible to use pricing for both demand management and revenue generation. Al-Najjar et al. (2011) and Tabieh et al. (2012) find price elasticities between −0.5 and −0.7 for Amman, confirming that Nauges' and Whittington's conclusion is transferable to our case. Pricing, therefore, has the potential to be an effective rationing mechanism, and to contribute to the efficiency and financial sustainability of the water supply system.

4.2.2 Pricing and Equity Considerations

Even an effective and efficient price rationing mechanism might be undesirable, however, if it requires prices to be raised in a way that jeopardizes the affordability of water for some members of society. As a solution to this problem, the structure of a water tariff can be adjusted to distribute the cost burden according to different user groups' ability to pay, allowing for equity and affordability considerations to be taken into account at any price level. Currently, an increasing block tariff is used in Jordan to promote an equitable distribution of the cost burden. Under this tariff structure, the price per cubic meter rises in a series of steps as increasing total quantities of water are consumed during a given time period. Thereby, luxury uses of water are intended to cross-subsidize an essential minimum consumption. Nevertheless, increasing block tariffs can also inadvertently increase the burden on low-income households, if these support relatively more people or share a connection with another household (OECD 2009). Appropriately implemented increasing block tariffs are, however, seen as a viable option for reconciling full cost recovery with affordability, due to their potential to provide a minimum endowment of water for little or no charge (ibid.. Dalhuisen et al. 2000). In a survey conducted in Amman, Potter and Darmame (2010) found that the piped water bill expenditure of households interviewed in low-income areas of the city was, on average, only slightly more than a quarter of that of households in high-income areas of the city, implying that cross-subsidization does occur. At the same time, the average income of high-income households in the survey was, however, eight times as high as that of low-income households. This demonstrates that there might still be scope for improving the equitability of the tariff structure. Even among low-income households in the survey, however, the average piped water bill value was only 2.1% of the average income, suggesting that the current tariff structure allows them to obtain piped water at an affordable rate.

However, the assessment of affordability changes when the expenditures necessary for households' coping strategies are taken into account. Among the low-income households interviewed by Potter and Darmame (2010), only 20% regularly bought bottled water, but among this subgroup, the average monthly expenditure for this supply source was 35.6 Jordanian Dinars (JD). This seems substantial, in view of the fact that the average monthly income across all low-income households (including those that did not buy bottled water) was only 235 JD. In addition, 84% of low-income and 64% of high-income households reported adjusting some their weekly activities to the temporal availability of piped water, which clearly imposes constraints on the discretionary use of their time. These observations indicate that increasing the piped water price to provide a better service could lead to substantial cost savings, if this reduces the need for households to rely on coping strategies. It might, counter-intuitively, even lower the overall cost burden, if the savings are substantial enough to offset the piped water price increase. As long as the basic water endowment is supplied at a sufficiently low rate, it would very probably improve the overall affordability of water for low-income households currently relying on costly coping strategies.

4.2.3 Acceptance: The Willingness to Pay for Higher Quality

Apart from the questions of whether piped water prices influence consumption and whether an equitable use of prices is possible, the implementation of a new piped water tariff also requires its acceptance by the respective users. In a situation similar to the one analyzed here, Whittington et al. (1991) interpreted high tanker water expenditures as an indicator of households' willingness to pay for improved public water supply. In Amman, the prices charged for tanker water are actually often higher than the estimated full cost price of piped water would be (Gerlach and Franceys 2009; Zawahri 2012), indicating a potentially sufficient willingness to pay for water. Behavioral pricing theory has identified the so-called reference price phenomenon in other contexts, which describes the fact that customers often use predefined expectations about the price of a product to decide about the acceptability of a newly encountered price (Gawel 2016; Kalyanaram and Winer 1995). In this sense, two additional, competing factors might influence water users' willingness to pay for piped water: On the one hand, the current low level of piped water tariffs in Amman would be expected to reduce the willingness to pay. This might further be exacerbated by the so-called fiscal illusion, the tendency of citizens to underestimate the true costs of publicly financed services (Gawel 2016). On the other hand, the emergence of the private tanker water market might have primed customers to accept somewhat higher prices for piped water as well, thereby contributing to the acceptability of a tariff increase.

Supplying piped water more continuously could both improve this water source's temporal availability and reduce intermittency-induced quality deteriorations (see above, Sect. 2), making it an attractive substitute for any of the other sources. As a confirmation of this, an extensive survey of Amman's water users in the year 2000 revealed that 97% of the households relying on tanker water and 96% of those using bottled water would replace these expensive sources with piped water, if it was supplied in sufficiently high quality and frequency (Theodory 2000). Although not all respondents explicitly reported a positive willingness to pay for such an improved service, 47% of high-income and 36% of upper-middle-income households did so. Across all income groups, 31% reported a willingness to pay for better water quality and 28% for a larger supply quantity. Rosenberg et al. (2007) simulated the willingness to pay for greater supply continuity in Amman on the basis of avoided coping costs. They validated their model with data from the Theodory (2000) survey, but even found that approximately 50% of households might be willing to pay a higher price for piped water, to avoid supply interruptions. These findings indicate that there seems to be some acceptance for a tariff increase, at least in parts of Amman's society. An adequately structured tariff could try to target these and to minimize resistance by, e.g., focusing on luxury uses, in particular.

An additional acceptance of higher tariffs for the sake of sustainability might hinge on whether water is generally perceived to be handled diligently. Experimental research in behavioral economics indicates that peoples' willingness to contribute to public goods (e.g., to conserve water) depends on whether free-riding is punished, because that reduces the fear of being exploited (Gächter et al. 2004). Currently, the existence of illegal water extractions from the piped network and from illegal wells is well-known in Jordan (Namrouqa 2016). Residents also observe wastage of water through leakages and quality deteriorations. Besides improving the piped water service, fighting water theft and wastage might, therefore, be important for raising the acceptance of piped water tariff increases.

4.2.4 Political Feasibility

Even if higher prices can allocate piped water in an effective, efficient, and equitable way, it might, nevertheless, be politically impossible to implement them, as the above-mentioned repeal of the 2009 tariff increase has shown. Jordan's regional neighborhood in the Middle East is characterized by political crises and, despite its relatively stable domestic situation, the country experiences regular protests for reforms that address deteriorating socio-economic conditions, water shortages, and corruption; therefore, avoiding public discontent is a high priority in Jordanian politics (Zawahri 2012). In such a situation, keeping water prices low is seen as a "small price" to pay for political stability, compared to attempting to solve more fundamental political and economic problems. In contrast to claims about the political infeasibility of raising water tariffs, Zawahri (2012), however, makes the case that a tariff increase could actually have a politically stabilizing effect on society, if the main financial burden was targeted at higher-income groups. It is questionable, though, whether price increases would be compatible with the societal influence of the well-established tribal system (Gao 2011). Moreover, Zeitoun et al. (2012) have argued that other stakeholders (e.g., owners of large farms) also have great influence on political decisions about water demand management and they, also, might be dissatisfied with such a solution.

Assuming that some progress could be made on increasing the price level for piped water, the transformation conundrum would, however, still not be solved.

Tackling the sustainability problems caused by intermittency and by the private tanker water market still requires improvements to the network infrastructure, reductions of NRW, and the enforcement of rules against illegal water extractions. A key challenge for decision makers is to actually make credible commitments to implement these improvements.

4.3 *Fighting Illegal Abstractions*

As long as prices are too low to equilibrate water demands and supplies, excess demands will increase pressure on switching to illegal alternatives, in order to overcome the prevailing shortage. In Amman, this can be observed, in particular, in the use of tanker water from unlicensed sources, but also in activities categorized as "administrative NRW", such as manipulations of water meters or direct illegal abstractions from water pipes (Namrouqa 2016).

With respect to fighting illegal and unsustainable tanker water abstractions, the question arises of whether increases in the piped water price level might actually undermine these efforts because they could boost demand for tanker water as a substitute. In this regard, a simulation experiment undertaken by Klassert et al. (2015) reveals that piped water tariff increases in Amman might actually reduce, rather than encourage, tanker water consumption. Even though the two sources appear to be substitutes, they do not behave as such. The authors explain this with the fact that intermittency rationing in Amman creates a highly artificial allocation of water with little correspondence to the differences in demands. In this situation, some households can rely fully on affordable piped water, whilst others have to complement their supply with much more expensive tanker water to obtain the same quantity. Increasing piped water tariffs can, therefore, help to distribute water more evenly and reduce the need for additional tanker water purchases, especially among households with very limited piped water access. In this way, a more uniform distribution of water is not only more equitable; it can also contribute to mitigating the negative sustainability impacts of the private tanker water market. This means that raising the price level for piped water would facilitate efforts to fight illegal tanker water abstractions by lowering the demand for tanker water, rather than undermine them. Most of the demand for illegal tanker water and other illegal abstractions will, however, only disappear when piped water is provided with a higher service level and greater reliability. In contrast, purely repressive approaches to combatting illegal water abstractions (e.g., monitoring and sanctions) that ignore their economic causes are unlikely to be successful. If a sensible demand-side policy is introduced, however, additional sanctions against water theft might strengthen its acceptability (see above, Sect. 4.2.3).

4.4 *Funding Infrastructure Improvements: Curing Revenue Deficits*

Demand patterns are interlinked in various ways with the performance of the technical supply infrastructure. On the one hand, one can expect increases in the willingness to pay for network-based water services when the technical system is improved. On the other hand, the possibility of implementing such improvements may be restricted by revenue constraints, especially due to tariffs not covering full cost, and difficulties in reducing administrative NRW. These revenue constraints conflict with the concept of the financial sustainability of water services (OECD 2010; see Section 2). A further challenge for the implementation of infrastructure improvements stems from the risk of water losses: As long as leakages and water theft remain prevalent, increasing supply durations will cause an amplification of water losses, which do not create revenue. This implies that there is little incentive for decision-makers to reduce intermittency.

Improving this situation will require investments both in the water supply infrastructure and in organizational efficiency. Some authors argue that organizational deficits could be mitigated by awarding a private management contract (Schomaker 2010; Oelmann 2010). Between 1999 and 2007, Amman's water utility was actually operated under a management contract with the private company Lyonnaise des Eaux-Montgomery Watson-Arabtech Jardanesh (LEMA). This, indeed, led to some improvements with regard to investing in infrastructure, combatting NRW, and increasing weekly supply durations to 66 h in 2005, but it also required substantial external funding by the World Bank (Potter and Darmame 2010; Humpal et al. 2012). This indicates that, no matter whether infrastructure improvements were to be implemented under the auspices of the publicly owned private company *Miyahuna*, or whether another, fully private management contract were to be negotiated, tariff increases alone will probably be insufficient to finance organizational and infrastructure improvements for a transformation to a piped water system with largely continuous supply. Initially, at least, external financial support might be required to fund these investments.

4.5 *Implementation Strategy*

The neglect of demand-side policies in Amman has led to a reliance on intermittency rationing, causing residents to adopt inefficient coping strategies and making the current public water supply system unsustainable in the long run. As we have seen, increasing the piped water price level could both reduce the need for intermittency rationing and contribute to financing an acceptable piped water service level. A higher price level could not only distribute water more equitably and efficiently, but also might even help to mitigate the indirect sustainability consequences that have resulted from the emergence of the private tanker water market. Of course, in order to ensure the affordability of water for households in all income categories, price increases should be implemented as an increasing block tariff, similar to the one currently in place, with an endowment for basic water needs provided at a very low or no charge. Although we have presented several arguments for the political feasibility of a water tariff increase, it is still far from clear whether such a policy initiative could be successful in practice. There is, however, relatively reliable evidence regarding one aspect: A tariff increase would encounter higher acceptance if it was introduced in exchange for service improvements with regard to the temporal availability and quality of piped water. Mitigating the shortcomings of the current network would, in any case, be necessary to increase resource efficiency. Judging from the experiences with the LEMA management contract, this might, however, at least initially, require investments larger than the potential revenue from water tariffs. In this regard, a credible commitment to improve the piped water system on both the supply and the demand side might be more successful in winning the support of the donor organizations involved in Jordan's water sector than projects that focus purely on the costly development of limited stocks of additional supply. Current efforts to increasingly combat illegal water extractions from the network and from private wells (Andrews 2015; Namrouqa 2016) could be a promising signal towards all stakeholder groups in the water supply system (residents, water regulators, industrial enterprises, farms, etc.) that water resources should be handled diligently and that deviating behavior will be sanctioned. This could enhance stakeholders' motivation to cooperate in taking further transformation steps (see above, Sect. 4.2.3). Even if investments in network improvements, tariff increases, and an effective regulation of the private tanker water market are subsequently initiated, however, substantial efforts will still be required to achieve a complete transformation to a more sustainable urban water supply system in Amman.

5 Conclusion

Amman's urban water supply system provides a clear example of the various coping mechanisms that can be prompted by quantity rationing schemes and of the subsequent inefficiencies in terms of meeting demands and ensuring water resource sustainability. Several lessons can be derived from this example:

A price level that reflects the scarcity value and costs of the public water supply relatively well will usually be the best option for maintaining a sustainable infrastructure for water supply services, because it creates efficient incentives for resource use and protection, market clearing, and adequate financing of the infrastructure. However, when the price level is below the true scarcity and equilibrium value, market clearing fails; consequently, non-price access barriers, such as supply intermittency, may act as substitutes with respect to the inevitable societal rationing mechanism needed in this case. This creates demand pressure, pushing for the use of (possibly) illegal water supply sources (e.g., tanker water from illegal sources, "administrative" NRW that is not paid for), and political pressure to provide more water intake into the system. The suboptimally large size of Amman's tanker water market indicates, e.g., how inefficient non-market quantity rationing mechanisms can be. Moreover, combatting illegal water abstraction with repressive measures not only requires well-established institutions of good governance, but it also does not cure the underlying economic cause for excess water demand being on the lookout for (illegal) alternatives, such as tanker water. Therefore, a repressive strategy will not help to mitigate the negative sustainability impacts of illegal water extractions, unless good water sector governance explicitly addresses the deficits in the public water supply system and tries to meet water demand, at least partly, through market-clearing. This would mean that each remaining abstinence from water use would be voluntary, thereby fulfilling the equilibrium condition. The example of Amman also demonstrates that the overall willingness to pay for public water supply may be higher than expected, especially if this supply is safe and reliable, thus increasing the service level and lowering non-pecuniary barriers to access to water. At the same time, equity and the affordability of the public water supply should be ensured by a tariff structure that distributes the cost burden in correspondence with different user groups' ability to pay.

Initiating a transformation to a more sustainable urban water supply system can be complicated by the fact that mechanisms compensating for the absence of an adequate price level may be highly interdependent, as is the case with the need to obtain additional funding for infrastructure improvements and the difficulty of raising tariffs without an improved infrastructure. These interdependencies can only be resolved if trade-offs between different sustainability dimensions, especially the three guiding principles for a sustainability-oriented urban transformation, are taken into consideration. As the example of Amman shows, the partially illegal private tanker water market could be seen as an unambiguous problem if it was viewed purely from the perspectives of resource efficiency and environmental sustainability. This would, however, neglect its important role in balancing the shortcomings of the piped water network and, thus, its contribution to residents' quality of life. As we have shown, an attempt to fully appreciate these trade-offs cannot only reveal the complexity of the governance challenge, but also identify steps towards resolving it.

Acknowledgments The chapter is part of a larger research undertaking in the Belmont Forum project "Integrated Analysis of Freshwater Resources Sustainability in Jordan", or Jordan Water Project (JWP) (see: https://pangea.stanford.edu/researchgroups/jordan/, accessed 25 January 2015). We would like to thank the entire JWP team for their valuable support.

This work was conducted as part of the Belmont Forum water security theme for which coordination was supported by the US National Science Foundation under grant GEO/OAD-1342869 to Stanford University. The authors of this work would also like to acknowledge support from the German Research Foundation. Any opinions, findings, and conclusions or recommendations expressed in this material do not necessarily reflect the views of the funding organizations.

References

Abu Amra B, Al-A'raj A, Al Atti N, Dersieh T, Hanieh R, Hendi S et al (2011) Amman water supply planning report, volume 3, appendices G & H. Report prepared by CDM for the United States Agency for International Development (USAID), the Ministry of Water and Irrigation (MWI), and the Water Authority of Jordan (WAJ). USAID, Amman, Jordan

Al-Najjar FO, Al-Karablieh EK, Salman A (2011) Residential water demand elasticity in greater Amman area. *Jordan*. J Agric Sci 7(1):93–103

Altaf MA, Whittington D, Jamal H, Smith VK (1993) Rethinking rural water supply policy in the Punjab, Pakistan. Water Resour Res 29(7):1943–1954

Andrews J (2015) Dr Hazim El-Naser – minister of water and irrigation, Jordan. The International Water Association's the Source. http://www.thesourcemagazine.org/dr-hazim-el-naser-minister-of-water-and-irrigation-jordan/#. Accessed 14 Sept 2016

Bedtke N (2015) Transformationsprozesse und Institutionen – eine theoretische Perspektive für die Wasserwirtschaft. In: Gawel E (ed) Die Governance der Wasserinfrastruktur, Band 2: Nachhaltigkeitsinstitutionen zur Steuerung von Wasserinfrastruktursystemen. Duncker & Humblot, Berlin, pp 9–76

Bonn T (2013) On the political sideline? The institutional isolation of donor organizations in Jordanian hydropolitics. Water Policy 15:728–737

Dalhuisen JM, de Groot HLF, Nijkamp P (2000) The economics of water: a survey of issues. Int J Dev Plan Lit 15(1):3–20

Dandy GC, McBean EA, Hutchinson BG (1984) A model for constrained optimum water pricing and capacity expansion. Water Resour Res 20(5):511–520

DOS – Department of Statistics (2014) JorInfo 1.0 online database. DOS, Amman. http://jorinfo.dos.gov.jo/jordaninfo7.0/libraries/aspx/Home.aspx. Accessed 25 Jan 2016

FDA/AFD – French Development Agency (2011) Jordan water demand management study. Report prepared for the Ministry of Water and Irrigation of Jordan (MWI). FDA/AFD, Paris, France

Gächter S, Herrmann B, Thöni C (2004) Trust, voluntary cooperation, and socio-economic background: survey and experimental evidence. J Econ Behav Organ 55:505–531

Gao E (2011) Do the buses run on time? Tribal diversity and public goods in Jordan. Paper presented at the 2011 Annual Meeting of the American Political Science Association, Seattle, USA, 1–4 September 2011

Gawel E (2016) Road pricing in Germany: a behavioral economics perspective. In: Beckenbach F, Kahlenborn W (eds) New perspectives for environmental policies through behavioral economics. Springer, Berlin, pp 237–251

Gawel E, Bretschneider W (2016) Sustainable access to water for all: how to conceptualize and to implement the human right to water. J Eur Environ Plann Law 13:190–217

Gerlach E, Franceys R (2009) Regulating water services for the poor: the case of Amman. Geoforum 40:431–441

Goldstein J (1986) Full cost water pricing. J Am Water Works Assoc 78(2):52–61

Grafton RQ, Ward MB (2008) Prices versus rationing: Marshallian surplus and mandatory water restrictions. Econ Rec 84:S57–S65

Green C (2003) Handbook of water economics: principles and practice. Wiley, Chichester

Griffin RC (2001) Effective water pricing. J Am Water Resour Assoc 37(5):1335–1347

Griffin RC (2016) Water resource economics, 2nd edn. MIT Press, Cambridge, MA/London

Griffin RC, Mjelde JW (2000) Valuing water supply reliability. Am J Agric Econ 82(2):414–426

Hashwa F, Tokajian S (2004) Intermittent water supply and domestic water quality in the Middle East. In: Zereini F, Jaeschke W (eds) Water in the Middle East and in North Africa: resources, protection and management. Springer, Berlin, pp 157–166

Humpal D, El-Naser H, Irani K, Sitton J, Renshaw K, Gleitsmann B (2012) Review of water policies in Jordan and recommendations for strategic priorities: final report. Report prepared for the Institutional Support and Strengthening Program (ISSP). United States Agency for International Development (USAID), Washington, DC, USA

IRG – International Resources Group (2015) Strategic master plan for municipal water infrastructure: final report. Report prepared for the Institutional Support and Strengthening Program (ISSP). United States Agency for International Development (USAID), Washington, DC, USA
Kabisch S, Kuhlicke C (2014) Urban transformations and the idea of resource efficiency, quality of life and resilience. Built Environ 40(4):475–485
Kalyanaram G, Winer RS (1995) Empirical generalizations from reference price research. Mark Sci 14(3):G161–G169
Klassert C, Sigel K, Gawel E, Klauer B (2015) Modeling residential water consumption in Amman: the role of intermittency, storage, and pricing for piped and tanker water. Water 7:3643–3670 doi:10.3390/w7073643
Kubursi A, Grover V, Darwish AR, Deutsch E (2011) Water scarcity in Jordan: economic instruments, issues and options, The Economic Research Forum (ERF) working paper no. 599. ERF, Cairo
Lancaster KJ (1966) A new approach to consumer theory. J Polit Econ 74(2):132–157
Madanat S, Humplick F (1993) A model of household choice of water supply systems in developing countries. Water Resour Res 29(5):1353–1358
MWI – Ministry of Water and Irrigation (2009) Water for life—Jordan's water strategy 2008–2022. MWI, Amman
Namrouqa H (2016) Authorities uncover theft of water equal to daily share of 48,000 people. The Jordan Times, April 18. http://jordantimes.com/news/local/authorities-uncover-theft-water-equal-daily-share-48000-people?platform=hootsuite. Accessed 14 Sept 2016
Nauges C, Whittington D (2010) Estimation of water demand in developing countries: an overview. World Bank Res Obs 25:263–294
OECD – Organization for Economic Co-operation and Development (2009) Managing water for all: an OECD perspective on pricing and financing. OECD, Paris
OECD – Organization for Economic Co-operation and Development (2010) Pricing water resources and water and sanitation services. OECD, Paris
Oelmann M (2010) Privatsektorbeteiligung als Lösung für unterfinanzierte Wassermärkte in Schwellen- und Entwicklungsländern? Korreferat zu Rahel Schomaker. In: Aufderheide D, Dabrowski M (eds) Effizienz und Gerechtigkeit bei der Nutzung natürlicher Ressourcen. Wirtschaftsethische und moralökonomische Perspektiven der Rohstoff-, Energie- und Wasserwirtschaft. Volkswirtschaftliche Schriften Heft 560. Berlin, Duncker&Humblot, pp 191–215
Potter RB, Darmame K (2010) Contemporary social variations in household water use, management strategies and awareness under conditions of "water stress": the case of greater Amman, Jordan. Habitat Int 34:115–124
Potter RB, Darmame K, Nortcliff S (2010) Issues of water supply and contemporary urban society: the case of Greater Amman, Jordan. Philos Trans R Soc A Math Phys Eng Sci 368:5299–5313. doi:10.1098/rsta.2010.0182
Rosenberg DE, Tarawneh T, Abdel-Khaleq R, Lund JR (2007) Modeling integrated water user decisions in intermittent supply systems. Water Resour Res 43:W07425 doi:10.1029/2006WR005340
Rosenberg DE, Talozi S, Lund JR (2008) Intermittent water supplies: challenges and opportunities for residential water users in Jordan. Water Int 33:488–504. doi:10.1080/02508060802474574
Schomaker R (2010) Wasserversorgung zwischen Staat und Markt – Das Beispiel Naher Osten. In: Aufderheide D, Dabrowski M (eds) Effizienz und Gerechtigkeit bei der Nutzung natürlicher Ressourcen. Wirtschaftsethische und moralökonomische Perspektiven der Rohstoff-, Energie- und Wasserwirtschaft. Volkswirtschaftliche Schriften Heft 560. Berlin, Duncker&Humblot, pp 191–215
Sigel K, Klassert C, Zozmann H, Klauer B, Gawel E (2017) Impacts of private tanker water markets on sustainable urban water supply: an empirical study of Amman, Jordan. UFZ Bericht. UFZ, Leipzig

Sima LC, Kelner-Levine E, Eckelman MJ, McCarty KM, Elimelech M (2013) Water flows, energy demand, and market analysis of the informal water sector in Kisumu, Kenya. Ecol Econ 87:137–144

Tabieh M, Salman A, Al-Karablieh E, Al-Qudah H, Al-Khatib H (2012) The residential water demand function in Amman-Zarka Basin in Jordan. Wulfenia J 19:324–333

Theodory G (2000) Willingness and ability of residential and non-residential subscribers in Greater Amman to pay more for water: a study conducted for the water authority of Jordan, Prepared for the United States Agency for International Development (USAID). Amman, Jordan: Development Alternatives, Inc. (DAI). http://pdf.usaid.gov/pdf_docs/pnacq616.pdf. Accessed 25 Jan 2016

UN – United Nations General Assembly, 60th Session (2005) World summit outcome. Resolution A/RES/60/1, New York, USA, 24 October 2005

WBWDRT – World Bank Water Demand Research Team (1993) The demand for water in rural areas: determinants and policy implications. World Bank Res Obs 8(1):47–70

WCED – World Commission on Environment and Development (1987) Report of the world commission on environment and development: our common future. Oxford University Press, New York

Whittington D, Lauria DT, Mu X (1991) A study of water vending and willingness to pay for water in Onitsha, Nigeria. World Dev 19(2/3):179–198

Yorke V (2013) Politics matter: Jordan's path to water security lies through political reforms and regional cooperation. Swiss National Centre of Competence in Research (NCCR) working paper no. 2013/19. NCCR, Bern, Switzerland

Zawahri NA (2012) Popular protests and the governance of scarce fresh water in Jordan. Arab World Geographer 15(4):267–301

Zeitoun M, Allan T, Al Aulaqi N, Jabarin A, Laamrani H (2012) Water demand management in Yemen and Jordan: addressing power and interests. Geogr J 178:54–66

The TEEB Approach Towards Sustainable Urban Transformations: Demonstrating and Capturing Ecosystem Service Values

Bernd Hansjürgens, Miriam Brenck, Robert Bartz, and Ingo Kowarik

1 Introduction

Green spaces are an important urban resource. They provide a bundle of ecosystem services that benefit humans (Elmqvist et al. 2013; Gómez-Baggethun and Barton 2013; Haase et al. 2014). For the future prospects of cities they are not only important for the health and well-being of their inhabitants, but also as a direct economic factor, e.g., to attract investors, companies, and employees. Thus, green (and blue) spaces are becoming increasingly decisive for providing a high quality of life in urban areas.[1] At the same time, urban green infrastructure contributes to more effective and efficient solutions for urban challenges, compared to traditional technical ("gray") solutions. There are many examples of their ecological and social superiority, although this has rarely been demonstrated in economic terms, i.e., in terms of a higher cost-benefit-ratio (see Naturkapital Deutschland – TEEB DE 2016). As a consequence, urban green infrastructures have not or only insufficiently been taken into account by local decision-makers. Accordingly, they have only been implemented occasionally in urban planning processes.

This chapter relies on major results of the Natural Capital Germany – TEEB DE – Report on cities (Naturkapital Deutschland – TEEB DE 2016).

[1]While focusing on urban green spaces and urban green infrastructure, we are fully aware that many of our arguments also hold for water-related ecosystem services ("blue" spaces; see Nivala et al., "Green Infrastructure for Increased Resource Efficiency in Urban Water Management", in this volume).

B. Hansjürgens (✉) • M. Brenck
Department of Economics, Helmholtz Centre for Environmental Research-UFZ, Leipzig, Germany
e-mail: bernd.hansjuergens@ufz.de; miriam.brenck@ufz.de

R. Bartz • I. Kowarik
Department of Ecology, Technical University of Berlin, Berlin, Germany
e-mail: robert.bartz@tu-berlin.de; kowarik@tu-berlin.de

S. Kabisch et al. (eds.), *Urban Transformations*, Future City 10,
DOI 10.1007/978-3-319-59324-1_7

The underlying economic reason is that many services provided by nature are public goods with widespread benefits for all humans. The benefits of these services, in the form of regulation of water bodies, protection against flooding, reduction of heat stress, etc., are therefore not mirrored in market activities (with resulting market prices). They also are not or only insufficiently taken into account by public decision-makers. In contrast, what is taken into account are the opportunity costs of urban green spaces, which increase over time with the urban dynamics (i.e. the dramatic increase of population, migration, economic activities, higher densities, etc.,). Both effects – the non-counting of benefits of urban green and the counting of its opportunity costs – lead to a neglect of green infrastructure in many planning and investment decisions and to a loss of urban green spaces. This is a result of existing pressures and interests that are of particular concern for growing municipalities. Because municipal revenues strongly depend on the settlement of private households and companies in cities and communities – due to corresponding revenues from income tax, business tax, and property tax – many open areas are released for building development and company settlement.

Against this background, two questions arise: How can "green" in cities be sustained and promoted? And, how can an urban transformation be initiated that follows the path of greening und sustainable urban development?

The TEEB study, "The Economics of Ecosystems and Biodiversity", offers an innovative answer to these questions. The central message is: Sustainable urban transformations can be achieved by recognizing and demonstrating the values of ecosystem services and by incorporating them into decision-making. *Recognizing* and *demonstrating* ecosystem services refers to the necessity that individuals and communities are aware of such services and that mechanisms exist that make ecosystem services' values visible (we call this "demonstrating" values of ecosystem services). *Incorporating* values into decision-making refers to specific instruments and measures to include ecosystem services in decisions (we call this "capturing" values of ecosystem services) (TEEB 2010; Naturkapital Deutschland – TEEB DE 2016). By taking such an economic (TEEB) approach, the prerequisites for "real" transformations, understood as "fundamental forms of change" (see Kabisch et al., "Introduction: Urban Transformations – Sustainable Urban Development Through Resource Efficiency, Quality of Life, and Resilience", in this volume) can be improved significantly.

This chapter refers to the role of ecosystem services for human health and human well-being (Sect. 2), how ecosystem services can be made economically visible (Sect. 3) and how they can be captured in decision-making (Sect. 4). We conclude with some comments on the importance of this perspective for sustainable urban transformations (Sect. 5).

2 Urban Green and Its Services for Human Health and Well-Being

Urban green, in its diverse manifestations, increases the attractiveness of cities as places to live and work. It serves all three goals that characterize the framework conditions for successful urban transformations (see Kabisch et al., "Introduction: Urban Transformations – Sustainable Urban Development Through Resource Efficiency, Quality of Life, and Resilience", in this volume): It contributes to resource efficiency by focusing on the natural basis of our lives and by emphasizing that resources should be used wisely (i.e., effectively and efficiently). It contributes to the quality of life because the benefits obtained from green infrastructures directly or indirectly improve humans' well-being. And it contributes to resilience by strengthening nature's capacity to protect against harm, e.g., in the case of flooding.

This section outlines essential ecosystem services that support human health, strengthen social cohesion, maintain citizens' contacts to nature, and increase, overall, the attractiveness of cities.

Human Well-Being in the City

Cities usually are more affected by overheating, high concentrations of particulate matter and other air pollutants, as well as noise, compared to their surroundings. These stress factors, individually and, especially, combined, can lead to serious health-related impairments and result in increased morbidity and mortality rates (Gabriel and Endlicher 2011; Heudorf and Meyer 2005; Hoffmann et al. 2008; Schneider et al. 2009). Furthermore, climate change contributes to greater stresses, which will increase considerably in the future. Heat waves will become more frequent, more intensive, and will last longer (IPCC 2013). Urban heat islands will expand, causing more people to be directly affected by overheating and heat waves (Li and Bou-Zeid 2013). Additionally, environmental risks will increase as a result of heavy rainfalls.

In particular, regulating ecosystem services can reduce these stresses. Trees and other vegetation elements, as well as rivers and lakes, reduce heat stress through shading and evaporative cooling effects. Vegetation also binds particulate matter and other air pollutants. Along roadsides, it can reduce the concentration of particulate matter by 15% (Kuypers et al. 2007). Water resources, forests, and parks are "cooling islands" that significantly contribute to a reduction of heat stress for neighboring city quarters during the night. Even smaller green spaces can reduce temperature by three to four degrees, compared to a built environment (Mathey et al. 2011; Bruse 2003; Ermer et al. 1996).

Incentives for humans' physical activity are created by proximity to nature and designed open spaces. Medical studies demonstrate the beneficial effects of physical activity on human health, for example, strengthening of the cardiovascular system as well as the immune system (Richardson and Mitchell 2010). Epidemiological studies show that an increasing distance to urban green spaces correlates with higher risks of disease and a lower life expectancy (Maas et al. 2009). Contact with nature increases the ability to reduce stress, aggression, or fear and positively influences concentration and performance (Hartig et al. 2003; Roe et al. 2013).

Strengthening Social Cohesion

Public open spaces such as parks, water body shores, communal gardens, and playgrounds, offer places for people to meet, relax, and communicate. Moreover, they contribute to environmental justice because access is free of cost.

Traditional allotment gardens are used by approximately 5 million people in Germany. Community gardens, such as neighborhood or intercultural gardens – often set up by residents' initiatives – are experimental spaces in cities that link social and ecological aims. They offer room for interaction and for creative cultural and political exchange. A garden project can strengthen the sense of community, which also contributes to the identification with the neighborhood or the city (HMWVL 2012; Jirku 2013; Wendorf 2011).

Children and Young People Experience Nature

The opportunity to play and roam freely in natural surroundings and to interact with nature's elements (water, soil, plants, animals) supports healthy physical and mental development of children and young people. Individual responsibility, creativity, risk competence, and social competence, as well as linguistic, motoric and scientific abilities benefit from these experiences (Dadvand et al. 2015; Gebhard 2003; Maller et al. 2006).

Green learning places and areas where nature can be experienced (so-called natural experience areas – "Naturerfahrungsräume") contribute to the experience of nature and to environmental education. This includes forest schools as well as urban agricultural projects. Urban wilderness, which often emerges on fallow land, creates opportunities to experience nature in a residential environment.

Growing Food – More than Just Nutrition

Urban agriculture and growing food in urban gardens play an important role – as provisioning services and for sustainable health promotion. Urban gardening and urban agriculture offer children the chance to find out about the origins of natural food. They offer experience in cultivating vegetables, fruit, and herbs. Awareness about local and regional products and about a healthy diet increases through gardening experiences, e.g., in community or school gardens (Lobstein et al. 2015).

Urban agriculture profits from better regional marketing and the direct sale of high-quality products. A partial self-supply with food from urban gardens or agricultural allotments also has economic relevance for some households (Neu and Nikolic 2014). In allotment gardens in the German Rhein-Ruhr area, the degree of self-supply with fruit and vegetables (excluding tropical fruits) was estimated to be around 50% (LUA NRW 2001). Nevertheless, experiencing the pleasures of gardening and working in nature are the main reasons for gardening (Neu and Nikolic 2014).

Urban Nature as a Location Factor

Municipalities compete against each other when it comes to tax revenues, production sites, workplaces and attractive places of residence. Companies usually make decisions about suitable location according to hard location factors such as transport links or proximity to sales markets. Nevertheless, if different areas are considered, "soft", location factors can be crucial. These soft factors involve environmental quality, valuable recreation areas, and an attractive environment for living and working.

A higher availability of urban green areas is reflected in higher real estate prices. The example of Cologne demonstrates this: The effect of shorter distances to parks on real estate prices is smaller than the impact of other factors such as size and age of properties. However, in view of the large number of real estate transactions, small effects also lead to a considerable capitalization (Kolbe and Wüstemann 2014).

This example illustrates the fact that urban green has substantial economic importance. It contributes to health, social cohesion, and healthy development of children, and also serves as a relevant location factor for municipalities.

3 Demonstrating the Value of Urban Green

In order to bring green spaces and green infrastructures to the tables of decision-makers, their advantages have to be made visible, particularly in economic terms. There are examples of green infrastructures that seem to be superior to "gray" (technical) infrastructure solutions; however, their economic superiority (in terms of better cost-benefit ratios) has rarely been demonstrated. Here, an economic perspective, with its own methods and indicators, is a helpful tool; it can help to develop arguments about the benefits of green solutions that can influence stakeholders' awareness and political decision-making.

To reveal, demonstrate, and assess ecosystem services, various methods can be applied, ranging from qualitative descriptions over quantitative approaches up to monetary evaluation (Gómez-Baggethun and Barton 2013; Gómez-Baggethun et al. 2015). Monetary evaluation methods are only one approach, amongst others, to assess urban green and its ecosystem services. In this section, we briefly describe suitable economic valuation methods for urban green spaces and infrastructures.

The economic perspective aims at uncovering the value of nature. Urban green can be seen as a valuable capital good – as "natural capital". This natural capital is comparable to human and physical capital: it is a stock – urban nature – that yields profit in the form of ecosystem services. By applying an economic perspective, awareness of the importance of urban green can be promoted. The economic perspective can reveal the "hidden values" of nature, which are considered insufficiently in decision-making processes related to urban planning, because they are overseen or underestimated.

3.1 *Which Values Are Covered by the Economic Approach?*

The term "value" can be interpreted in two directions (Oxford University Press 1989): Firstly, it can be seen as the monetary worth of something, e.g., price as the equivalent of a commodity, which is shown in money or other means of payment. In its second meaning, it can be determined as validity, importance, or significance of an object, person, or circumstance. This is also linked to the word's Latin origin (valor = validity).

It should be noted that every valuation is context-specific and depends on its basic conditions: ecological, social, and cultural circumstances, individual preferences, basic values, views and opinions, perceptions of the society, individuals' or societies' prosperity, economic situation, etc. (Vatn 2009; Brondizio et al. 2010). These factors are situational, as well as space- and time-related. There is, thus, not one "correct" value; instead, there are diverse values of nature, based on different moral concepts (or value paradigms) (Spangenberg and Settele 2016).

Several values of nature can be distinguished (see Fig. 1). Here, we take an *anthropocentric view* on nature's values because we focus on the benefits humans obtain from nature.

The *physiocentric* view highlights the "intrinsic" value of nature (Krebs 1999; WBGU 1999; EPA 2009). The recognition of the inherent value of nature is independent of an appreciation by humans from their usability perspective (Eser and Potthast 1999). This intrinsic value thus exists without direct or indirect benefits for people. In contrast, the economic view is based on anthropocentric paradigms and therefore excludes this inherent value of nature in its approach.

Social values are located between individual (anthropocentric) values and the inherent (physio-centric) values of nature. Some authors place social values among economic values, because they are based on aggregated individual preferences. Others consider them to be a category of their own (for details see Kenter et al. 2015; Hansjürgens et al. 2016). In this chapter, the term "social values" assumes that a person has certain preferences, as a member of society, that differ from individual preferences. For instance, if individuals act as citizens, they might make other decisions than they would if they followed their individual interests (Gómez-

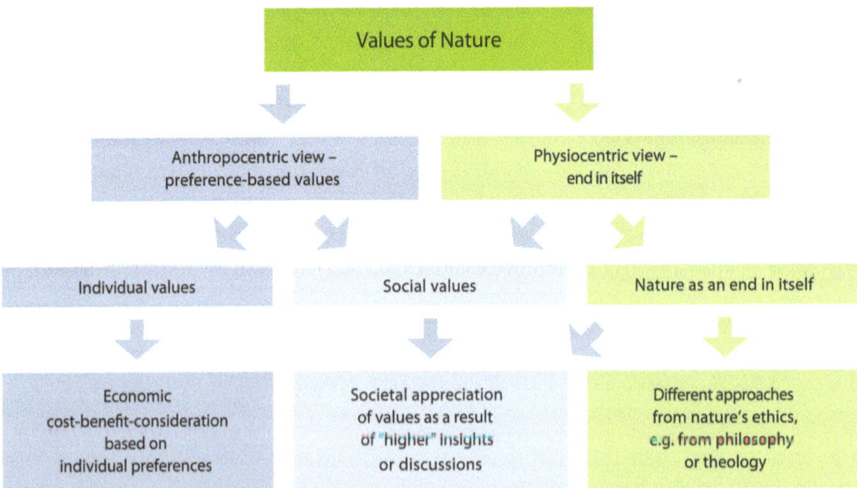

Fig. 1 Different values of nature (Source: Naturkapital Deutschland – TEEB DE 2016, p. 31)

Baggethun et al. 2015; Kenter et al. 2015, Chan et al. 2016). This phenomenon is described by Vatn (2009) as "We-rationality", in contrast to "I-rationality".

3.2 Methodological Approaches to Demonstrating Values of Urban Green

Various methodological approaches can be applied to identify and demonstrate urban ecosystem services with regard to health and quality of life. Economic assessment methods are based on individual preferences. Here, basically, two approaches are available: (1) the use of market prices of goods, which are related to the demand for environmental goods, for example by their complementarity; and (2) the direct inquiry about the willingness to pay for the environmental goods at issue.

1. In the first group of valuation approaches, actors have made purchase decisions for goods that are related to various environmental qualities. In this way, users indirectly express a demand for environmental quality. In the terminology of economics, they reveal their preferences – described as "Revealed Preferences Approaches" (for an overview, see Freeman 2003). For the urban scope, these include, in particular, decisions on the place of residence or travel costs to attractive green cities with recreational value (Travel Cost Method).
2. The second group of approaches deals with direct survey-based inquiries of the willingness of demanders to pay for nature (or, to be more precise: for possible improvements of particular components of environmental quality). For this purpose, the components of environmental quality (e.g., the extension of an urban park) are described in all their dimensions, potential scopes are differentiated (e.g., changes of the area of the urban park), and the individual willingness to pay is investigated. Because the willingness to pay is stated directly, these approaches are described as "Stated Preferences Approaches".

Participatory and deliberative methods (processes of participation or "negotiation") are particularly suitable for capturing the social values of urban green. Here, groups of stakeholders can exchange information about the need for urban green and its societal relevance (for details, see Lienhoop et al. 2015).

4 Capturing Urban Green in Decision-Making

As shown above, urban green provides many services for humans – especially for health and wellbeing – with strong effects on social costs. How can these green spaces and their multi-functionality be sustained and developed? Which strategies and measures exist, which are suitable and which are to be pursued? At this point, we highlight four aspects (see Fig. 2):

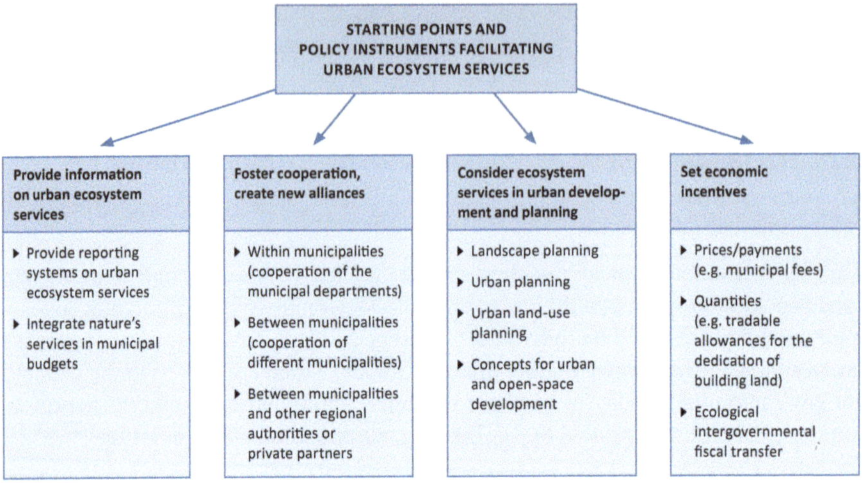

Fig. 2 Policy Instruments for strengthening urban green (Source: Naturkapital Deutschland – TEEB DE Naturkapital Deutschland – TEEB DE 2016, p. 280)

1. Delivering information on nature and ecosystem services in cities;
2. Enhancing cooperation between sectors and areas within a municipality as well as between municipalities and other levels of regional authorities;
3. Considering urban green in integrative planning approaches; and
4. Providing economic incentives.

4.1 Delivering Information

An essential basis for a stronger consideration of urban green in decisions is information about the various services of nature, as they are presented in the Natural Capital Germany report on cities (Naturkapital Deutschland – TEEB DE 2016). Information on urban ecosystem services can constitute important support for decision-makers. So far, such information is only used occasionally in decision contexts. Municipalities might ask for information about urban ecosystem services, but they often lack the necessary budgetary resources for a local assessment. But even general information on urban ecosystem services can be very helpful for pointing out the relevance of multifunctional green spaces. Based on this information, synergies and trade-offs between different land-use options or alternative design options can be revealed and discussed.

To strengthen the promotion and development of urban green, it is important that the "tangible green assets" – the natural capital – are visualized in budgetary decisions of cities and municipalities. One possible approach is the implementation of a "green" double-accounting budget system, where urban green can be assessed in terms of financial assets (Hilgers and Burth 2011; Güse et al. 2009). In this system, urban natural capital can complement municipal assets. The green double-accounting makes the value of urban green areas explicit: they bind significant public investments, and they also are subject to ongoing investments for maintenance and design measures. It needs to be kept in mind that a monetary accounting of urban green can only be an indication of its minimal values, because relevant cultural ecosystem services such as social cohesion, environmental education, or effects on well-being cannot adequately be captured in monetary values.

4.2 Enhancing Cooperation and Creating New Alliances

Another strategic approach to strengthen urban green and ecosystem services is the enhancement of collaboration. This concerns alliances between (1) different sectors within a municipality and the comprehensive coordination between them, (2) different municipalities, and (3) the municipality and other governmental levels.

Cooperation Between City Sectors
With regard to cities and communities, it is especially important to connect diverse sectors within the local community that benefit from urban green spaces, such as infrastructure, health, education, climate protection and adaptation, and economic development.

Some examples: It was shown that urban green promotes health. In addition, positive effects arise for social cohesion in neighborhoods and for children and adolescents who can experience positive effects due to urban green (Röhner 2013). In this context, the responsible authorities in the community, such as the departments for urban green, health, youth welfare, and social security, should have a common interest in promoting and facilitating access to urban green for various social and demographic groups. In many cases, the objectives of the individual administrative departments are not aligned with each other because they compete for small public budgets. Improved coordination and comprehensive planning can contribute to identifying synergy effects, in order to reach common aims and to push for appropriate measures. Hence, new paths for decision-making and coordination are needed. This involves, in particular, making the multi-functionality of urban green visible, making use of the given synergies, collectively discussing trade-offs and common interests, and developing cooperation to benefit from urban green spaces and green infrastructure.

A German example for a comprehensive urban development concept that integrates the promotion of ecosystem services as a target criterion in urban planning is provided by the city of Augsburg. In July 2015, the city council adopted the "Future Guidelines for Augsburg" as a basis for the sustainable development of the city (Stadt Augsburg 2015). The concept integrates various administrative bodies and departments and formulates common and comprehensive objectives. It differs from other communities in the comprehensive inclusion of numerous urban departments and sectors to deal with issues related to green urban infrastructures.

Cooperation Among Cities

An important field for inter-communal cooperation is the promotion of municipal economic development. Until now, municipalities see themselves in competition for attracting investors and businesses. This competition led to the designation of new commercial and industrial areas for the settlement of companies. Spaces were designated but were not used, due to a lack of demand (see, for Europe: EEA 2006; for Germany: Schröter-Schlaack 2013).

This land consumption is alarming, with regard to, e.g., the German sustainability goal of limiting land consumption to 30 hectare per day until 2020 (Bundesregierung 2016). In addition, it represents a considerable fiscal burden for municipalities. Instead of additional tax income from property, business, and income taxes, deficits that are the result of the development and maintenance of oversized infrastructure projects have emerged.

A stronger collaboration for greater harmonization of land-use policies between municipalities is urgently needed. Inter-communal cooperation cannot only save space. It can also reduce costs, without forcing individual municipalities to give up the advantages of attracting new business. The goal is a joint marketing of regional commercial premises and can be implemented by the following approach: The municipalities gather these industrial and commercial areas into a common pool. The share of each municipality is allocated according to the value of the properties within this pool. The incorporated commercial premises are jointly marketed by an agency and the collected business tax is distributed according to the shares in the common pool (Naturkapital Deutschland – TEEB DE 2016). In doing so, price competition among the municipalities is avoided and full utilization of already assigned commercial areas is promoted.

Cooperation Between Cities and Other Governmental Levels

The multifunctional advantages of nature-based solutions can also be used in the collaboration between municipalities and superordinate levels of regional authorities (e.g., federal government and federal states). For example, a program dealing with adaptation to climate change, which is initiated by the federal government, will, in many cases, refer to concrete implementation projects that must be implemented locally. One example of the facilitation of nature-based solutions is the "Green Roof Strategy Hamburg": the aim of this statewide strategy for green roofs is the greening of 100 ha of roof area until 2020. The project is financed with 3 Mio. Euro by the German federal environmental ministry's program "Measures for adaptation to climate change". In cooperation with the HafenCity University of Hamburg, it works towards a legally binding requirement for installing green roofs (Stadt Hamburg 2014). The design of such programs in a way that includes urban green and also takes municipal decisions into consideration can generate considerable synergies that exceed adaptation to climate change. Effects can then, in addition to the actually intended effects on climate adaptation, also relate to health, social coherence, and life quality, as described above.

4.3 *Strengthening Ecosystem Service Perspectives in Urban and Land Use Planning*

Urban planning is a key component of urban development. This is where urban green needs to be taken into account. The planning instruments should consider the multi-functionality of green infrastructure. On the one hand, landscape planning is based on a weighing-up of all relevant impacts, including environmental impacts, which can be referred to as the "supply side" of ecosystem services. However, a focus of the ecosystem services concept is also on the "demand side", i.e., on the stakeholders in a certain area (inhabitants and visitors) and how they benefit. This aspect is, to date, not or only implicitly taken into account in planning processes (Hansen et al. 2015). Initial ideas about how the ecosystem service approach can be integrated into this planning system, by identifying stakeholders and their preferences, exist. Within the framework of specific planning decisions, multi-criteria analyses can be applied that include qualitative arguments in addition to monetary costs and benefits. In this way, the value of nature-based solutions can be considered appropriately in the evaluation processes.

4.4 *Providing Economic Incentives*

Finally, an additional area of activity for strengthening urban green and urban ecosystem services is provided by economic instruments. They are supposed to increase the price for using nature for decision-makers, or to make the nature-friendly variant cheaper. Most importantly, if such instruments are implemented and the resulting price signal is transferred to producers and consumers via market prices, private decision-makers will decide whether they take up the offered incentive or not. The result is to reach the environmental aim in an efficient way.

In the municipal sector, however, options for introducing economic incentives are rather limited. There usually is no sovereignty, on the municipal level, for levying taxes or other new price-based instruments (Droste et al. 2017). The municipal level is, basically, only allowed to regulate those areas in which it has sovereignty on its own. And, with respect to quantity-based economic instruments, although there are proposals for introducing tradable development rights, their implementation is, so far, restricted to a few cases in a pilot testing phase (Schröter-Schlaack 2013; Umweltbundesamt 2016).

A possible area of action is to use charges, either through an incentive-oriented design of existing charges (e.g., municipal fees for water services) or by levying new ones (e.g., water- or land-use charges). Charges represent a solution in terms of price-indicating opportunity costs of using ESS. Fees for water services usually have to conform to the principle of cost recovery (Gawel 1995). However, margins exist here, because the determination of underlying costs also allows – at least in principle – the inclusion of environmental and resource costs (Gawel 2016). Apart from the degree of cost recovery, setting environmental incentives can be achieved through appropriate levy design. For example, a separate rainwater run-off charge can offer incentives to avoid soil sealing (Rüger et al. 2015). The same is true for wastewater fees that are oriented to the area of sealed natural ground (Geyler et al. 2014). Fee-based policies may also contribute to abating disincentives to shrink green spaces in cities, as is the case for cemetery charges in Germany (TEEB DE 2016). Here, surface-oriented user pricing keeps the public green function of these areas free of charge (overcoming the lack of money problem for municipalities) and sets disincentives for users, in order to reduce the demand for large graves. This leads to a significant trend to reduce this kind of green areas in cities. Incentives to curtail land consumption may also be provided by tradable land-use designation rights, creating a genuine market price for "consuming" land-use (Schröter-Schlaack 2013).

Another instrument consists of the communal fiscal equalization scheme, as it exists in Germany (Ring 2008). This instrument regulates fiscal compensations between municipalities (district-independent cities, districts, district municipalities) within a federal state, while the federal state provides specified compensation payments in accordance with the financial revenue power and the expenditure requirements of the municipalities. The municipal financial needs are based on the number of inhabitants as well as some other indicators that essentially reflect the centrality of a municipality and the interconnected financial need (Ring and Mewes 2013). Examples of indicators of financial needs are central institutions such as opera houses but also the number of school students. The financial needs for nature conservation issues or the safeguarding of natural capital are currently not taken into account within the communal financial compensation in Germany.

This is where reform proposals start, but they have, so far, only been discussed in the literature and have not yet been integrated into on debates political reforms. According to these proposals, ecological indicators that represent the provision of environmental public goods and services are to be additionally considered as indicators to calculate financial needs and analogous allocations. This would be an important step towards providing economic incentives for public actors to promote ecosystem services.

5 Concluding Remarks

The promotion of urban ecosystem services is a promising strategy for achieving comprehensive and sustainable urban transformation. The TEEB approach can play a decisive role in implementing such a strategy because it helps to demonstrate and capture values of ecosystem services. Based on this focus on economic aspects, several recommendations for urban transformation can be developed. They include, inter alia, a focus on the benefits of nature's services and the identification of beneficiaries, and an inclusion of these benefits in cost-benefit ratios. If these issues are taken into account, they can foster urban transformation in the direction of a more resource-efficient and resilient municipality, with increased quality of life.

With respect to instruments that support this overarching task, several come into play, and many of them are already available: strengthening of cross-sectoral communication within the different sectors and departments of cities (particularly between sectors such as planning, infrastructure, mobility, health, or education), inter-communal collaboration between municipalities, as well as collaboration between the local municipal level and upper governmental levels. Proposed measures also include financial incentives or improved planning tools. Key elements are cross-sectoral thinking, alliances of various stakeholders, identifying synergies, defining common goals of urban transformation, knowledge transfer, and nature-based solutions.

References

Brondizio ES, Gatzweiler FW, Zografos C, Kumar M (2010) The socio-cultural context of ecosystem and biodiversity valuation. In: Kumar P (ed) TEEB – The Economics of Ecosystems and Biodiversity Ecological and Economic Foundations. Routledge, London, pp 149–176

Bruse M (2003) Stadtgrün und Stadtklima. Wie sich Grünflächen auf das Mikroklima in Städten auswirken. LÖBF-Mitteilungen 1:66–70. http://www.helix-pflanzensysteme.de/media/block_downloads/213/21-stadtgruen-und-stadtklima-bruse-loebf-mitteilungen-1-03.pdf. Accessed 11 Nov 2016

Bundesregierung (2016) Deutsche Nachhaltigkeitsstrategie (German Sustainability Strategy). New Edition 2016. Draft of 30. May 2016. https://www.bundesregierung.de/Content/DE/StatischeSeiten/Breg/Nachhaltigkeit/0-Buehne/2016-05-31-download-nachhaltigkeitsstrategie-entwurf.pdf?__blob=publicationFile&. Accessed 11 Nov 2016

Chan KMA, Balvanera P, Benessaiah K, Chapman M, Díaz S, Gómez-Baggethun E et al (2016) Opinion: why protect nature? rethinking values and the environment. Proc Natl Acad Sci 113(6):1462–1465

Dadvand P, Nieuwenhuijsen MJ, Esnaola M, Forns J, Basagaña X, Alvarez-Pedrerol M et al (2015) Green spaces and cognitive development in primary schoolchildren. Proc Natl Acad Sci 112(26):7937–7942

Droste N, Schröter-Schlaack C, Hansjürgens B, Zimmermann H (2017) Implementing nature-based solutions in urban areas: financing and governance aspects. In: Kabisch N, Korn H, Stadler J, Bonn A (eds) Nature-based solutions to climate change adaptation in urban areas. Linkages between science, policy and practice. Springer, Cham, pp 307–321

EEA - European Environment Agency (2006). Urban Sprawl in Europe the ignored challenge EEA Report No 10/2006. http://www.eea.europa.eu/publications/eea_report_2006_10/eea_report_10_2006.pdf. Accessed 11 Nov 2016

Elmqvist T, Fragkias M, Goodness J, Güneralp B, Marcotullio PJ, McDonald RI, et al (Eds.) (2013) Urbanization, biodiversity and ecosystem services: challenges and opportunities – a global assessment Springer, Dordrecht, Heidelberg, New York, London

EPA – Environmental Protection Agency (2009) Valuing the protection of ecological systems and services. A report of the EPA advisory board. EPA, Washington, DC

Ermer K, Hoff R, Mohrmann R (1996) Landschaftsplanung in der Stadt. Ulmer, Stuttgart

Eser U, Potthast T (1999) Naturschutzethik. Eine Einführung in die Praxis. Nomos, Baden-Baden

Freeman AM (2003) The measurement of environmental and resource values, theory and methods, 2nd ed. Resources for the Future Press, Washington, DC

Gabriel K, Endlicher W (2011) Urban and rural mortality rates during heat waves in Berlin and Brandenburg, Germany. Environ Pollut 159(8):2044–2055

Gawel E (1995) Die kommunalen Gebühren. Duncker & Humblot, Berlin

Gawel E (2016) Environmental and resource costs under article 9 water framework directive. Challenges for the implementation of the principle of cost recovery for water services. Duncker & Humblot, Berlin

Gebhard U (2003) Kind und Natur. Die Bedeutung der Natur für die psychische Entwicklung. VS Verlag für Sozialwissenschaften, Wiesbaden

Geyler S, Bedtke N, Gawel E (2014) Sustainable rainwater management in existing settlements. Gwf-Wasser, Abwasser 155(1):96–102. 155(2): 214–222

Gómez-Baggethun E, Barton DN (2013) Classifying and valuing ecosystem services for urban planning. Ecol Econ 86:235–245. doi:10.1016/j.ecolecon.2012.08.019

Gómez-Baggethun E, Martín-López B, Barton D, Braat L, Saarikoski H, Kelemen E et al. (2015). State-of-the-art report on integrated valuation of ecosystem services. EU FP7 OpenNESS project. Deliverable 4.1, European Commission FP7. http://www.openness-project.eu/sites/default/files/Deliverable%204%201_Integrated-Valuation-Of-Ecosystem-Services.pdf. Accessed 17 Nov 2016

Güse, E, Thieme-Hack, M, Thomas, J (2009) Grüne Doppik - Vermögensbewertung von Vegetation für die Doppelte Buchführung der öffentlichen Hand. Kommunalnachrichten Sachsen-Anhalt - KNSA - im Städte- und Gemeindebund Sachsen-Anhalt. Magdeburg, Ausgabe 8/9, Nr. 476, pp. 9–20

Haase D, Larondell N, Andersson E, Artmann M, Borgstrom S, Breuste J et al (2014) A quantitative review of urban ecosystem service assessments: concepts, models, and implementation. Ambio 43(4):413–433

Hansen R, Frantzeskaki N, McPherson T, Rall E, Kabisch N, Kaczorowska A (2015) The uptake of the ecosystem services concept in planning discourses of European and American cities. Ecosyst Serv 12:228–246. doi:10.1016/j.ecoser.2014.11.013

Hansjürgens B, Schröter-Schlaack C, Berghöfer A, Lienhoop N (2016) Justifying social values of nature: economic reasoning beyond self-interested preferences. Ecosyst Serv 22:228–237

Hartig T, Evans GW, Jamner LD, Davis DS, Garling T (2003) Tracking restoration in natural and urban field settings. J Environ Psychol 23(2):109–123

Heudorf U, Meyer C (2005) Gesundheitliche Auswirkungen extremer Hitze am Beispiel der Hitzewelle und der Mortalität in Frankfurt am Main im August 2003. Gesundheitswesen 67(5):369–374

Hilgers D, Burth A (2011) Konzept einer doppischen Kommunalschuldenbremse für das Land Hessen. Verwaltung Manag (5):242–251

HMWVL (2012) Freiräume entwickeln, Lebensräume schaffen. HMWVL, Wiesbaden

Hoffmann B, Hertel S, Boes T, Weiland D, Jockel KH (2008) Increased cause-specific mortality associated with 2003 heat wave in Essen, Germany. J Toxicol Environ Health, 71(11–12), 759–765. http://www.lanuv.nrw.de/uploads/tx_commercedownloads/malbo14_web.pdf. Accessed 17 Nov 2016

IPCC - Intergovernmental Panel for Climate Change (2013) Summary for policymakers. In: Stocker TF, Qin D, Plattner G-K, Tignor M, Allen SK, Boschung J et al (eds) Climate change 2013: the physical science basis. Contribution of Working Group I to the fifth assessment report of the Intergovernmental Panel on Climate Change. Cambridge University Press, Cambridge/New York

Jirku A (ed) (2013) StadtGrün. Fraunhofer IRB, Stuttgart

Kenter JO, O'Brien L, Hockley N, Ravenscroft N, Fazey I, Irvine KN et al (2015) What are shared and social values of ecosystem services? Ecol Econ 111:86–99

Kolbe J, Wüstemann H (2014) Estimating the value of urban green spaces: a hedonic pricing analysis of the housing market in cologne, Germany. Folia Oeconomica 5(307):45–61

Krebs A (1999) Ethics of Nature: A Map. De Gruyter, New York

Kuypers VH, De Vries EA, Tonneijck F, Hofschreuder P (2007) Grüne Maßnahmen für saubere Luft. Ein neuer Blick auf die Luftqualität in den Niederlanden. In: Endlicher W, Gorbachevskaya O, Kappis C, Langner M (eds) Tagungsband zum Workshop über den wissenschaftlichen

Erkenntnisstand über das Filterungspotenzial (qualitativ und quantitativ) von Pflanzen am 1. July and in German/Adlershof (Berliner Geographische Arbeiten, 109), pp 35–40
Li D, Bou-Zeid E (2013) Synergistic interactions between urban heat islands and heat waves: the impact in cities is larger than the sum of its parts. J Appl Meteorol Climatol 52:2051–2064. doi:10.1175/JAMC-D-13-02.1
Lienhoop N, Bartkowski B, Hansjürgens B (2015) Informing biodiversity policy: the role of economic valuation, deliberative institutions, and deliberative monetary valuation. Environ Sci Pol 54:522–532. doi:10.1016/j.envsci.2015.01.007
Lobstein T, Jackson-Leach R, Moodie M, Hall KD, Gortmaker SL, Swinburn BA et al (2015) Child and adolescent obesity: part of a bigger picture. Lancet 385(9986):2510–2520
LUA NRW - Landesumweltamt Nordrhein-Westfalen (2001) Verzehrstudie in Kleingärten im Rhein-Ruhrgebiet. LUA-Materialien zur Altlastensanierung und zum Bodenschutz, Band Nr 14
Maas J, Verheij RA, De Vries S, Spreeuwenberg S, Schellevis FG, Groenenwegen PP (2009) Morbidity is related to a green living environment. J Epidemiol Community Health 63(12):967–973
Maller C, Townsend M, Pryor A, Brown P, Leger LS (2006) Healthy nature healthy people: »Contact with nature« as an upstream health promotion intervention for populations. Health Promot Int 21(1): 45–54
Mathey J, Rössler S, Lehmann I, Bräuer A, Goldberg V, Kurbjuhn J, et al. (2011) Noch wärmer, noch trockener? Stadtnatur und Freiraumstrukturen im Klimawandel Naturschutz und Biologische Vielfalt, 111. Münster: Landwirtschaftsverlag
Naturkapital Deutschland – TEEB DE (2016) In: Kowarik I, Bartz R, Brenck M (eds) Ökosystemleistungen in der Stadt – Gesundheit schützen und Lebensqualität erhöhen. Leipzig, Berlin
Neu C, Nikolic L (2014) Die (neuen) Selbstversorger – Zwischen Not und Weltanschauung? In: Berger PA, Keller C, Klärner A, Neef R (eds) Urbane Ungleichheiten: neue Entwicklungen zwischen Zentrum und Peripherie. Wiesbaden, Springer VS
Oxford University Press (Ed.) (1989). Oxford English dictionary. Ed. by John Simpson and Edmund Weiner. 2nd edn. Oxford University Press, Oxford
Richardson EA, Mitchell R (2010) Gender differences in relationships between urban green space and health in the United Kingdom. Soc Sci Med 71(3):568–575
Ring I (2008) Compensating municipalities for protected areas: fiscal transfers for biodiversity conservation in Saxony, Germany. GAIA – Ecol Perspect Sci Soc 17:143–151
Ring I, Mewes M (2013) Ausgewählte Finanzmechanismen: Zahlungen für Ökosystemleistungen und ökologischer Finanzausgleich. In K Grunewald, O Bastian (Eds.), Ökosystemdienstleistungen Konzept, Methoden und Fallbeispiele (pp. 167–177). Berlin/Heidelberg: Springer.
Roe JJ, Ward Thompson C, Aspinall PA, Brewer MJ, Duff EI, Miller D et al (2013) Green space and stress: evidence from cortisol measures in deprived urban communities. Int J Environ Res Public Health 10(9):4086–4103
Röhner C (2013) Kinder und Natur – Zur Bedeutung der natürlichen Umgebung für die kindliche Entwicklung. In: Becher A, Miller S, Oldenburg I, Detlef P, Schomaker C (eds) Kommunikativer Sachunterricht. Schneider Verlag Hohengehren, Baltmannsweiler, pp 167–181
Rüger J, Gawel E, Kern K (2015) Reforming the German rain water charge – approaches for an incentive-oriented but still workable design of the charge. Gwf – Wasser, Abwasser 156(3):364–372
Schneider A, Breitner S, Wolf K, Hampel R, Peters A, Wichmann H-E (2009) Ursachenspezifische Mortalität, Herzinfarkt und das Auftreten von Beschwerden bei Herzinfarktüberlebenden in Abhängigkeit von der Lufttemperatur in Bayern (MOHIT). Mwünchen: Helmholtz-Zentrum München – Deutsches Forschungszentrum für Gesundheit und Umwelt, Institut für Epidemiologie
Schröter-Schlaack C (2013) Steuerung der Flächeninanspruchnahme durch Planung und handelbare Flächenausweisungsrechte. Dissertation, University of Halle-Wittenberg

Spangenberg J, Settele J (2016) Value pluralism and economic valuation – defendable if well done. Ecosyst Serv 18:100–109. doi:10.1016/j.ecoser.2016.02.008

Stadt Augsburg (2015). Zukunftsleitlinien für Augsburg. Stadt Augsburg. http://www.nachhaltigkeit.augsburg.de/zukunftsleitlinien.html. Accessed 7 Aug 2015

Stadt Hamburg (2014) Bürgerschaft der Freien und Hansestadt Hamburg. Gründachstrategie Hamburg. Drucksache 20/11432

TEEB - The Economics of Ecosystems and Biodiversity (2010) Mainstreaming the economics of nature: a synthesis of the approach, conclusions and recommendations of TEEB http://www.teebweb.org/publication/mainstreaming-the-economics-of-nature-a-synthesis-of-the-approach-conclusions-and-recommendations-of-teeb. Accessed 17 Nov 2016

UBA – Umweltbundesamt (2016) Planspiel Flächenhandel. http://www.flaechenhandel.de. Accessed 17 Nov 2016

Vatn A (2009) An institutional analysis of methods for environmental appraisal. Ecol Econ 68:2207–2215

WBGU – German Advisory Council on Global Change (1999) Welt im Wandel. Umwelt und Ethik. Sondergutachten 1999. Metropolis-Verlag, Marburg

Wendorf G (ed) (2011) Wohnsiedlungen im Umbruch: Impulse inter- und transdisziplinärer Forschung zur Gestaltung von Nachkriegssiedlungen. München, Oekom

Green Infrastructure for Increased Resource Efficiency in Urban Water Management

Jaime Nivala, Andreas Zehnsdorf, Manfred van Afferden, and Roland A. Müller

1 Introduction

Green Infrastructure can be defined as a network of natural and engineered areas that are designed and managed to provide an increased breadth and depth of ecosystem services in both urban and rural areas. It is based on the concept that a specific area of land that has healthy ecosystems can actively mitigate challenges such as urban growth and climate change. Green Infrastructure offers a myriad of environmental, economic, and social benefits, as well as valuable ecosystem goods and services. This chapter addresses resource efficiency as it relates to water, and specifically to resilience and quality of life in urban areas.

One key aspect of Green Infrastructure is that it is comprised of many individual components, all of which function simultaneously in order to improve overall ecosystem health in a given area. Previous approaches, commonly referred to as 'Gray Infrastructure', often performed a singular function and did not take into account any environmental or ecological considerations. The multifunctional approach of Green Infrastructure is attractive because it offers simultaneous benefits to both nature and society.

Green Infrastructure includes components that span a wide range of scales, and the combined impact is much more than simply the sum of the individual elements. Current approaches include technologies such as green roofs, rain gardens, constructed wetlands, and vegetated swales, and are often implemented in parallel with approaches such as pervious pavement (for increased infiltration) as well as rainwater harvesting systems and storage tanks (for collection, storage, and eventual re-use of water). Each component in the greater network of a Green Infrastructure initia-

J. Nivala (✉) • A. Zehnsdorf • M. van Afferden • R.A. Müller
Centre for Environmental Biotechnology, Helmholtz Centre for Environmental Research-UFZ Leipzig, Germany
e-mail: jaime.nivala@ufz.de; andreas.zehnsdorf@ufz.de; manfred.afferden@ufz.de; roland.mueller@ufz.de

S. Kabisch et al. (eds.), *Urban Transformations*, Future City 10,
DOI 10.1007/978-3-319-59324-1_8

tive contributes to overall improvements for nature and society. Many urban Green Infrastructure approaches focus on retention and treatment of water on a local scale, by reducing shock loads to sewage treatment plants, thereby decreasing the discharge of untreated stormwater to receiving waters.

Green Infrastructure initiatives also challenge the common misperception that there is not enough land available in urban landscapes, or that the associated costs are prohibitive. In fact, although installation costs for Green Infrastructure may be higher, compared to conventional approaches, they can be much more cost-effective when evaluated on a cost-per-unit-volume of managed water.

Not everything that is physically green can be automatically considered to be a component of Green Infrastructure. In order to be integrated into a Green Infrastructure approach, it must be interconnected with other components in the network and provide benefits for more than one kind of ecosystem service; these include, but are not limited to, increased biodiversity, restored natural habitat, rainwater retention, flood alleviation, reduced stress on local water supplies, lower energy demands for building, increased food production, increased protection of coastal areas, and more recreation opportunities, among many others.

2 Graywater

The separation of graywater and blackwater as wastewater component streams, combined with specially adapted processes for treating these streams, makes it possible to almost fully close local water and material cycles. In addition, energy and nutrients can be recovered significantly more efficiently from wastewater if separation into component streams with differing loading levels is carried out at the location where the wastewater arises (Kjerstadius et al. 2015). In urban areas, the localised treatment of graywater leads to a significant reduction in the residual wastewater (blackwater) to be treated in a centralised manner, because this blackwater then only comes from toilets and accounts for around 30% of the total wastewater volume.

Depending on its origin, graywater can be divided into light graywater (e.g., wash basins, showers, and bath tubs) and dark graywater (e.g., wastewater from washing machines, dishwashers, and kitchen sinks). Sievers et al. (2014) quote the following typical concentration ranges for dark graywater in Europe, based on a literature review with 130 references: chemical oxygen demand 102–1583 mg/L, biochemical oxygen demand 56–427 mg/L, total nitrogen 3–48 mg/L, and total phosphorous 0.5–15 mg/L.

The degree of engineering sophistication required for the processes used to treat graywater differs, depending on both the intended subsequent use and the quantity of the water to be treated (Ghaitidak and Yadav 2015; Teh et al. 2015). Thus, there are systems for installation in single-family homes as well as engineering solutions for residential developments with several hundred residents. Fig. 1 shows an urban graywater treatment in Oslo, Norway, where the treatment system is integrated into

Fig. 1 Klosterenga urban graywater treatment system in Oslo, Norway (Source: Jaime Nivala)

the front courtyard of a block of residential apartment buildings. The system is comprised of a septic tank, a wetland constructed for vertical flow, followed by a wetland constructed for horizontal flow. Both wetlands contain porous filter media and are planted with flowering wetland plants. The system footprint is 1 m^2 per person equivalent. The system provides high levels of treatment that meet European summing water standards with respect to pathogens as well as standards for drinking water set by the World Health Organization (Jenssen 2005), making the treated effluent suitable for discharge to nearby surface waters.

Particularly strong interest in the treatment and re-use of graywater exists in areas of the world where little water is available (Al-Zu'bi et al. 2015; de Gois et al. 2015; Lam et al. 2015). Systems such as sand filters (Ochoa et al. 2015) and constructed wetlands (Laaffat et al. 2015; Saumya et al. 2015) are simple in nature, yet highly effective.

Combinations of conventional wastewater treatment processes can also be used, particularly when large flows are to be treated in confined spaces. For example, the combination of a membrane bioreactor for the removal of biodegradable substances, ozonisation, UV disinfection and subsequent adsorption and filtration have been used to treat graywater up to drinking-water quality, (Knerr et al. 2009).

Good purification performance has also been achieved with biological systems such as membrane bioreactors and biological aerated filters that include a reliable barrier for the removal of solids (Jefferson et al. 2004). However, a comparison of the potential of pilot wastewater treatment processes for in-building graywater recycling has revealed that the various treatment processes (membrane bioreactors and biological aerated filters) differ with regard to the microbiological water quality criteria they achieve (Laine 2001).

3 Rainwater and Surface Runoff

Managing rainwater and surface runoff from roofs, parking lots, and other impervious surfaces in urban areas is a challenge because the timing and quantity of rainwater and runoff is highly variable throughout the year, as well as from 1 year to the next. Extended dry periods can be followed by heavy rain events, where large quantities of water fall within a short period of time, leading to hazards such as flash flooding. Moreover, this water often contains pollutants such as nitrogen and phosphorus, particulate matter, salts, and heavy metals, as well as compounds such as pesticides and hydrocarbons. Green infrastructure can help increase resiliency in urban landscapes by providing management, storage, and treatment of storm water, as well as improving local micro climates via natural cooling effects. Such systems can also contribute to improving the quality of life in urban areas, when designs also include ancillary benefits. Nature-based technologies can often be integrated directly into urban landscapes, providing additional ecosystem services such as more habitats for plant and animal species as well as the potential for increased biodiversity within the urban realm.

In principle, constructed wetlands and rainwater gardens are suitable for the localised treatment and storage of rainwater and surface runoff because they provide storage, treatment, and controlled release, thereby reducing the quantity of water and pollutants sent to local sewage treatment plants. A case study in Burnsville,

Minnesota, (USA) showed a 93% reduction in runoff after the installation of 17 rain gardens in a neighborhood with a watershed area of approximately 2 ha (City of Burnsville 2016).

Rainwater can also be treated using wetland plant mats made of fleece materials or other engineered textiles that are intertwined with the roots of helophytes. Because this structure does not use gravel or sand, which otherwise generally serve as the substrate matrix in constructed wetlands, these systems are comparatively lightweight and enable rainwater treatment to be shifted to the roofs of buildings, leaving ground-level areas available for other uses. This type of helophyte roof is a form of extensive roof greenery; the plant mats are installed on the surface of the roof and watered daily at regular intervals.

In general, green roofs can be used both for rainwater treatment and as "natural air-conditioning systems" (Blumberg 2011). When installed properly, they also have a positive impact on the building climate by acting as an insulation layer that effectively reduces the solar energy transferred to the building.

Green roofs have also been shown to protect roof structures, thereby increasing their service life, as well as improving the microclimate by reducing heat island effects. Green roofs also help to retain rainwater and serve as habitats for a range of organisms (Song et al. 2013). Moving rainwater treatment to green roofs with lush vegetation covers – such as wetland roofs – provides positive ecosystem services, such as filtering of pollutants, air humidification, carbon dioxide absorption, and oxygen production (Fig. 2).

The helophyte vegetation is markedly photosynthetically active under summer conditions. If rainwater is collected and dosed to the helophtye mat on a controlled and regular basis, it allows for significantly higher levels of evaporation and transpiration (evapotranspiration) than green roofs planted with, for example, water-efficient plants. The choice of green roof vegetation depends strongly on project aims. Projects aiming to reduce the volume of water would utilize a helophyte mat, whereas projects aiming to maximize the volume of water available for re-use would use water-efficient vegetation, such as succulent plants.

4 Wastewater

Wastewater is another component of the urban water cycle that can be managed with ecologically engineered systems. New construction in urban areas can be planned in a way that separates household (domestic) wastewater from the centralized sewer system, so that it can be managed and treated separately. Many urban areas, however, utilize what is known as combined sewers, which convey both household sewage and storm water runoff. Such combined sewers have a limited capacity to carry storm water, which can result in an overflow of the untreated wastewater-storm water mix to the surrounding environment. Domestic wastewater and combined sewer overflows can both be effectively treated in urban areas with technologies such as constructed wetlands.

Fig. 2 Wetland roof with a surface area of 120 m² for the treatment of rainwater in an office building in Jena, Germany (Source: Michael Blumberg)

Constructed wetlands, also known as treatment wetlands, are one of the most common kinds of ecologically engineered systems for onsite treatment of domestic wastewater. Treatment wetlands provide many advantages, compared to other wastewater treatment systems. Unlike many conventional technologies, wetland systems are simple to operate, can be constructed from locally available materials, and can be easily adapted to local topographical and climate conditions. Treatment wetlands can also be integrated into the landscape, thus adding aesthetic appeal to the surrounding environment.

Treatment wetlands that receive only domestic (household) wastewater, such as septic tank effluent, can be designed in many different ways. The most basic design is a horizontal subsurface flow wetland, which consists of a water-tight liner, 50 cm of coarse gravel, and wetland plants. Water in this system is kept below the surface of the gravel (thus avoiding the risk of human exposure to untreated wastewater). The wastewater flows horizontally, by gravity, from one end of the wetland to the other, and is treated by physical, chemical, and biological processes. This type of treatment system can operate without any external energy inputs, but requires a relatively large surface area; approximately 5 m^2 of area is required per person equivalent.

Another kind of treatment wetland that is used for domestic wastewater is an unsaturated vertical subsurface flow wetland. This type of wetland still has a liner and wetland plants, but the main media is sand or fine gravel, and the water is intermittently dosed throughout the day with a small pump. The water percolates downward through the sand or gravel, and is treated as it passes through the unsaturated media. Unsaturated vertical subsurface flow wetlands, however, require a small amount of energy for dosing the water to the wetland bed, but the trade-off is a smaller area requirement. Vertical flow constructed wetlands generally require an area of 3–4 m^2 per person equivalent.

Treatment wetlands can also provide wastewater solutions in situations where land availability is extremely limited. Design modifications exist, whereby an incremental increase in energy requirements (for example, additional pumps for moving water, or for providing additional air for increased microbial degradation) can result in a significantly reduced system footprint. Intensified treatment wetland designs require much smaller areas, generally 1–2 m^2 per person equivalent, but they also require slightly more sophisticated components and, inevitably, need access to the energy grid.

Treatment wetlands can also be customized to fit specific site conditions. An excellent example is the Constructed WetRoof in Tilburg, The Netherlands, where a horizontal subsurface flow wetland design was modified to fit the roof of an office building (Fig. 3).

The system treats 500 L/d of wastewater that is generated onsite. The wastewater undergoes primary treatment in a septic tank, and is pumped twice every day to the Constructed WetRoof system, where it flows through a 9 cm thick layer of porous media (a mix of local and volcanic sands and a lightweight expanded clay aggregate) covered with sod (Zapater-Pereyra et al. 2012). Approximately 25% of the treated effluent is re-used onsite for toilet flushing, whereas the remaining treated effluent is discharged to the soil for groundwater recharge.

Sewer systems that receive large amounts of rainwater during storm events generally have a bypass installed, so that, when the total amount of sewage plus rainwater exceeds the capacity of the sewage treatment plant, it is discharged directly (without treatment) to the nearest receiving watercourse. This is termed *Combined Sewer Overflow*. The annual pollutant load from combined sewer overflow discharges can often exceed the pollutant loads discharged by conventional sewage treatment plants, simply due to the sheer volume of water that is effectively "washed through" the sewer network and discharged during rain events (Meyer et al. 2013).

In many urban areas, the sewer lines and sewage treatment plants are decades old and running at or near capacity. The cost of new sewer lines and new treatment plants is often prohibitively high. Treatment wetlands for CSO treatment can be installed (even retroactively) between the sewage treatment plant and the receiving water body. In such cases, the wetland only receives flow during rain events for which the sewage treatment plant is overloaded.

The sewage treatment plant of Lyon, France, was confronted with this problem and began to search for alternative solutions to constructing new sewer lines. A pilot CSO treatment wetland was installed in a community at the periphery of the Lyon urban area, as a demonstration of a Green Infrastructure solution that could be adopted throughout the peri-urban areas of Lyon (Fig. 4).

The system is designed to accept combined sewer overflow for a sub-watershed in Lyon. The system covers an area of 530 m^2 and is partitioned into two wetland

Fig. 3 A 300 m^2 constructed wetland for domestic wastewater treatment on a rooftop in the Netherlands. A portion of the treated effluent is re-used for toilet flushing (Source: Frank van Dien, ECOFYT)

cells that are fed alternatingly. The system has a storage capacity of 1200 m^3. Overflow is sent to the wetland for buffering, and receives treatment during the days after the rain event. Discharge from the treatment system is controlled, ensuring that the water has received adequate treatment before it is released to the nearby watercourses over a timespan of days to weeks.

5 Outlook

This chapter demonstrates that a broad range of Green Infrastructures for urban water management with high treatment efficiencies is available for a variety of applications related to urban development. However, the national and international perspective reveals that there is substantial opportunity for increased implementations of green infrastructure for water management. Decision makers and investors often choose well established and well-known technologies based on existing knowledge about investments and costs. The value of environmental protection with public protection targets (e.g., reducing air pollution, mitigating microclimate effects, closing local water balances, and improving urban landscapes) often plays

Fig. 4 The Marcy Etoile constructed wetland for combined sewer overflow treatment in Lyon, France (Source: Stephane Troesch, EpurNature)

only a secondary role. Communities are now aware of the benefits of green technologies and use these innovative approaches for minimizing the negative effects of urbanization.

In this framework, future activities require a systematic approach in order to bring together innovative technologies with adapted and corresponding framing conditions. Infrastructural and urban planners should be aware of the possibilities and advantages that green technologies offer at an early stage of urban development. Pilot projects that demonstrate not only the technological performance but also the economic, ecological, and social relevance of this approach are needed, to bring together different segments of the decision-making process. Finally, an operation and maintenance system is required, in order to prepare clear management responsibilities for the growing market.

Making Green Infrastructure a real alternative to traditional, gray infrastructure harmonization across infrastructure sectors is required to replace isolated planning and operation approaches. The sectors of water, building, transport, energy and agriculture apply their own standards for construction and performance; these treat Green Infrastructure almost as an afterthought and make cross-sectoral public tendering of Green Infrastructures complicated.

Furthermore, the implementation of Green Infrastructure requires adequate financing schemes. The focus of many institutional lenders and investment banks does not actively promote the deployment of Green Infrastructures because their substantial benefits have, up till now, been largely undervalued.

In order to leverage investments and reduce regulative and administrative barriers, the European Commission recently developed the Green Infrastructure Strategy and initiated an EU Research & Innovation agenda for nature-based solutions. The strategy aims to ensure that the implementation of Green Infrastructure becomes an integral part of spatial planning and territorial development and it promotes EU leadership in the new global market for nature-based solutions with new economic opportunities, products, services and new local green jobs (EU 2015).

References

Al-Zu'bi Y, Ammari TG, Al-Balawneh A, Al-Dabbas M, Ta'any R, Abu-Harb R (2015) Ablution graywater treatment with the modified re-circulated vertical flow bioreactor for landscape irrigation. Desalin Water Treat 54(1):59–68

Blumberg M (2011) Sumpfpflanzendächer als Variante der Dachbegrünung. In: Ziegler C (ed) Regenwasserbewirtschaftung - GWF Praxiswissen, 1. Deutscher Industrieverlag, München, pp 189–196

City of Burnsville, Minnesota. (2016) Burnsville Rainwater Gardens: a Nationally Significant Demonstration Project. http://www.ci.burnsville.mn.us/DocumentCenter/View/450. Accessed 26 Nov 2016

de Gois EHB, Rios CAS, Costanzi RN (2015) Evaluation of water conservation and reuse: a case study of a shopping mall in southern Brazil. J Clean Prod 96:263–271. doi:10.1016/j.jclepro.2014.08.097

EU - European Union (2015) Towards an EU Research and Innovation policy agenda for Nature-Based Solutions & Re-Naturing Cities. Final Report of the Horizon 2020 Expert group on nature-based solutions and re-naturing cities. Luxembourg: Publications Office of the European Union

Ghaitidak DM, Yadav KD (2015) Reuse of graywater: effect of coagulant. Desalin Water Treat 54(9):2410–2421

Jefferson B, Palmer A, Jeffrey P, Stuetz R, Judd S (2004) Grey water characterization and its impact on the selection and operation of technologies for urban reuse. Water Sci Technol 50(2):157–164

Jenssen PD (2005) Decentralized urban greywater treatment at Klosterenga Oslo. In: v. Bohemen H (ed) Ecological engineering – bridging between ecology and civil engineering. Æneas Technical Publishers, Maastricht, pp 84–86

Kneer H, Eindfeld K, Engelhart M, Heck A, Legrix J-C, Rechenburg A, Wölle J (2009) Modulare Gesamtlösung für dezentrales Wasserrecycling. Gwf – Wasser, Abwasser 150(1):64–71

Kjerstadius H, Haghighatafshar S, Davidsson A (2015) Potential for nutrient recovery and biogas production from blackwater, food waste and graywater in urban source control systems. Environ Technol 36(13):1707–1720

Laaffat J, Ouazzani N, Mandi L (2015) The evaluation of potential purification of a horizontal subsurface flow constructed wetland treating graywater in semi-arid environment. Process Saf Environ Prot 95:86–92. doi:10.1016/j.psep.2015.02.016

Laine, A. (2001) Technologies for gray water recycling in buildings. Thesis (PhD) Cranfield University

Lam L, Kurisu K, Hanaki K (2015) Comparative environmental impacts of source-separation systems for domestic wastewater management in rural China. J Clean Prod 104:185–198. doi:10.1016/j.jclepro.2015.04.126

Meyer D, Molle P, Esser D, Troesch S, Masi F, Dittmer U (2013) Constructed wetlands for combined sewer overflow treatment – comparison of German, French and Italian approaches. Water 5(1):1–12

Ochoa SIC, Ushijima K, Hijikata N, Funamizu N (2015) Treatment of domestic graywater by geo-textile filter and intermittent sand filtration bioreactor. J Water Reuse Desal 5(1):39–49. doi:10.2166/wrd.2014.042

Saumya S, Akansha S, Rinaldo J, Jayasri MA, Suthindhiran K (2015) Construction and evaluation of prototype subsurface flow wetland planted with Heliconia angusta for the treatment of synthetic graywater. J Clean Prod 91:235–240. doi:10.1016/j.jclepro.2014.12.019

Sievers JC, Londong J, Albold A, Oldenburg M (2014). Characterisation of Greywater – estimation of design values. In J. Lohaus (Ed.), Proceedings of 17th international EWA symposium "WatEnergyResources – water, energy and resources: innovative options and sustainable solutions" during IFAT, 5–9 May 2014, Munich, Gemany. European Water Association, Hennef

Song U, Kim E, Bang JH, Son DJ, Waldmann B, Lee EJ (2013) Wetlands are an effective green roof system. Build Environ 66:141–147. doi:10.1016/j.buildenv.2013.04.024

Teh XY, Poha PE, Gouwandaa D, Chonga MN (2015) Decentralized light graywater treatment using aerobic digestion and hydrogen peroxide disinfection for non-potable reuse. J Clean Prod 99:305–311. doi:10.1016/j.jclepro.2015.03.015

Zapater-Pereyra M, van Dien F, van Bruggen JJA, Lens PNL (2012) Material selection for a constructed wetroof treating wastewater. In: Proceedings of the 13th international conference on wetland systems for water pollution control, 25–29 November 2012, Perth, Australia. IWA Publishing, the Australian Water Association, and Murdoch University, pp 464–475

Sustainable Urban Water Governance – Main Aims, Challenges and Institutional Approaches in Germany and Beyond

Moritz Reese and Erik Gawel

1 Introduction

In the pursuit of a liveable urban environment, water infrastructures have always played an essential role. The development of safe sewage and drinking water supply systems since the mid-nineteenth century has ended the ages of severe pestilence and created the hygienic preconditions for the exponential growth of cities in the decades of industrialization and urbanization. In the future quest for sustainable urban transformation, too, water infrastructures will remain an essential factor, even though the factual challenges have changed and transformation of the historic infrastructure systems seems to be necessary, in multiple regards: In many cities around the globe, water infrastructures are facing major challenges. Demographic change, climate change, aging infrastructure, and more stringent water pollution control standards are placing the "blue infrastructure" under pressure. Scenarios involving re-orientation and system transformation – e.g., shifting the emphasis towards semi-central and closed-loop technologies – are being discussed (see, e.g., the contribution of Nivala et al., "Green Infrastructure for Increased Resource Efficiency in Urban Water Management", in this volume). In this context, there is also a growing focus on the governance of urban water infrastructures, and it is questioned whether existing laws and organizational settings are adequate for meeting the above-mentioned challenges.

M. Reese (✉)
Department Environmental and Planning Law, Helmholtz Centre for Environmental Research-UFZ, Leipzig, Germany
e-mail: moritz.reese@ufz.de

E. Gawel
Department of Economics, Helmholtz Centre for Environmental Research-UFZ, Leipzig, Germany
e-mail: erik.gawel@ufz.de

S. Kabisch et al. (eds.), *Urban Transformations*, Future City 10,
DOI 10.1007/978-3-319-59324-1_9

However, the debate in this area tends to be fragmentary and to oscillate between the various individual objectives and problems of a sustainable water supply and sewage systems. In the light of this, this contribution provides a *comprehensive* review of the challenges, objectives, and institutional priorities that should underlie the development of *sustainable* water infrastructures. Firstly, we briefly review the factual challenges that (many) agglomerations are facing with regard to water infrastructures. Then, drawing on the general and sector-specific sustainability debate, 15 key objectives and preconditions for sustainable water infrastructure development are formulated and discussed as "imperatives" to be observed in the interests of sustainability. Lastly, conclusions will be drawn as to how these imperatives can be accounted for within the governance framework of water infrastructure development (Sect. 4).

At the outset, it should be noted that this paper refers to *German* and *European* examples with regard to both the factual situation and the governance framework. It is apparent, however, that similar challenges exist in many cities around the world. Moreover, the general key objectives and requirements of sustainable water infrastructure development discussed in Sect. 3 are of a global nature and are applicable, regardless of the specific local situation. The same holds true with regard to the basic institutional implications that we analyse in Sect. 4. As the reader will hopefully agree, these implications are, similarly, of a general nature and equally relevant beyond Germany and Europe.

Lastly, we wish to clarify that the sustainability challenges, criteria, and approaches analysed below are not generally confined to urban agglomerations but are, in many regards, equally relevant to smaller and even rural settlements. Nevertheless, they are essential factors for sustainable *urban* development, therefore, need to be duly considered and analysed in the context of urban transformation.

2 Seminal Challenges for Urban Water Infrastructure Development

Little research is needed to identify the challenges facing urban water infrastructure development and governance: the key problems and development issues are more or less obvious and are widely discussed among experts (for concise summaries see Pinnekamp 2014; Bauer 2014; for the legal aspects see Hanke and Libbe 2012). The following survey relates mainly to the situation in Germany and Central Europe. However, as mentioned above, comparable problems also exist – to varying degrees – in other countries and regions (Grafton et al. 2015; Wong and Brown 2009).

2.1 Widespread Reinvestment Gaps and Obsolescence of Existing Waste Water Structures

In many parts of Germany, the water infrastructure suffers from serious underinvestment (Dohmann 2014; Grabow and Schneider 2014). It is estimated that the annual amount that needs to be spent on the country's sewage infrastructure alone is €8.5 billion (Dohmann 2014, p. 17), yet, in 2013, the actual investment totalled a mere €4.8 billion (Leptin et al. 2014). Many systems are out-dated and in need of replacement. Hence, reform of the financing and investment policies is apparently necessary if the infrastructure is to be not only appropriately maintained but also upgraded and transformed in accordance with technical developments.

2.2 Demographic Change (Shrinkage/Growth/Aging)

Demographic change poses a twofold challenge for water infrastructure development. Firstly, the population density is decreasing in many areas. Shrinkage is occurring especially in rural areas but also in many urban neighbourhoods (Statistisches Bundesamt 2015). The central water supply and sewage infrastructure in these shrinking areas is increasingly underutilized, resulting in significant inefficiencies and, in some cases, in steeply increased costs for operators and users (private households, businesses). Underutilization can also place the functioning of the infrastructure at risk (Londong et al. 2011, p. 153ff.; Siedentop et al. 2015, p. 15). Secondly, the average age of the German population is rising steadily (Statistisches Bundesamt 2015), with attendant changes in user behaviour. In particular, the discharge of pharmaceuticals into domestic sewage is increasing (Tränckner and Koegst 2011; Hillenbrand et al. 2010). The human and environmental risks and impacts of this remain largely uninvestigated and consideration needs to be given to the strategies and technologies (e.g., active carbon adsorption or ozonation as a fourth stage of sewage treatment) that should be deployed to reduce the concentration of pharmaceuticals in sewage (Pinnekamp 2014, p. 5). The higher water purity standards that currently apply are an important factor here.

2.3 New and Stricter Environmental Standards

In Europe, wastewater infrastructures must comply with increasingly stringent water quality standards, largely as a result of the European Union's water policy. The Water Framework Directive 2000/60, which entered into force in 2000, makes it compulsory for member states to achieve a good status of surface and ground water bodies, normally by 2015, and, at the latest, by 2027 (Article 4.1 WFD).

Compared to the status quo, the Directive sets high standards for both the ecological and chemical components of these objectives.

Good ecological status of surface waters is defined as a situation in which 'the values of the biological quality elements ... deviate only slightly from those normally associated with the surface water body type under undisturbed conditions' (1.2 of Annex V to the Water Framework Directive). For highly modified and artificial water bodies, values are permitted to deviate only slightly from those found at the 'maximum ecological potential' consistent with the necessary modifications. Good chemical status of surface waters is defined by the environmental quality standards of Directives 2008/105/EC and 2013/39/EU. Some of these quality objectives – especially those for ubiquitous pollutants – are still significantly exceeded in many water bodies. In addition, Directive 2013/39 introduces a 'watch list' that should be monitored with a view to limiting permitted levels at a later, if this proves necessary. Pharmaceuticals are included in this watch list for the first time. This means that further standards might be imposed, posing an additional challenge for municipal sewage management and rendering additional treatment stages necessary (Kern 2014; UBA 2015). Water quality targets also present challenges for rainwater management, for example, with regard to combined sewer overflows, which often enable untreated sewage to drain into receiving water bodies.

2.4 *Increasing Risk of Urban Flash Floods as a Result of Land-Use Change and Climate Change*

Climate change may cause an increase in heavy rain events in Central Europe and elsewhere. Increasingly extreme precipitation increases the risk of urban flooding, as a result of rainwater volumes that overload the drainage system. This risk of sewer overflow and flooding is, of course, both caused and exacerbated by the fact that, in cities, increasingly more land is being sealed and not being developed in a 'water-sensitive' way that enables extreme precipitation to be temporarily stored or channelled off without causing damage. Under the banner of terms such as 'water-sensitive development' and 'sponge cities', a wide range of technical and planning-related development options is now being discussed (Bunster-Ossa 2013).

2.5 *Increasingly Dry Summer Seasons and Drought Periods as a Result of Climate Change*

As part of regional variations in climate change patterns, periods of drought could increase and make it more difficult to secure adequate supplies of water, especially for the high demand of large agglomerations (Zebisch et al. 2005; Castell-Exner and Zenz 2010; Karthe 2015). Drier summers are likely to become the norm in some parts of Germany, making it necessary either to restrict water withdrawal (e.g., for power-plant cooling or agricultural irrigation) or to put expensive, alternative procurement systems (long-distance water supply) into place.

2.6 Strong Call for More Energy and Resource Efficiency

For reasons of climate change mitigation, environmental conservation, and cost efficiency, there is constant pressure on the water service sector to reduce the use of energy and resources (Flasbarth 2012) – particularly in the energy-intensive field of wastewater treatment. In Germany, for example, wastewater treatment uses between 15 and 80 kWh of energy per person per year (Hansen 2014); a significant energy-saving potential seems to exist here (DWA 2010). The main contribution of wastewater management to resource conservation is via the recovery of valuable substances from sewage. The discussion focuses currently on phosphorus, because this element is in particularly short supply (Cordell et al. 2011; Roskosch and Rechenberg 2015). However, other nutrients, energy, and the water itself also need to be considered (Mo and Zhang 2013).

2.7 Increasing and Highly Diverging Water Prices

As already mentioned, coping with the challenges that have been described requires significant investment; financing this investment is, itself, a key challenge for water infrastructure development. A major problem in this area is that local authorities do not necessarily spend the revenue from water and sewage charges on infrastructure development but, instead, use it for other municipal purposes (Gawel 2011). Linked to this, one of the reasons why costs and financing are regarded as a key challenge is that there are large regional differences in water and sewage charges in Germany and, in some cases, these charges are rising markedly (Gawel and Bedtke 2013). This leads one to suspect that, in some places, management is inefficient or that consumers are being overcharged.

2.8 The Liberalization Question

With regard to cost efficiency, the finger is often pointed at the state's monopoly of activities in this field and the frequently competition-free regulatory framework of the water management system: these two issues are often seen as causes of inefficient management and excessively high charges (see, for example, SRU 2002, p. 295ff.). In England, for example, competition in this field has already been significantly extended, whereas Germany – partly as a result of the negative experiences of some other countries (Lobina and Hall 2007; Bakker 2010) – has virtually avoided this path (Gawel 2016b). Nevertheless, the issue of an appropriate competitive regulatory framework remains on the agenda (Gawel and Bedtke 2015), especially from the European perspective (Finger et al. 2007).

3 Towards Sustainability: Essential Aims and Requirements of Sustainable Water Infrastructure Development

The summary of challenges, in itself, reveals some of the key objectives of the development of urban water infrastructures, such as increasing the resilience of water supply and sewage systems to extreme weather conditions, improving environmental performance in terms of water quality, and boosting energy efficiency. These are also some of the core objectives of *sustainable* development. As will be shown below, however, sustainable development of urban water infrastructures– at least if it is based on the prevailing understanding of sustainable development, in general, and of sustainable water governance, in particular – also needs to satisfy a number of other core objectives and basic conditions.

In the professional debate on the above-mentioned challenges, there is frequent reference to the topos of *sustainability* and the ideal of *sustainable* water management (e.g., Hellström et al. 2000; Makropoulos et al. 2008; Novotny 2008). Various, very valuable concepts for implementing the requirement for sustainability in water governance have also been developed (Kahlenborn and Kraemer 1999; Brackemann et al. 2001).[1] We have analysed this sustainability debate and the relevant concepts in detail elsewhere, as a basis for arriving at a comprehensive requirement profile for sustainable water supply and sewage disposal systems (Reese and Bedtke 2015) that:

- is based on a broad definition of sustainability, involving *long-term* optimization of basic ecological, economic, and social conditions for wellbeing and, in particular, permanent and ubiquitous compliance with minimum standards for a life compatible with human dignity and minimum conditions for the conservation of the natural foundations of life.
- should specify only 'pure' objectives and basic conditions that are essential for sustainable management in the above sense, and should not include strategic or instrumental decisions about ways and means. The initial aim is to clarify the sustainability standards needed for further strategic, technical, and institutional development.

[1] Key documents are the *Dublin Statement on Water and Sustainable Development*, International Conference on Water and the Environment, Dublin, January 1992; Agenda 21: United Nations Conference on Environment and Development in June 1992 in Rio de Janeiro, Chapter 18 – *Protection of the quality and supply of freshwater resources.*

On the basis of these premises, we have identified the 15 core objectives and basic conditions listed in the table below (Reese et al. 2015).

Each one of these "sustainability imperatives" will be further explained and analysed below. Initially, however, some important general aspects will be detailed.

All the sustainability imperatives apply to the entire urban water management system – i.e., on both the supply and the disposal side. The two sides are considered here in parallel – with particular attention not only to their connections and synergies but also to differences and specific challenges.

There are many trade-offs between the various sustainability imperatives; the aim is, therefore, optimization – i.e., realization of the combined objectives in the best way possible. The core requirement of the sustainability maxims is the achievement of the optimum and, in the long term, most viable balance between the ecological, economic, and social issues that the imperatives cover. However, how conflicts/trade-offs between the individual sustainability imperatives are to be resolved in practice – e.g., whether affordability should be reduced in order to achieve a higher level of environmental protection – is clearly a question of weighting, for which the concept of sustainability does not provide a definitive answer. Instead, there is considerable room for political manoeuvring to achieving a balance between the different imperatives: various 'sustainable' solutions are possible.

Nevertheless, there are also limits to this room for manoeuvring; to completely ignore one of the imperatives or to neglect one in an obviously inappropriate way oversteps these limits. In particular, the boundaries of sustainability are transgressed if a compromise solution is *not globally applicable* or *not permanently sustainable* and the problem is shifted to other regions or to future generations (SRU 2012, p. 31ff.). With regard to security of resources and environmental sustainability, this applies, above all, to practices that do not meet the requirement for long-term conservation and lead, instead, to continuous degradation of the natural foundations of economic activity. It is true that the level of long term conservation needs to be determined politically (Reese and Bedtke 2015; Kopfmüller et al. 2001, p. 64 ff.). On this basis, however, the conservation principle and the 'sustainability management rules' that arise from this principle can well be regarded as defining the 'outer' limits of sustainable resource management, also in the field of urban water management (see SRU 2002 for a more detailed account).

Within these ultimate sustainability boundaries, the sustainability imperatives provide an indicator of the sustainability effects of various management options. The better an option scores, in terms of meeting the objectives of the individual imperatives, the more sustainable it is in the sense used in these imperatives. If the option involves conflicting sustainability requirements and trade-offs are therefore unavoidable, measuring its (positive or negative) sustainability benefit is, normally, partly a question of weighting. Some initial thoughts on these issues of weighting and 'room for manoeuvring' in connection with sustainability on an urban scale will now be presented.

It is not the authors' intention to put forward their own preferences as a basis for a particular sustainability (preference) model. The aim is, rather, to highlight the various areas of tension and scope for value judgements, and the limits of these

Table 1 Sustainability imperatives: Core objectives and basic requirements of sustainable urban water governance

1. Demand orientation	9. Polluter/user pays
2. Secure resource availability	10. Acceptance
3. Health protection	11. Knowledge orientation
4. Security of supply	12. Precaution
5. Environmental soundness	13. Flexibility
6. Resource efficiency	14. Integration
7. Economic efficiency	15. Participation
8. Affordability	

areas. One of the purposes of the list of objectives in Table 1 is to specify the key criteria for assessing the technical and institutional aspects of sustainable urban water management.

(1) Demand orientation

A demand-oriented water supply and sewage disposal system is the main aim of residential water services. The actual need on the demand side (domestic use, industry, trade, and agriculture) in quantitative and qualitative terms must form the basis and the benchmark of water governance activities. Demand-orientation has fundamental points in common with economically efficient water management systems that provide water and sewerage services geared to the preferences of consumers (see imperative no. 7). From the social perspective, though, one aspect of this demand orientation is that water and sewage services must also take account of the other sustainability requirements.

A maximum level of supply based on what consumers demand and in the absence of any cost and other limitations (the 'saturation quantity') is clearly not compatible with the objective of sustainability. The sustainability objective, in the narrow sense, must instead be a – sustainable – level of supply that is also compatible with the other sustainability imperatives. However, the hypothetical maximum demand remains a key guiding criterion, although it will only be met if this can be achieved *sustainably*; the other sustainability imperatives may impose significant restrictions on this. In particular, if supplies are scarce, the need to secure resource availability may mean that the demand from user groups cannot be met in full.

Even if scarcity is not an issue, it makes economic sense to regard demand as relevant only if is articulated in awareness of and with due consideration to the full economic costs of providing water services. Restrictions may also be required for cost effectiveness and efficiency. For example, the costs of providing infrastructure may exceed the utility derived from it, or cost-effectiveness calculations may require the scope of provision to be reduced, or render consideration of alternative technological options necessary.

It should be borne in mind here that key social priorities in connection with demand-oriented supply are a basic supply of clean water and sewage hygiene. However, with regard to the relationship between the precedence of basic supply, on the one hand, and competing uses and sustainability goals on the other, it is necessary to clarify how this precedence relates to the 'impositions' of payment, procurement, or change of location. These 'impositions' can, after all, (partly) serve to steer the demand for basic provision into more sustainable channels. Two fundamental aspects should be considered in this connection:

- Firstly, impositions do not obstruct basic provision if they are within the means of those affected (Gawel and Bretschneider 2016). Obviously, no one should find it impossible to obtain access to clean water but, provided that the impositions are tolerable, basic provision per se is not in question. The issue, rather, is the interest of the recipients of such provision in avoiding expenditure. Against this interest, however, the competing interests of sustainability, such as the interest in securing resource availability and the interest in environmental soundness must inevitably be set. Balanced solutions are therefore required.
- Secondly, it is not necessarily the task of urban water services to ensure that everyone can cope with the impositions. In particular, if poverty is an obstacle, responsibility for addressing this lies with social policy institutions (Bretschneider 2016). Consequently, the primacy of basic provision does not mean that major environmental, resource-related, or efficiency problems should be accepted unconditionally, in order to ensure basic provision in even the most unfavourable location.

In Germany and Central Europe, it is, in fact, often decreasing demand that poses challenges for water governance. If population shrinkage and falling consumption result in systems being underused, the imperative about demand-oriented water supply and sewage services becomes relevant because it requires supply and disposal systems to be protected against critical under-utilization. The demand orientation benchmark, thus, also includes the need to avoid functionally critical over-capacity. Avoidance of over-capacity is also required under the imperatives relating to economic and resource efficiency, and indeed long before the under-utilization reaches an extent that puts functionality at risk.

(2) Secure resource availability

Ensuring the long-term adequacy of water resources is a crucial requirement for sustainable supply and one that involves both the supply and the demand side. On the supply side, sources of raw water in sufficient quantity and quality must be secured, especially with regard to the high demand of large agglomerations. Water pollution control, including, in particular, the harmless disposal of wastewater, is clearly of central importance in this respect.

In connection with securing water resources, it is always worth considering the use of sources outside the local area; the exchange of resources between water-rich export regions and demand-heavy import regions can be economically beneficial to both sides. However, long-distance water supply systems are widely criticized – partly on account of the need to conserve resources for the future – and because German water law establishes the primacy of 'water supplies from local water resources' as standard practice, provided that these supplies can be secured in sufficient quantity and quality and at reasonable expense (Article 50.2, German Federal Water Act). One of the cited justifications for this is that obtaining supplies locally is in keeping with the 'polluter pays' principle and provides incentives for conserving local water resources that might be lost if water is obtained from a distance (Kahlenborn and Kraemer 1999). In addition, transporting water over long distances may involve accepting reductions in quality and incur additional costs that conflict with the requirements for economic efficiency and affordability. The conflict between local and long-distance supply involves a number of advantages and disadvantages that must be weighed against each other (Kampe 1987; Kriener 2004).

On the demand side, a wide range of technical and regulatory measures can be used to encourage users to reduce the amount of water consumed. This is relevant in agriculture and for industrial plants and equipment that use water. It may also be possible to manage water consumption by limiting allocation to different uses. Finally, the consumption side and the supply side can be linked through processing and re-use, to form a resource-saving circular management system. However, demand management and circular solutions have a significant impact on central supply and disposal systems, which need to be adjusted to reduced procurement volumes and wastewater quantities. The costs of the necessary system adjustments may be considerable; in the short term they may impose limits on resource-saving for efficiency reasons. In the long term, however, system adjustments can result in significant cost savings, especially when abandonment of (oversize) pipe networks goes hand in hand with closed-loop management.

(3) Health protection

Health is a central precept of sustainable water supply and sewage disposal. *On the supply side,* this imperative requires making the available water have a quality appropriate to the purpose for which it is used and to be harmless to health. For drinking water supplies, this means maintaining adequate hygiene and avoiding problematic pollutant content. Shortage of supplies may make it necessary to compromise on quality or to use less pure sources of water or treated wastewater (graywater, rainwater). In the interests of affordability, it may be appropriate to dispense with costly stages of treatment and to differentiate quality standards according to the purpose for which the water will be used. When attempting to strike a balance between health protection and resource conservation, drinking water quality standards clearly play a key role.

On the disposal side, from the point of view of the health protection objective, it is important to ensure that wastewater is collected, drained away, and treated in ways that preserve municipal hygiene, that there is a physical barrier impervious to human pathogens between wastewater and water supplies, and that any planned use of surface waters, as swimming areas or sources of drinking water, is unaffected by the chosen disposal option.

(4) Security of supply

The goal of security of supply encapsulates the need to secure supply and disposal systems, which are part of the 'critical infrastructure' (BMI 2011), against extreme events and to ensure uninterrupted provision of water services. In the public debate, the concept of security of supply often goes further than this and is regarded as including long-term resource security (Scheele 2006). When considering the specific risks and challenges, it seems appropriate to distinguish between these objectives and to recognize security of supply in the narrower sense – security against interruption and breakdown – as a separate sustainability condition.

The causes of risks to the security of the water supply, in this narrower sense, are varied. They include worst-case scenarios, in particular:

- technical failure of the supply infrastructure as a result of age- or obsolescence-related defects, inadequate maintenance, or over- or under-use (BMI 2011),
- extreme rainfall and flooding,
- periods of extreme drought, and, also
- hostile events ranging from terrorist attacks (Gleick 2006) to defence-related situations;
- if services are provided by the private sector, security of supply may also be jeopardized when a supply company ceases operation as a result of bankruptcy.

During times of stable development, these unusual sustainability risks tend to remain in the background. Nevertheless – and not least because of the historical experience of two world wars – security risks are an important political consideration connected to the development of 'critical' infrastructure. National security of supply, including supply to the energy industry and to agriculture, continues to be a key driver of state infrastructure policy.[2] The challenges that arise from this include making supply and disposal systems as invulnerable as possible and maintaining emergency systems for basic supply purposes. In organizational terms, steps should be taken to enable those enterprises and administrative systems responsible for water supply and disposal services to remain functional and to ensure effective crisis management.

(5) Environmental soundness

Sustainable water infrastructures must be environmentally sound in the long term. According to sustainability management rules, this means that steps must be taken to minimize the environmental burdens arising from water procurement, treatment, and transport – and, in particular, those arising from waste water disposal – and that the capacity limits of the environmental media and ecosystems must not be exceeded (Enquete-Kommission 1994; Kahlenborn and Kraemer 1999). From the point of view of urban water management, these are primarily issues of protecting

[2] See, in particular, the publications of the German Federal Office of Civil Protection and Disaster Assistance (BBK) on the protection of critical infrastructure: www.bbk.bund.de/DE/AufgabenundAusstattung/KritischeInfrastrukturen/Publikationen/publikationen_node.html

water bodies and maintaining the hydrological regime (groundwater recharge). The requirements of water law and the resulting need for action have already been referred to in the introduction of Sect. 1 above. Reference has also been made to the opportunities for climate change mitigation within urban water management, which, in large measure, also form part of the objectives of energy and resource efficiency.

(6) Water resource efficiency

The sparing, efficient, and future-oriented use of water resources is another fundamental requirement of sustainability. This applies, in particular, to natural resources that are exhaustible and difficult to replace. However, renewable resources must also be considered if they are being overused or if their capacity for regeneration is being irreversibly impaired. The objective of water resource efficiency is, thus, about contributing to the maintenance of society's production opportunities. Furthermore, resource efficiency is also required for environmental protection, because the extraction, use, and disposal of resources often entail significant environmental burdens.

'Resource efficiency' is only indirectly linked to 'securing resource availability'. If securing resource availability is interpreted as imposing an absolute limit on (or reduction in) human access to resources, this cannot be ensured through economically efficient resource use alone; the best possible use of inputs does not necessarily imply the restrained use of these resources, in the sense of absolute moderation. This is the 'ecological trap of economic resource efficiency' – a subject that has been discussed often and at length (e.g., Gawel 2016c).

Nevertheless, economic resource efficiency can make an important contribution to securing resource availability by weeding out uses that are 'economically inefficient' or, in other words, wasteful. Economic resource efficiency, as a usage rule, results in a specific form of rationing of scarce resources that is unavoidable in a world of scarcity, namely rationing on the basis of relative urgency: scarce resources are used where – according to the benchmark of value that is used – they produce the highest yield, i.e., where they can be deployed with the greatest utility (the economic aspect). At the same time, this rule condenses resource use, confining it to efficient uses (the ecological aspect). This means that economic resource efficiency can be simultaneously seen as a kind of ecological precautionary principle that conserves the resource base by reducing (inefficient) usage and, at the same time, as a rule of economic wisdom that seeks to maximize the utility of the remaining resource appropriation.

(7) Economic efficiency of water services

To ensure sustainable urban water management, it is essential that the provision of water services is based on economic criteria and that economic resources are used *without waste*. At the microeconomic level, this means that service provision must be as *cost-efficient* as possible, whilst, at the macroeconomic level, it requires *efficient allocation* of supply and disposal tasks/services.

In the name of cost efficiency it is necessary to realize the efficiency potential of the sector, which is likely to reside mainly in technological and administrative economies of scale. Efficiency potential also exists in the synergies that can be achieved through linking and harmonization of supply and disposal services at the operational level. However, one of the reasons why boosting economic efficiency in the water sector poses a challenge is that services are largely provided by regional territorial monopolies and, moreover, usually in a vertically integrated form and without competitive control of prices. This is why service provision is widely regarded as inefficient (e.g., Monopolkommission 2010; see Gawel 2016b for an overview).

In addition, the efficiency imperative renders both static and dynamic consideration necessary. In the light of the climatic, demographic and settlement-structure-related changes described earlier, solutions must be found that remain efficient under the changed circumstances or can be adapted at low cost.

(8) Affordability

The objective of sustainable urban water services – like the objectives of other public services – is often linked to the social policy requirement that these services be offered to users at affordable and roughly equal prices (Bretschneider 2016; Gawel and Bretschneider 2014). This requirement is based on the conviction – which is also widespread outside Germany – that the public sector has a responsibility to intervene to ensure a minimum level of supply of basic infrastructure services to the entire population (Reese and Koch 2010; Britz 2004). It is the task of state and local authorities to make sure – by providing services themselves or through appropriate regulation – that the necessary structures are created and maintained and that basic provision is also available to the entire population at affordable and approximately equal prices. In connection with the general objective of meeting demand, it has already been stated that the human right to water gives rise to a requirement to provide basic water supply and waste water disposal services at affordable prices (Laskowski 2010; from an economic perspective: Bretschneider 2016). The question of which barriers to access are nevertheless appropriate – especially in the interests of sustainable water supply and sewage services – must be considered in connection with the affordability objective, with a particular focus on the price.

There are questions to be asked about what pricing impositions can be regarded as 'affordable' and the extent to which offsetting should take place between central locations, to which services can be provided at low cost, and decentral locations where, because of higher fixed costs and smaller effects of scale, connection to supply and disposal systems may prove to be significantly more expensive. It is doubtful whether it should be the task of public water services to guarantee equal affordability across the board, even in the most remote areas. Finally, moving on from the core guarantees of the human right to water, it is necessary to clarify whether the affordability postulate should apply beyond the area of basic provision and whether it should also protect water users' interest in merely avoiding added expense.

The affordability objective raises significant conflict issues, depending on how these questions are answered. For example, if the answers mean that cost-recovery prices cannot be imposed and (cross-) subsidies are therefore required, this would clearly conflict with the 'user pays' principle (see imperative no. 9 below) and with the objectives of securing resource availability and resource efficiency (nos. 2 and 6 above), because subsidizing water services tends to militate against sparing use and sustainable technological and structural change. These other sustainability requirements therefore appear to argue, prima facie, for the affordability objective to be taken into account via social 'poverty assistance' that is specifically targeted at the needy and, as far as possible, does not affect the elementary allocational function of prices for water services.

(9) Polluter/User pays

Fair distribution of the burdens associated with supply and disposal services is another common objective of sustainability. It applies, in the first instance, both to the distribution of the burden within the present generation of users/polluters (intra-generational) and to fair distribution of the burden between this generation and future ones (inter-generational) (Rogall 2008). Intra- and inter-generationally fair distribution of the burden needs to be guaranteed institutionally, and systematic fairness gaps must be avoided.

Full cost recovery under the 'user pays' principle not only accords with the general sense of justice but is also a significant incentive for sustainable consumption behaviour and a basic requirement of an economic management system that remains viable in the long term. An infrastructure management system that systematically transfers costs to future user groups or to the general public is clearly not sustainable; it is not economically viable, in the long term, and it creates incentives to waste resources. In the interests of economic sustainability, it is therefore desirable for each user and each generation of users to pay the share of the costs that reflects their usage.

This must also apply, in principle, to the external costs of water services, especially the costs arising from consumption and pollution of environmental assets (Gawel 2016a). From the point of view of 'sustainable' equivalence of burden, the costs of rehabilitating these environmental assets and the costs of lost environmental utility must also be recovered from the service recipients, in order to avoid cost-shifting, to create sustainable avoidance incentives, and to ensure restoration – at least on a compensatory basis – of the consumed assets.

The call for cost recovery that fully reflects the usage made and the benefit obtained therefore seems appropriate for a number of reasons. However, implementing this approach raises significant conflict, attribution, and calculation issues. As already mentioned above, there is, in particular, a fundamental conflict with the objective of basic supply and affordability. From the perspective of the 'user pays' principle, the redistribution of costs required by the affordability principle seems unfair and reduces the incentive for beneficiaries and service providers to adapt to real scarcity and cost considerations. Calculation and attribution problems arise, for example, in connection with reliable credit financing, assessment of depreciation and risk, and assessment and attribution of environmental costs. Another significant challenge involves ensuring that funds are used in accordance with the purpose for which they were obtained. This is also a necessary condition of the 'user pays' principle and one that deserves special attention.

An essential requirement, at least for inter-generational equivalence of burden, is that the cost-equivalent funds collected from one generation of users are actually used to offset the consumption of assets that has taken place and to conserve the substance of the infrastructure. Otherwise, cost-equivalent prices will be calculated but the burdens will still be transferred to future generations of users. This aspect of equivalence of burden is by no means trivial, especially because providers of municipal water services have a strong tendency to transfer some of the revenue from water and sewage charges to the general municipal budget and use it for entirely different purposes, to the detriment of the infrastructure stock (Gawel 2011). A key challenge in accordance with the imperative of 'the user pays' is, therefore, to ensure full purpose-appropriate use of the cost-covering refinancing funds that are collected.

With regard to water supply, in particular, the cost generators are, primarily, the connected customers or water users. As the beneficiaries, they give rise both to the provision of the supply and to the factor costs, environmental costs, and resource costs that may result from the withdrawal of water. If the factor cost of supplying water is further increased by external influences, these cost sources (farmers, plant operators) would also need to be included as causes of costs. This gives rise to attribution and distribution questions related to the proportion of costs attributable to the connected customer (as the originator of the purpose) and the 'external' cost generator. Difficult assessment problems also arise in connection with the 'true' environmental costs (Gawel 2014) for further references to the Common Implementation Strategy (CIS) in connection with Article 9 of the Water Framework Directive).

The above comments apply analogously to the sewage disposal system. The environmental costs of wastewater disposal are the adverse effects on water bodies caused by the waste water. It is precisely because the 'user pays' principle has been ignored for so long and standing practices have been implemented at the expense of asset depletion that sewage disposal is, in many places, confronted with significant sustainability issues.

(10) Acceptance

Technological and institutional solutions cannot be sustainable unless they are sufficiently accepted by users, those responsible for supply, and any third parties involved. Acceptance is an asset in itself – as an expression of satisfaction and agreement with a particular supply situation or standard. Acceptance is also a key implementation condition; in particular, it is essential in order to ensure that the specific solutions are supported by the stakeholders in the long term and not obstructed. Conflicts of objectives therefore arise in connection with supply and disposal concepts and solutions that – at least in the eyes of those affected – are associated with significant cost burdens, interference with private property, or loss of convenience (see, for an overview, Po et al. 2003; Mankad and Tapsuwan 2011). Acceptance problems also arise in connection with the production of drinking water from treated wastewater, because although the resulting water may be hygienically safe, the 'yuck' factor may still predominate (Russel and Lux 2009). In Germany, the traditional idea prevails that drinking water, as 'food no. 1' must, as far as possible, be free of non-natural ingredients. Supply solutions that do not ensure this therefore meet with acceptance problems. This applies, in particular, to competitive solutions involving long-distance transport and piping systems, in which adequate hygiene can only be ensured by adding chemicals (Besche 2004). Another area in which acceptance issues may arise relates to connection with decentralized or innovative solutions – such as small-scale sewage treatment systems for individual properties or rainwater harvesting systems – that require greater involvement of the users of the infrastructure services in setting them up (Parsons et al. 2010; Geyler et al. 2014). However, acceptance of a new water management option, including its disadvantages, will be significantly enhanced if the option is recognizably in keeping with the sustainability imperatives elaborated here and it is, therefore, clear that it is in the common interest. As experience with major infrastructure projects has shown, another key factor is involvement of the affected population in the decision-making process at an early stage, so that people are given an opportunity to state their concerns and these concerns are at least considered during decision-making. A related point is that acceptance can only be expected if possible alternatives are included and examined with an open mind as to the outcome, and if the decision-making process and the reasoning behind the decision are made transparent and comprehensible (Renn et al. 2014).

(11) Knowledge orientation

A basic requirement for broad implementation of all the above-mentioned core objectives is a knowledge base that is as extensive as possible and that is continually expanded through research and innovation. The knowledge base should include, firstly, technical knowledge relating to the options for advanced water extraction, treatment and distribution, and wastewater disposal. Secondly, it should cover knowledge of relevant natural and social conditions. This must include reliable projections and detailed knowledge of the driving forces behind climatic and demographic developments and the causal relationships between these developments. Knowledge in predictive situations also involves knowing what we don't know and dealing with this lack of knowledge. When important infrastructure decisions are to be made, uncertainties must be identified and disclosed, different possible courses of development must, if possible, be illustrated by scenarios, and action strategies must be devised that have the prospect of proving successful in all the possible courses of development.

The technical challenge extends into the realm of science. It entails improving our incomplete knowledge of the regional consequences of climate change and their impacts on water governance through research (Hasse et al. 2012; Keßler et al. 2012). Research into demographic, social, technical and settlement-structure-related developments is also needed. Here, detailed regional monitoring programmes and advanced, continuously improved modelling are required.

(12) Precaution

Precaution is a key condition for sustainability. It is particularly important in connection with sustainable development of water supply and sewage systems.

Precaution means, above all, that future developments and risks must be assessed as clearly as possible and that supply and disposal systems and usage structures that are capable of meeting changed requirements in the future, or of being adapted, must be developed in that manner at an early stage.

Forward-looking action is crucial for sustainable development of urban water infrastructures because central supply and disposal systems have a long service life. The demand side is also characterized by 'sluggish' developments in settlement structures – developments that can only be influenced in the long term and through advance planning. From the point of view of the precautionary approach, this suggests that it is necessary to think well ahead and act on a long-term basis.

Rational precaution will, where possible, select 'no regret' options and develop systems that will remain functional across a range of development scenarios (Reese 2010). To this end, it can be useful to establish flexible, decentral supply and disposal solutions that can more easily be adapted to unforeseen developments. However, given the current situation, this requires a system transformation that would itself need extensive preparation. Another, potentially complementary strategy could involve creating particularly 'stable' systems that perform adequately under a wide range of conditions.

Economic precaution is also essential for sustainability. In tandem with the objectives of demand orientation and equivalence of burden, economic precaution requires that constant attention be paid to making the investment and to building up the reserves that are necessary in order to maintain the functionality of supply and disposal systems in a cost-efficient way and to finance necessary adaptations and innovations.

(13) Flexibility

Flexibility is now widely regarded as an important factor in the sustainability of water infrastructure, especially in light of the uncertainty that exists with regard to future demographic, economic, and ecological developments (Spiller et al. 2015; Laskowski 2012). As described above, flexibility that enables adaptation to various developments is, in principle, a key precautionary strategy. The more dynamic the way in which conditions develop and the more uncertain and variable the developments that occur, the more important flexibility becomes.

However, flexibility comes at a price and flexible technologies may have disadvantages in terms of other sustainability objectives. In particular, there are likely to be economic disadvantages if a wide range of flexible solutions results in fewer benefits of scale, and if some sustainable technologies that are currently only practicable and affordable when used on a large scale cannot be used in decentral structures (e.g., fourth treatment stage, some energy-saving measures, or methods of recovering resources such as phosphorus). Overall, therefore, decentral solutions cannot categorically take precedence. Instead, the advantages and disadvantages of all relevant sustainability aspects must be weighed up, taking into account the adaptation risks of less flexible solutions (Bedtke and Gawel 2015). It is therefore advisable to aim for as much flexibility as is possible without incurring overriding disadvantages.

(14) Integration

In the discussion of the sustainability imperatives above, the key components of the objectives of the sustainability postulate have been described and the main synergies and conflicts between them have been set out. It has become clear that sustainability-oriented development that aims at optimum achievement of the ecological, economic, and social objectives that have been discussed must, above all, integrate these objectives while also considering the various synergies and conflicts. Institutionally, this requires a framework that is not merely limited to sectoral aspects and individual sustainability issues but that takes account of developments and their economic, ecological, and social consequences and steers them in a way that optimizes the overall outcome. With regard to urban water management, one must therefore ask whether suitable institutions exist for this complex integration task that can, in particular, promote the following:

- Investigation of the specific manifestations and circumstances of the above-mentioned economic, ecological, and social sustainability objectives, as they relate to the specific supply and disposal situation.
- On the basis of this, creation of balanced strategies/plans for sustainable infrastructure development that take account of the above-mentioned circumstances.
- Linking and harmonization of water supply and sewage disposal services with particular attention to the synergies described in relation to the above-mentioned sustainability imperatives, and
- Linking and harmonization of these development plans with relevant project, sectoral, and overall plans, especially spatial plans and water resources management plans.

In particular, aligning supply systems closely with spatial planning and development is an important condition for sustainable urban water infrastructures and one that is already implicit in the other sustainability objectives. It is clear that a demand-oriented, resource-appropriate, environmentally sound, secure, and at the same time efficient and economical water supply and sewage disposal system can only be guaranteed if water infrastructure and urban development are optimally coordinated (Bahri 2012; Wickel 2015; Wong and Brown 2009). Equally evident is also the need for coordination with water body and river basin management, which is important in connection with securing the necessary raw water reserves. However, river basin management also sets important parameters for the location and quantity of permissible water withdrawal and wastewater discharge and for opportunities for decentral water extraction and closed-loop management.

(15) Participation

Participation, understood as informing and listening to/consulting members of the public and organizations potentially affected by administrative decisions is, today, viewed as an indispensable condition for sustainability for several reasons (see, on the state of debate, SRU 2008, p. 52ff, and also the papers in Jonuschat et al. 2007). Involving the people and organizations affected by a decision can play a valuable role in establishing the knowledge base for sustainable decisions (Bulkeley and Mol 2003). It may yield not only the particular factual knowledge of those affected but also information about their interests and priorities, which represent important points of view in a decision that is based on the most satisfactory balance of interests that can be achieved. Above and beyond its function in providing decision-related knowledge, participation is also recognized as an element of democratic decision-making and as an important basis for acceptance (Habermas 1992; Mason 1999). In representative democratic systems, the executive obtains its legitimation primarily from the election and influence of the representative bodies that govern it; at the local government level, these are often the mayor and the community council. However, this does not in itself provide any opportunity for discussing sectoral decisions with those affected in terms of effects, stakeholder interests, and alternatives. This requires plan- and project-related participatory instruments. Matching the indirect sustainability purposes of participation, it has already been made clear, in the exposition of the relevant substantive sustainability objectives, that the participation of affected people and organizations is a key requirement in the development of water infrastructure, in order to:

- identify or estimate present and future supply and disposal needs as a benchmark for demand-oriented development of supply structures,
- obtain a complete picture of key interests and impacts, as well as of technical options and alternatives,
- develop water supply and wastewater disposal solutions that users and executing organizations will accept and be prepared to implement to an adequate extent.

Existing and potential expanded forms of participation have been discussed in recent years, mainly in connection with problems in acceptance of major infrastructure projects (see Renn et al. 2014 for an overview). However, this debate has only touched on water infrastructure, perhaps because water infrastructure projects seldom involve construction or environmental impacts on a scale that would attract attention outside the region (but see Laskowski 2010, p. 889 ff.; Laskowski 2012, p. 606). Nevertheless, the construction and operation of urban water management facilities can undoubtedly have significant adverse impacts on the neighbourhood. This is particularly true of sewage treatment plants but it also applies to reservoirs, drains, and large-scale sewer construction work. In addition, development of water infrastructure systems can also affect the interests of citizens, companies, and associations, for example through decisions on:

- which systems plots of land are to be connected to,
- which decentral facilities may be self-built and self-operated,
- which charges and prices are to be levied for water services,
- to which extent expensive connections to decentral properties are to be maintained/expanded and potentially co-financed by the community of charge payers,
- which protection standards are to be ensured for regional water bodies, and

– how secure against extreme events the supply and disposal systems are designed to be.

The above-mentioned sustainability arguments for participation apply to all these decisions.

4 Institutional Foundations for Sustainable Urban Water Infrastructure

Various aspects of the necessary institutional conditions have already been mentioned in the above exposition of the challenges and core objectives of the sustainable development of urban water infrastructure; in each case, in relation to a particular sustainability objective or problem. In this section, the institutional framework will be considered from an *overarching* perspective and the main approaches and organizational issues that arise from the challenges and objectives set out above will be identified. These issues include the question of the division of responsibility between the market and the state, the regulatory framework, economic incentives as a governance tool, the organization of water supply and sewage disposal services, and, finally, the use of planning instruments as a basis for cognitive preparation, integration, coordination, participation, and long-term orientation.

4.1 State-Sponsored vs. Competitive Fulfilment of Infrastructure Tasks

The division of responsibility for various tasks between the state and the market is of fundamental importance for fulfilment of the sustainability imperatives, for a number of reasons. The issue of an appropriate regulatory framework that includes competitive elements was identified in the summary at the start of this paper as a key issue in the development of urban water infrastructures (on the relevance of market competition as to transformation of the water sector see also the contribution of Bedtke and Gawel, "Linking Transition Theories with Theories of Institutions – Implications for Sustainable Urban Infrastructures Between Flexibility and Stability", in this volume). A glance at the complete package of sustainability objectives reveals that this issue is relevant to most of them – e.g., to demand orientation, health protection, environmental soundness, affordability, efficiency, flexibility, and so on. In each case, it is necessary to ask to what extent implementation of the sustainability objective is dependent on or benefits from the public monopoly and whether equally good or even better results could be achieved through competitive fulfilment of the task within a suitably regulated framework.

It is not possible to give clear priority to one side of the argument or the other on the basis of the above consideration of the package of sustainability aspects. Instead, various advantages and disadvantages, as well as opportunities and risks, can be identified on both sides. Securing sustainable service provision by private stakeholders, especially in terms of health protection and environmental soundness, usually requires specific regulation. The extent to which (one-off) competition among private companies in a tendering process (for the market) or possibly new and revised price regulation of existing territorial monopolies can be designed in a way that it boosts efficiency – in particular by cutting prices – while at the same time securing sustainability, clearly depends to a large extent on how concessions, tendering processes, prices, and competition are regulated. In Germany and Europe, the relevant regulations continue to be in flux. At all events, there is no suggestion that the water sector should be granted a general exemption from cost-effectiveness postulates.

Moreover, in view of the special features of water, classical liberalization, as in other network-tied sectors (rail transport, telecommunications), is only possible to a very limited extent – and from the point of view of other sustainability considerations, especially security of supply, health protection, environmental soundness, affordability, and acceptance, it is not particularly useful. By contrast, competitive elements, as part of the regulatory framework, are certainly worthy of consideration (Gawel 2016b). However, in connection with competition for the market, which may involve subdividing infrastructure services into tendering periods, areas, and performance segments, it should be borne in mind that – as has been described here – many of the sustainability imperatives can only be effectively and efficiently fulfilled through systematic integrated development and, in particular, through close coordination with associated public infrastructure and spatial development activities, especially those relating to water supply, sewage disposal, water resources management, and urban development. Prima facie, this integration requirement supports the primacy of public, municipal responsibility for such tasks, although this does not imply any exemption from competitive and efficiency-oriented regulatory elements.

With regard to the possible sustainability advantages of decentral solutions, further consideration needs to be given to whether, and to what extent, these options should be promoted through more rigid application of sustainability criteria to connection and usage dictates, thereby providing more scope for private initiatives and decentral solutions. In this respect, however, it is also necessary to consider the whole package of sustainability conditions from a long-term perspective and to ensure that the infrastructure system *as a whole* is developed coherently and sustainably.

4.2 *The Regulatory Framework*

A regulatory framework that provides mandatory instructions for action is clearly important for almost all the sustainability conditions – in particular, for health protection and environmental soundness, but also for the efficiency (rules on public contracts, benchmarks) and affordability (price regulation) of infrastructure services. In Germany, the regulatory framework for all these sustainability aspects is already well developed, but there is still scope for further optimization, or at least further consideration, as is discussed elsewhere (Reese et al. 2015).

4.3 Incentives

Economic incentive instruments direct stakeholders' behaviour by using the price mechanism to influence their economic decision-making calculations. Charges for water services play a key role, because they need to address a large number of sustainability considerations in a single variable. Sewage and water withdrawal levies represent another important method of creating incentives for environmentally friendly, resource saving, and efficient user behaviour and generating funds for the financing of infrastructure development. In this respect, the precept of cost recovery on the basis of the 'polluter pays' or 'user pays' principle, which has been classified here as an independent sustainability condition, is clearly pivotal. This imperative is also directly enshrined in the EU Water Framework Directive (Article 9 WFD – Gawel 2016a). There are fundamental conflicts of objective that need to be resolved between the objectives of resource efficiency and environmental soundness, on the one hand, and affordability and cost recovery on the other. In addition, improving the efficiency of supply and disposal services remains a challenge – all the more so because it must not be pursued in isolation, as in the past, but needs to be reconciled with the other sustainability requirements.

4.4 Organization

The organization of urban water services must be examined in the light of the above-mentioned sustainability objectives and conditions and consideration must be given to the forms of organization best suited to the pursuit of the sustainability objectives. Organization may obstruct realization of the sustainability objectives and the associated integration requirements, in particular, through sectoralization and a lack of coordination instruments. Important aspects of coordination are coordination of water supply and wastewater disposal, coordination of overall water infrastructure development with urban planning, and coordination between water infrastructure development and water resources management. The more strongly decentral structural elements are included in water supply and disposal provision, the greater the importance of coordination between the officials responsible for central structures and the operators of decentral structures. Organizational obstacles to sustainability can also arise if the areas for which officials have spatial responsibility are too small, with the result that they lose the capacity to pursue demanding and complex sustainability objectives effectively.

4.5 Planning

Planning is 'a systematic process for defining the objectives and sequences of actions over a longer period' (Fürst 1995, p. 11). One of the key purposes of the planning process is to achieve optimum coordination of substantively related individual decisions in the light of shared objectives. Planning law instruments therefore aim to ensure such systematic, long-term oriented, cognitively and predictively prepared, and adequately coordinated processes. The corresponding planning institutions are, naturally, mainly involved in situations related to the development of complex land-use, settlement and infrastructure elements.

The fact that infrastructure development is essentially a complex planning task involving investigation, coordination, and optimization is of major importance in connection with the realization of almost all the sustainability conditions and, in particular, the necessary coordination and overall optimization of the frequently conflicting sustainability requirements. It is difficult to meet the multiple optimization requirements of the sustainability principle effectively without proper planning that develops both the internal and external aspects of the infrastructure system in the light of its relationship to urban development and water resources management, that builds on thorough investigation of the relevant facts, forecasts and interests, and that arrives at a well-founded development strategy by involving relevant bodies and other stakeholders. However, in Germany, corresponding specialist planning law relating to urban water infrastructure exists only in a more or less rudimentary form in the law of the *Länder*, i.e., the national states (Wickel 2015). The development of an upgraded planning system geared to sustainability criteria as a basis for both environmental impact assessment and the involvement of the general public could therefore be an important driver of the sustainable orientation of water infrastructure (Wickel 2015).

5 Conclusion

In the context of urban transformations, urban water management is facing major challenges and corresponding pressure to change. These changes should be systematically based on the principle of sustainable development. Future organization of urban water infrastructure, as a pillar of sustainable urban development, will – as we have set out to show here – depend on a number of core objectives and conditions of sustainability that, in some cases, conflict with each other. For future co-development of technical and institutional solutions for blue infrastructure in the city, it will be important to consider *all* these key conditions of sustainability and to 'optimize' them by adopting an integrating perspective. This applies, of course, both within the water sector and to the linkages with other urban/spatial change processes and infrastructures.

References

Bahri A (2012) Integrated urban water management. TEC background papers no.16. Global Water Partnership. http://www.mynanbo.org/public/documents/outils/uploaded/lts46ngv.pdf. Accessed 24 Nov 2016

Bakker K (2010) Commons versus commodities: political ecologies of water privatization. In: Peet R, Robbins P, Watts M (eds) Global political ecology. Routledge, London, pp 345–368

Bauer W (2014) Morgenstadt – Städte der Zukunft. In: Pinnekamp J (ed) 47. Essener Tagung für Wasser- und Abfallwirtschaft – Ist unsere Wasserwirtschaft zukunftsfähig? Gewässerschutz, Wasser, Abwasser 234. Ges. zur Förderung der Siedlungswasserwirtschaft an der RWTH, Aachen

Bedtke N, Gawel E (2015) Flexible Wasserinfrastruktursysteme – Konzepte und institutionelle Ansatzpunkte. In: Gawel E (ed) Die Governance der Wasserinfrastruktur, Band 2: Nachhaltigkeitsinstitutionen zur Steuerung von Wasserinfrastruktursystemen. Duncker & Humblot, Berlin, pp 77–123

Besche B (2004) Wasser und Wettbewerb, Möglichkeiten und Grenzen einer Öffnung des Wassermarktes. Lang, Frankfurt

BMI – Bundesministerium des Inneren (2011) Schutz Kritischer Infrastrukturen – Risiko- und Krisenmanagement – Leitfaden für Unternehmen und Behörden http://www.bbk.bund.de/SharedDocs/Downloads/BBK/DE/Publikationen/PublikationenKritis/Leitfaden_Schutz-Kritis.pdf?__blob=publicationFile. Accessed 24 Nov 2016

Brackemann H, Bartel H, Höring H, Kühleis C, Rechenberg J, Teuchert C (2001) Nachhaltige Wasserversorgung in Deutschland – Analyse und Vorschläge für eine zukunftsfähige Entwicklung. Erich Schmidt Verlag, Berlin

Bretschneider W (2016) Access, affordability and efficiency: an institutional economic approach of social concerns in potable water allocation. PhD dissertation, Leipzig University

Britz G (2004) Kommunale Gewährleistungsverantwortung – Ein allgemeines Element des Regulierungsrechts in Europa? Die Verwaltung 37:145–163

Bulkeley H, Mol APJ (2003) Participation and environmental governance: consensus, ambivalence and debate. Environ Values 12(2):143–154

Bunster-Ossa IF (2013) Sponge city. In: Pickett STA, Cadenasso ML, McGrath B (eds) Resilience in ecology and urban design: linking theory and practice for sustainable cities. Springer, Dordrecht, pp 301–306

Castell-Exner C, Zenz T (2010) Klimawandel und Wasserversorgung. Energie, Wasser-Praxis 61(3):20–23

Cordell D, Rosemarin A, Schröder JJ, Smit AL (2011) Towards global phosphorus security: a systems framework for phosphorus recovery and reuse options. Chemosphere 84(6):747–758

Dohmann M (2014) Voraussichtliche Investitionen für eine nachhaltige Abwasserinfrastruktur. In: Pinnekamp J (ed) 47. Essener Tagung für Wasser- und Abfallwirtschaft: Ist unsere Wasserwirtschaft zukunftsfähig? – Gewässerschutz, Wasser, Abwasser 234. Ges. zur Förderung der Siedlungswasserwirtschaft an der RWTH, Aachen, pp 4/1–4/18

DWA – Deutsche Vereinigung für Wasserwirtschaft (2010) Energiepotenziale in der deutschen Wasserwirtschaft – Schwerpunkt Abwasser. DWA, Hennef

Enquete-Kommission – Schutz des Menschen und der Umwelt (1994) Die Industriegesellschaft gestalten. (12.07.1994, BT-Drs. 12/8260). Bundestag, Bonn

Finger M, Allouche J, Luís-Manso P (2007) Water and liberalisation – European water scenarios. IWA Publishing, London

Flasbarth J (2012) Konsequenzen der Energiewende für die Wasserwirtschaft. Korrespondenz Abwasser, Abfall 6(59):560–562

Fürst D (1995) Planung. In: ARL (Akademie für Raumforschung und Landesplanung) (ed) Handwörterbuch der Raumordnung. Hannover, ARL, pp 708–711

Gawel E (2011) Kapitalentnahmen und Gebührenkalkulation. Zur Finanzierungswirkung kalkulatorischer Kosten. Der Gemeindehaushalt 112(10):217–222

Gawel E (2014) Umwelt- und Ressourcenkosten: Begriff und Stellung im Rahmen von Art. 9 WRRL. Die öffentliche Verwaltung 67(8):330–338
Gawel E (2016a) Environmental and resource costs under article 9 water framework directive. Challenges for the implementation of the principle of cost recovery for water services. Duncker & Humblot, Berlin
Gawel E (2016b) The regulation framework of the German water sector from an economic point of view – taking stock and perspektives. gwf –Wasser, Abwasser 157(5):538–556
Gawel E (2016c) Ressourceneffizienz als ökonomisches Konzept. In: Reimer F (ed) Ressourceneffizienz – Leitbild für das Umweltrecht? – Gießener Abhandlungen zum Umweltrecht, vol 26. Nomos, Baden-Baden, pp 31–62
Gawel E, Bedtke N (2013) Wasserpreise zwischen Kartellkontrolle und Nachhaltigkeit. Wirtschaftsdienst 93(2):94–102
Gawel E, Bedtke N (2015) Efficiency and competition in the German water sector between "Modernization" and "Regulation". J Public Nonprofit Services 38(2/3):97–132
Gawel E, Bretschneider W (2014) The affordability of water and energy pricing: the case of Germany. In: Fitzpatrick T (ed) International handbook on social policy and the environment. Edward Elgar, Cheltenham, pp 123–151
Gawel E, Bretschneider W (2016) Content and implementation of a right to water. An institutional economics approach. Metropolis, Marburg
Geyler S, Bedtke N, Gawel E (2014) Nachhaltige Regenwasserbewirtschaftung im Siedlungsbestand – Teil 1: Ziele, Optionen und Herausforderungen. gwf – Wasser, Abwasser 155(1):96–102
Gleick PH (2006) Water and terrorism. Water Policy 8(6):481–503
Grabow B, Schneider S (2014) KfW Kommunalpanel 2014. Frankfurt am Main, KfW
Grafton Q, Daniell KA, Nauges C, Rinaudo J-D, Wai Wah Chan N (2015) Understanding and managing urban water in transition. Springer, Dordrecht
Habermas J (1992) Faktizität und Geltung: Beiträge zur Diskurstheorie des Rechts und des demokratischen Rechtsstaats. Frankfurt a. M, Suhrkamp
Hanke S, Libbe J (2012) Ansätze und rechtliche Rahmenbedingungen für neuartige Lösungen in der Abwasserentsorgung. Zeitschrift für Umweltpolitik und Umweltrecht 2:151
Hansen J (2014) Chancen für eine Steigerung der Energieeffizienz kommunaler Abwasseranlagen. In: Pinnekamp J (ed) 47. Essener Tagung für Wasser- und Abfallwirtschaft: Ist unsere Wasserwirtschaft zukunftsfähig? Gewässerschutz, Wasser, Abwasser 234. Aachen, Ges. zur Förderung der Siedlungswasserwirtschaft an der RWTH, pp 60/1–60/15
Hasse JU, Wienert B, Bolle F-W (2012) Die Wasserwirtschaft gestaltet den Wandel in der Emscher-Lippe-Region – Erfolge des Netzwerk und Forschungsprojekts dynaklim. In: Pinnekamp J (ed) 45. Essener Fachtagung für Wasser- und Abfallwirtschaft: Wasserwirtschaft und Energiewende Gewässerschutz, Wasser, Abwasser 230. Aachen, Ges. zur Förderung der Siedlungswasserwirtschaft an der RWTH, pp 30/1–30/14
Hellström D, Jeppsson U, Karrman E (2000) A framework for systems analysis of sustainable urban water management. Environ Impact Assess Rev 20(3):311–321
Hillenbrand T, Niederste-Hollenberg J, Menger-Krug E, Klug S, Holländer R, Lautenschläger S, Geyler S (2010) Demografischer Wandel als Herausforderung für die Sicherung und Entwicklung einer kosten- und ressourceneffizienten Abwasserinfrastruktur. UBA-Texte; 36/2010. UBA, Dessau
Jonuschat H, Baranek E, Behrendt M, Dietz K, Schlußmeister B, Walk H, Zehm A (eds) (2007) Partizipation und Nachhaltigkeit. OEKOM, München
Kahlenborn W, Kraemer A (1999) Nachhaltige Wasserwirtschaft in Deutschland. Springer, Berlin
Kampe D (1987) Dezentrale Konzeptionen räumlicher Wasserversorgung. Informationen zur Raumentwicklung 5(6):323–333
Karthe D (2015) Bedeutung des Klimawandels (insbesondere hydrometeorologischer Extremereignisse) für die Trinkwasserhygiene in Deutschland. In: Evers M, Diekkrüger B

(eds) Aktuelle Herausforderungen im Flussgebiets- und Hochwassermanagement. Prozesse, Methoden, Konzepte. Hennef, FHGW, pp 33–43
Kern K (2014) New standards for the chemical quality of water in Europe under the new Directive 2013/39/EU. J Eur Environ Plan Law 11(1):31–48
Keßler S, Nilson E, Koflak S (2012) Ergebnisse des Forschungsprojekts KLIWAS und Nutzen für wasserwirtschaftliche Fragen der Anpassung an den Klimawandel. In: Pinnekamp J (ed) Wasserwirtschaft und Energiewende – 45. Essener Tagung für Wasser und Abfallwirtschaft, Aachen: Gesellschaft zur Förderung der Siedlungswasserwirtschaft an der RWTH Aachen. Aachen, RWTH, pp 31/1–31/11
Kopfmüller J, Brandl V, Jörissen J, Paetau M, Banse G, Coenen R, Grundwald A (2001) Nachhaltige Entwicklung integrativ betrachtet – Konstitutive Elemente, Regeln, Indikatoren. edition sigma 2001, Berlin
Kriener F (2004) Wasserversorgung von Ballungsräumen. Nomos, Baden-Baden
Laskowski SR (2010) Das Menschenrecht auf Wasser – Recht der nachhaltigen Entwicklung. Mohr Siebeck, Tübingen
Laskowski SR (2012) Flexibilisierung der kommunalen Abwasserentsorgung in Zeiten des klimatischen Wandels. Zeitschrift für Umweltrecht 23(11):597–606
Leptin C, Bellefontaine K, Breitenbach H, Graf P, Roosen C (2014) Wirtschaftsdaten der Abwasserbeseitigung. Korrespondenz Abwasser, Abfall 61(8):701–707
Lobina E, Hall D (2007) Experience with private sector participation in Grenoble, France, and lessons on strengthening public water operations. Util Policy 15(2):93–109
Londong J, Hillenbrand T, Niederste-Hollenberg J (2011) Demografischer Wandel: Anlass und Chance für Innovationen in der Wasserwirtschaft. Korrespondenz Abwasser, Abfall 58(2):152–158
Makropoulos CK, Memon FA, Shirley-Smith C, Butler D (2008) Futures: an exploration of scenarios for sustainable urban water management. Water Policy 10(4):345–373
Mankad A, Tapsuwan S (2011) Review of socio-economic drivers of community acceptance and adoption of decentralised water systems. J Environ Manag 92(3):380–391
Mason M (1999) Environmental democracy. Earthscan, London
Mo W, Zhang Q (2013) Energy–nutrients–water nexus: integrated resource recovery in municipal wastewater treatment plants. J Environ Manag 127:255–267
Monopolkommission (2010) Mehr Wettbewerb, wenig Ausnahmen. Achtzehntes Hauptgutachten der Monopolkommission gemäß § 44 Abs. 1 Satz 1 GWB (BT.-Drs. 17/2600). Monopolkommission, Bonn
Novotny V (2008) Sustainable urban water management. In: Feyen J, Shannon K, Neville M (eds) Water and urban development paradigms: towards an integration of engineering, design and management approaches. Taylor & Francis, London, pp 19–32
Parsons D, Goodhew S, Fewkes A, de Wilde P (2010) The perceived barriers to the inclusion of rainwater harvesting systems by UK house building companies. Urban Water J 7(4):257–265
Pinnekamp J (2014) Abwassertechnik der Zukunft. In: Pinnekamp J (ed) 47. Essener Tagung für Wasser- und Abfallwirtschaft Ist unsere Wasserwirtschaft zukunftsfähig? Gewässerschutz, Wasser, Abwasser 234. Ges. zur Förderung der Siedlungswasserwirtschaft an der RWTH, Aachen
Po M, Kaercher JD, Nancarrow BE (2003) Literature review of factors influencing public perception of water reuse, Technical report 54/03. CSIRO, Perth
Reese M (2010) Die Herausforderung der Klimaanpassung an Bürger, Politik und Recht. In: Reese M, Bovet J, Möckel S, Köck W (eds) Rechtlicher Handlungsbedarf für die Anpassung an die Folgen des Klimawandels – UBA-Berichte 1/2010. UBA, Dessau-Roßlau, pp 1–26
Reese M, Bedtke N (2015) Was ist "Nachhaltigkeit" und was ist "nachhaltige" Wasserwirtschaft? Allgemeine Nachhaltigkeitskonzeptionen und Ableitungen für den Bereich der Wasserver- und Abwasserentsorgung. In: Gawel E (ed) Die Governance der Wasserinfrastruktur. Band 1: Rahmenbedingungen, Herausforderungen und Optionen – Studien zu Umweltökonomie und Umweltpolitik, 11. Duncker & Humblot, Berlin, pp 198–289

Reese M, Koch HJ (2010) Abfallwirtschaftliche Daseinsvorsorge im Europäischen Binnenmarkt Deutsches Verwaltungsblatt, 1393–1403
Reese M, Gawel E, Geyler S (2015) Die Nachhaltigkeitsgebote der Siedlungswasserwirtschaft – Kernziele, Grundvoraussetzungen und institutionelle Ansatzpunkte nachhaltiger Wasserver- und Abwasserentsorgung. In: Gawel E (ed) Die Governance der Wasserinfrastruktur. Band 1: Rahmenbedingungen, Herausforderungen und Optionen – Studien zu Umweltökonomie und Umweltpolitik, 10. Duncker & Humblot, Berlin, pp 197–289
Renn O, Köck W, Schweizer P, Bovet J (2014) Öffentlichkeitsbeteiligung bei Vorhaben der Energiewende. Zeitschrift für Umweltrecht 25(1):281–287
Rogall H (2008) Ökologische Ökonomie. Verlag für Sozialwissenschaften, Wiesbaden
Roskosch A, Rechenberg B (2015) Phosphorrückgewinnung als Ressourcenschutz. gwf – Wasser, Abwasser 156(2):214–220
Russell S, Lux C (2009) Getting over yuck: moving from psychological to cultural and sociotechnical analyses of responses to water recycling. Water Policy 11(1):21–35
Scheele U (2006) Versorgungssicherheit und Qualitätsstandards in der Wasserversorgung – Neue Herausforderungen unter Veränderten Rahmenbedingungen, netWORKS Diskussionspapier Nr. 23. Berlin netWORKS
Siedentop S, Moritz H, Schulwitz M (2015) Kommunale Infrastrukturkosten und Demographie. https://www.wegweiser-kommune.de/documents/10184/16915/Infrastrukturkosten + und + Demographie/03a26f 08-b007-4c5c-9c59-7691e 326271b. Accessed 24 Nov 2016
Spiller M, Vreeburg J, Leusbrock I, Zeeman G (2015) Flexible design in water and wastewater engineering – definitions, literature and decision guide. J Environ Manag 149:271–281
SRU – Sachverständigenrat für Umweltfragen (2002) Umweltgutachten 2002: Für eine neue Vorreiterrolle. SRU, Berlin
SRU – Sachverständigenrat für Umweltfragen (2008) Umweltgutachten 2008 – Umweltschutz im Zeichen des Klimawandels. Erich-Schmidt, Berlin
SRU – Sachverständigenrat für Umweltfragen (2012) Umweltgutachten 2012: Verantwortung in einer begrenzten Welt. Erich-Schmidt, Berlin
Statistisches Bundesamt (2015) Bevölkerung Deutschlands bis 2060–13. koordinierte Bevölkerungsvorausberechnung. Statistisches Bundesamt, Wiesbaden
Tränckner J, Koegst T (2011) Entwicklung der Arzneimittelkonzentrationen im Abwasser durch demografischen Wandel – Konsequenzen für die Abwasserbehandlung. Korrespondenz Abwasser, Abfall 58(2):128–137
UBA – Umweltbundesamt (2015) Organische Mikroverunreinigungen in Gewässern – Vierte Reinigungsstufe für weniger Einträge – UBA-Positionspapier March 2015. UBA, Dessau-Roßlau
Wickel M (2015) Rechtliche Pflichten, Verfahren und Anreize zur besseren Vernetzung von Siedlungswasserwirtschaft und Stadtentwicklung. In: Gawel E (ed) Die Governance der Wasserinfrastruktur, Band 2: Nachhaltigkeitsinstitutionen zur Steuerung von Wasserinfrastruktursystemen. Speyer und Peters, Berlin, pp 399–459
Wong TH, Brown RR (2009) The water sensitive city: principles for practice. Water Sci Technol 60(3):673–682
Zebisch M, Grothmann T, Schröter D, Hasse C, Fritsch U, Cramer W (2005) Klimawandel in Deutschland: Vulnerabilität und Anpassungsstrategien klimasensitiver Systeme. UBA, Dessau-Roßlau

Transitioning the Heat Supply System – Challenges with Special Focus on Bioenergy in the Context of Urban Areas

Volker Lenz, Cornelia Rönsch, Kay Schaubach, Sebastian Bohnet, and Daniela Thrän

1 Background and Challenges

The reduction of greenhouse gas (GHG) emissions is a central goal of international climate policy agreements. Thus, Europe and, consequently Germany, as one of its major economies, have decided to reduce their GHG emissions by 80–95% by the year 2050. Therefore, the energy supply system needs substantial decarbonizing (Merkel 2015). The predominant form of energy consumed by end users in Europe and Germany is heat, which is mainly used for residential demands, i.e., space heating and hot water. The most important renewable source for heat is biomass, which accounts, for example in Germany, for almost 90% of the overall renewable heat (see Fig. 1).

The constantly increasing share of the human population living in towns and growing urban areas has reached about 50%, which corresponds to approx. 3% of the land surface worldwide (Reusswig et al. 2014). Not only in countries with increasing of the total population but also in those with even decreasing numbers of inhabitants, the urban areas are growing in space and population. In the years to come, it is expected that the energy demand of these densely populated areas will rise dynamically. This, in combination with the anticipated decarbonization of the energy sector, poses a significant challenge but also the opportunity to implement

V. Lenz (✉)
Department Thermo-chemical Conversion, Deutsches Biomasseforschungszentrum gemeinnützige GmbH (DBFZ), Leipzig, Germany
e-mail: Volker.Lenz@dbfz.de

C. Rönsch • K. Schaubach • D. Thrän
Department of Bioenergy Systems, Deutsches Biomasseforschungszentrum gemeinnützige GmbH (DBFZ), Leipzig, Germany
e-mail: cornelia.roensch@dbfz.de; kay.schaubach@dbfz.de; daniela.thraen@dbfz.de

S. Bohnet
Delitzscher Land e.V., Delitzsch, Germany
e-mail: sebastian.bohnet@delitzscherland.de

S. Kabisch et al. (eds.), *Urban Transformations*, Future City 10,
DOI 10.1007/978-3-319-59324-1_10

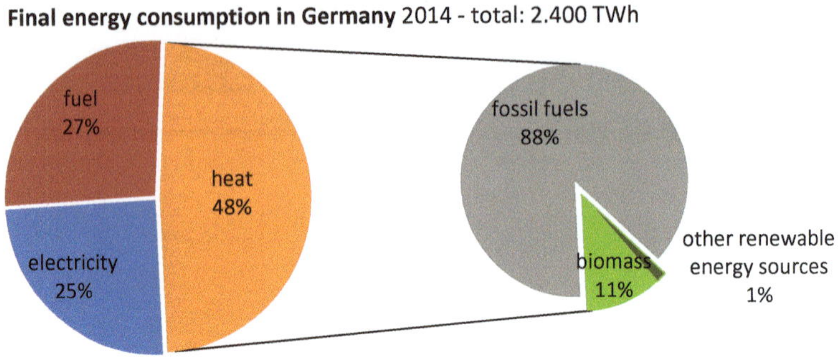

Fig. 1 Final energy consumption in Germany (Source: Based on BMWi 2015)

renewable heat on a large scale. Urban areas and their districts differ greatly in terms of, e.g., building characteristics and density, infrastructure, proportion of residential areas, public buildings, industries and businesses, which also results in diverse heating systems on levels from single edifices up to the urban area as a whole. Therefore, decarbonization strategies for these areas have to consider a broad range of key factors, such as available sustainable fuels, various conversion technologies at different scales, varying demands, integration with the sectors of waste management, power supply and, possibly, material use, as well as the parameters set by the existing infrastructure.

Using biomass as a source for renewable heating within urban areas involves specific challenges:

- Within growing cities, the transport distances for solid and liquid biofuels, as well as for biogenic waste materials, will expand, as long as there is no internal production, use, and recovery or disposal.
- Air pollution is becoming a severe problem in urban areas. Due to many emitters, a large number of parties are concerned and only a low natural regeneration potential is available. Urban areas have a share of 60–80% of worldwide emissions (Reusswig et al. 2014). Moreover, the combustion of solid biofuels in inefficient heating units (e.g., for cooking) is very often one of the major sources of air polluting emissions (see Kindler et al., "Socio-Spatial Distribution of Airborne Outdoor Exposures – An Indicator for Environmental Quality, Quality of Life, and Environmental Justice: The Case Study of Berlin", in this volume).
- Urban areas depend strongly on a secure energy supply. Black-outs can cause serious damage. Therefore, grid stability and independent back-up capacities for power and heat generation are essential, e.g., with biomass.
- Very often, the inhabitants of urban areas live with greater anonymity in the near neighborhood than in rural areas. But, at the same time, a greater interest in improving general living conditions of their own part of the city is possible, including the energy supply. Thus, changing the latter depends very much on a committed involvement of the inhabitants.

Europe and, here, especially Germany, is among the areas with the most ambitious plans to reduce greenhouse gas emissions. Whilst, for electricity supply, the

"Energiewende" is already in implementation, additional measures are currently being discussed for heat supply (Stryi-Hipp et al. 2015). Therefore, this article focusses mainly on German conditions in the largest urban areas (e.g., Berlin, Rhein-Main-area, or Ruhr-area), building on the experiences from the transformation process of the electricity sector. In a first step, the different heating systems and exemplary data on heat demand and percentage of renewable provision are elaborated. Secondly, the possible potentials of biogenic fuels, especially in urban areas, are listed. This is complemented by a definition of criteria for promising future biogenic energy systems and a chapter describing these new technology options. A brief discussion of change management is given thereafter. Taking the distinction between European member states and urban areas in other countries into account, generalized conclusions for future biogenic heat supply in urban areas are drawn as an impetus for further research and discussion.

Besides the technical, economic, and ecological aspects of the transformation of the heat supply in urban areas, there are legislative regulations allowing, blocking, or incentivizing transformation options. This article does not explicitly consider legislative impacts, so that the implementation of some of the described options may require changes in national and international legal frameworks.

2 Heat Supply in Urban Areas

Urban areas in Germany are characterized by heterogeneous structures in terms of their use of land and the type of buildings they contain. Depending on the specific building layout, they are classified as (1) concentrated districts with multi-storey, mostly non-residential buildings, (2) commercial and industrial areas, (3) apartment buildings, as well as (4) low-rise developments with garden structures. In other regions of the world, different historical styles exist. However, in general, the structures in urban areas are comparable, in spite of (5) slums in developing countries and, sometimes, (6) districts without perspectives, due to, e.g., economic decline or severe environmental problems.

The heat demand profiles of specific districts depend on the composition of building types (industrial, commercial, public, residential), the properties of the building stock such as age (usually covering a large range) and renovation status, resulting in a range of insulation standards, as well as the residential milieu. For example, office buildings and schools have a different daily heat load course, compared to apartment houses. The specific heat demand ranges from 15 W/m^2 to 250 W/m^2, resulting in, e.g., 5–25 kW heat demand for a typical detached apartment house (Erhorn et al. 2010). The heterogeneous nature of the building stock and the stakeholders (owners, renters, public sector) in the urban areas as well as the variety of fuels used, leads to a wide range of systems and technologies for heat generation. Basically, there is a distinction between the heat-only generation and the cogeneration of heat and electricity (Combined Heat and Power, CHP). Heat-only plants use the thermal energy of a fuel for the production of heat by means of combustion.

Systems are available for a wide range of installed capacity as well as energy sources.

- Heating systems for single rooms (e.g., stove, fireplace):
  - The generated heat is only used at the place of installation. Installed capacity depends on the size of the room, with a typical load of 6–8 kW_{th}, up to a maximum of 20 kW_{th}.
  - Solid fuels dominate, especially wood, occasionally coal. Nevertheless, some electrical heat-storing radiators or mobile radiators are in use.
- Central heating systems (e.g., (condensing) boilers, heat pumps, solar-thermal systems) with and without heating grid (district heating):
  - Generated heat is fed into a distribution system, supplying a single building, a group of a few buildings, a district, or even entire cities. Systems are available from 4 kW_{th} up to a few MW_{th}, depending on the energy source and application (e.g., detached houses, multi-storey dwelling, schools, hospital, industrial sites).
  - Fossil fuels, especially natural gas, dominate in urban areas due to the gas distribution system. In the spectrum of renewable sources, woody biomass is particularly important in central heating systems (primarily wood chips, logs, and pellets). Particularly in view of fuel logistics and the existing gas networks, bio-methane might gain in importance.
- Combined Heat and Power Plants (CHP plants)
  - Heat generated in conjunction with electricity production processes is fed into a local or area distribution system. Plants are available from 1 kW electric capacity (and 4 kW thermal capacity) up to a few MW_{el} and MW_{th}, depending on the energy source and application (e.g., detached houses, businesses, public supply).
  - Fossil fuels, i.e., natural gas, oil, and coal in regions with coal mining dominate. As a result of the German Renewable Energy Act, more than 9000 CHP plants based on biomass have been installed in Germany (2013). Biogas plants represent 65% of the installed capacity, wood-based CHP plants 31% and vegetable oil-based CHP plants 4% (Scheftelowitz et al. 2014).

The heat distribution systems listed above can be differentiated by the number of edifices they supply and, hence, the distances covered and the applied technologies. If the heat is used in close proximity to the conversion site, the system is categorized as object supply. In case large physical distances are covered, the term "grid-connected supply" is used in this paper. Depending on the actual transport distance, this is subdivided into local heating or long-distance heating. However, there is no exact quantitative differentiation between both terms (Fischedick et al. 2007). When it comes to the use of excess heat from industrial processes and CHP plants, the main focus is on the grid-connected supply.

As a result of the described variety of structures and stakeholders in urban areas, the whole range of heat supply and distribution systems is implemented to different degrees. Due to the close proximity of consumers in urban areas, grid-connected supply predominates very often and in many parts of the urban area, either as long-distance heating, gas networks, or even the power grid. The three most important urban areas in Germany, with the key figures for the heat supply, are described in Table 1.

For decentralized heating furnaces based on biomass, there is a greater need for fuels with superior properties for long-distance transport, especially energy density, to reduce logistics costs. For example, wood pellets are superior to log wood, which is found more often in rural areas. The use of log wood in urban areas is limited to single- or two-family houses in the sub-urban areas and single combustion units functioning as additional heating systems in the comfort segment. Other possible renewables, besides biomass, that can be used in single-family houses as well as

Table 1 The most important urban areas in Germany, compared to the national average

Area	Size	Inhabitants	Heat supply	Source
Ruhr-area (e.g., Essen)	4400 km^2 (210 km^2)	Approx. 5,100,000 (approx. 540,000)	Essen: 58%individual gas boilers, 17% long-distance heating, 10% oil heaters, 15% electric heaters; large biomass heating units with local heating system based on local wood, waste incineration plant feed in long-distance heating (55%)	Rat der Stadt Essen (2009)
Berlin	890 km^2	Approx. 3,500,000	44% individual gas boilers, 31% long-distance heating/ steam, 25% oil heating, renewable energies <0.5%	Reusswig et al. (2014)
Rhein-Main-area (e.g., Frankfurt/ Main)	4300 km^2 (250 km^2)	Approx. 2,600,000 (app. 720,000)	Frankfurt/Main: 50% individual gas boilers, 38% long distance heating/steam from gas plants, 7.5% oil heaters, 0.5% coal heaters, 3% renewables (one third of the renewable heat is produced by an organic waste fermentation plant, a large biomass CHP unit and a variety of smaller decentralized biomass CHP units; in total 130 GWh/a)	Raatz et al. (2014)
Germany	357,000 km^2	Approx. 82,200,000	11% long-distance heating (of this, 18% renewable energy sources), 36% industrial heating systems, 53% small heating systems (15% renewable energy sources)	AG Energiebilanzen (2016)

apartment houses and office buildings are heat pumps and solar thermal systems. For the city of Frankfurt, several projects have proven the feasibility beyond new buildings and single-family houses; large plants have been realized, even in the densely built financial district, as well as in commercial areas (Schumacher et al. 2015).

3 Biogenic Urban Sources of Heat Supply

Biomass can be differentiated into woody, herbaceous, and other biomass. Biomass for heat supply is mainly forest wood, generated in rural areas (BMWi 2014). But there are also biogenic resources within cities that are applicable for heat generation.

Woody biomass of urban origin includes waste wood. The assessments of (Thrän 2009) report that 70% of this wood comes from settlement, packaging, and construction waste. In addition to woody biomass, herbaceous biomass includes landscaping material and material that comes from the maintenance of roadside vegetation and green waste from the landscaping of parks and cemeteries, in particular. The biological recycling of brownfields also generates biomass that has to be dealt with (see Banzhaf et al., "What Really Matters in Green Infrastructure for the Urban Quality of Life? Santiago de Chile as a Showcase City", in this volume). These biomass streams are characterized by increased heavy metal content (roadside maintenance) and, often, increased ash content (landscape maintenance wood), which make it difficult to use them in small-scale combustion units. Herbaceous biomass is very inhomogeneous in composition and is characterized by high water content that requires drying to be considered for thermochemical conversion (see figures in Table 2).

In addition to woody and herbaceous biomass, urban areas also generate wet and damp residues, by-products and waste of biogenic origin from businesses and industries, as well as municipal waste and organically polluted wastewater. In urban areas, the most important form is the municipal waste, which consists of organic

Table 2 Urban biomass resources in t (dry mass) and the theoretical energy potential in PJ, each per year, in the three main urban areas in Germany

Urban area	Waste wood and landscape wood	Wet and damp residues/wastes (households and commercial)	Food production residues	Wastewater (sludge)	Energy potential in PJ
Ruhr-area	800,700	446,800	395,300	229,000	18.8
Berlin	549,500	306,600	271,300	157,200	12.9
Rhein-Main-area	408,200	227,800	201,500	116,700	9.6

Sources: Reusswig et al. (2014), Rat der Stadt Essen (2009), Raatz et al. (2014)

household waste and commercial waste similar to household waste (e.g., waste from canteens and markets, and residues from the retail sector). This extensive resource could and should be used for energetic purposes (e.g., in Germany, since 2005, only waste with an organic mass component of $\leq 5\%$ in the dry base may be deposited legally in landfills (BMJV 2009)). Separately collected biogenic waste (organic waste, market and canteen waste) can be used to generate energy in a biogas plant (see figures in Table 2).

Food production, as well as food processing, takes place in urban areas to fulfil the food demands of the inhabitants. Usually, such food industry plants are located at the periphery of the city center. However, the use of by-products for energy can be exploited to supply companies' sites and surrounding commercial units.

Biomass also includes organically polluted wastewater that occurs in households, as well as in industry and commerce. The regulations for the discharge of wastewater into receiving waters require pre-treatment of waste water by aerobic and anaerobic processes. Quantitatively, plants using exclusively aerobic treatment processes predominate. With increasing capacity and, thus, plant size, anaerobic processes are integrated into sewage treatment plants, so that an additional energy production is possible; it is often used to meet the energy needs of the sewage treatment plants.

Therefore, for the three urban areas – Ruhr, Berlin, and Rhein-Main – in Germany, approximately 7% of the heat demand in each region, based on the average national demand, could, theoretically, be supplied by the existing waste potentials, if all the resources are used for heat generation. If a significant reduction of heat demand by insulation and higher efficiency of commercial processes were to be realized (BMVI 2013), even up to 12% of the future heat demand could be supplied by waste biomass. It has to be noted that the utilization of these potentials is linked to special emission risks. Thus, special legal aspects have to be considered very often, e.g., special emission reduction measurements or safe distances to housing. In addition to the space required and concerns of the general public, there is the challenge of integrating such systems into urban areas from where the biomass flows and heat sinks originate. However, the City of Vienna impressively demonstrates how a complex system, such as an incineration plant, can be integrated attractively into the urban heart of the city (Wien Energie 2016).

Nevertheless, even more important biomass resources are located in rural areas; these include straw, forest wood residues and by-products, as well as agricultural biomass sources. To be used in the urban area, these fuels have to be transported either by truck (e.g., compressed solid fuels) or by existing grids (e.g., bio-methane via the gas grid, renewable electricity via the power grid). Already today, road transport in urban areas causes 10% of the global CO_2 emissions (BMBF 2010). Even if transport will have less specific CO_2-emissions in future, transport is time and space consuming. Therefore, not only the transport of energy carriers by road into the urban areas has to be avoided, but also the transport of urban by-products, residues, and waste. Furthermore, biomass can be provided in a local context to consumers by using urban brownfields for biomass production (Ponitka et al. 2010).

In general, the challenge is to make biomass flows, that cannot be used for energy purposes, available for energy recovery, by achieving suitable properties. This requires, on the one hand, adapting combustion technology to the requirements of the fuels. On the other hand, corresponding processes have to be developed and adapted; these include compaction (optionally in combination with additives/pre-treatment), torrefaction, or hydrothermal carbonization. Because urban biomass can be stored, at least for short periods, conditioning can be performed when surplus electricity from renewable sources is in the grid. Thus, additional CO_2 emissions do not occur. From a cost perspective, biogenic waste, residues, and by-products have to be dealt with anyway, because deposition in landfills represents a severe problem in terms of greenhouse gas emissions. For example, in Germany, the deposition of biomass in landfills is forbidden. Therefore, it is not a question of comparing the costs of fuels from rural areas according with those of urban sources but rather a matter of cost-effective conversion of urban waste, residues, and by-products.

4 Criteria for Promising Supply Systems in Urban Areas for Heat from Biomass

Currently, heat from biomass is the most important renewable heat source worldwide, as well as in most regions of the world (see also Fig. 1) (Thormann et al. 2016). Although, in some countries, there is still biomass potential and the political target to increase the use of biomass for energy, e.g., in Japan (JIHK 2014), the global discussion about food security, sustainability of land use and conservation of biodiversity has a more specific focus. Fulfilling global climate protection targets will lead to a supply system dominated by renewable energy, with variable contributions from wind and solar power, as well as from solar-thermal and geothermal heat, in combination with heat from waste sources and surplus electricity. It is expected that, despite heat storage, there will be heat supply gaps, both in total and also temporarily in individual city zones, that will need to be closed with renewable chemical energy storage.

Besides the growing variety of heat supply options, the specific heat demand of the buildings will also continue to differentiate. Thus, there will be many buildings that have very low specific heat requirements, due to extensive insulation work – sometimes even only selective heat demands for domestic hot water. At the same time, there will also be buildings for which an extensive insulation is not possible, for various reasons, so that they will need very substantial heat inputs. In conjunction with the integration of various, inexpensive renewable energy resources, biomass can be used, with high output in the short-term (peak loads), but with comparatively small amounts of heat production overall.

Therefore, the use of biomass for bioenergy will undergo a fundamental change in applications towards much more flexibility and integration into regional value

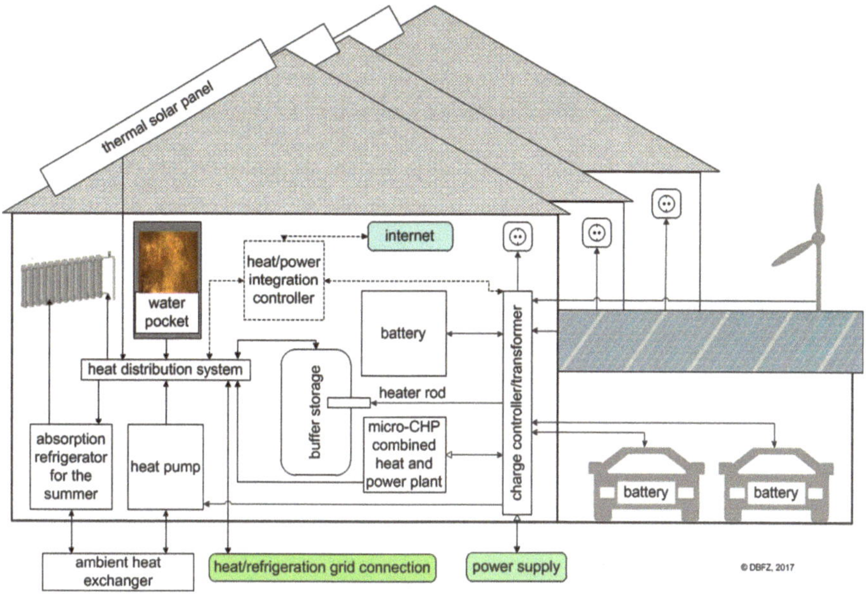

Fig. 2 Exemplary overview of the integration of different renewable heating systems into heat, power, and mobility supply in settlements (Source: Based on Lenz 2017)

chains and a more intensive integration of heat and power supply, as well as of mobility (Lenz and Thrän 2015) (Fig. 2).

Using all the different energy technologies shown in Fig. 2 in one and the same house is definitely not cost-effective. However, if we look at several buildings coupled by a heating grid, such multi-heating systems and multi-combined heating and power systems might be feasible. Additionally, even in single houses, systems with a heat pump and a stove with water heat exchanger (so called water pocket) could be economically feasible (Lenz et al. 2017). Thus, it is always a question of the optimal combination of the different systems, depending on existing technologies, space, available energy sources, housing conditions, and so on.

Additionally, the energetic use of biomass will have to focus much more on residues, by-products, and wastes instead of on high-quality biomass resources from tree wood or agricultural farming for primary use.

In terms of health aspects in urban areas, air quality is one of the greatest challenges for the future; therefore, emissions from thermal combustion processes will have to be minimized.

Both, fuel conditioning and storage, as well as the actual conversion, will need areas within the urban area. Odour and noise emissions are also possible; these have to be considered when selecting suitable sites. In addition, aesthetic design is important for the acceptance of systems whenever they are clearly visible. The last point, in particular, raises the question about how local residents will accept the systems.

In addition to an intensive on-site communication and involvement of residents, the general image of each technology and method is of considerable relevance.

Besides all the listed aspects, the costs are also extremely important. The disposal costs for fuels from waste and the provision of certain ancillary services (electricity grid stabilization, heat supply reliability) need to be considered.

A summary of the above-mentioned criteria is provided in Table 3.

Table 3 Criteria for promising supply systems for heat from biomass in urban areas

Criteria	Range/Unit	Comments
Operating flexibility – load change rate 30–100% and 100–30%	In minutes	Possibly under 5 min, high partial load efficiency
Operational flexibility – cold start and stop time	In minutes	Start possibly under an hour; stop under 5 min
Specific investment costs	€/kW$_{el}$ and €/kW$_{th}$	Competitiveness to alternative renewable heating technologies and feasible insulation/energy efficiency measures
Fuels starting material	Usable materials	Range of usable residual and waste materials
Fuels (fuel processing expenses)	High final quality	Quality criteria, possibly price
Fuels (storability)	Time and space requirements per unit of energy	Durability, water uptake, biodegradation, self-ignition, energy density and stack height (flowability, bridging), emissions during storage
Near-zero emission technology	Dust emissions below 1 mg/m^3, CO emissions below 0.1 g/m^3, org-C. less than 1 mg/m^3; NO_x is below 0.05 g/m^3 in each standard at 13% O_2	Dust, CO, organic carbon, NO_x
Local fuel source	Collection and preparation of local biomass resources	Options of collection and preparation of landscape biomass as well as residues and by-products
Noise emissions	dB (A) ... at the side of the plant	Reducible with better insulation
Odor emissions	Clearly recognizable disturbance or measured by olfactometry	Often connected to quality of combustion
Aesthetics/optics	Visibility, personal likes and dislikes	Very dependent on attitudes of the protagonists
Acceptance of technology	Unacceptable up to "sexy"	Very dependent on attitudes of the protagonists
Space requirements	In m^2	Value of used area

Source: Lenz and Thrän (2015)

5 Future Technologies for Heat from Biomass

Heating options used in urban areas that are based on biomass vary widely on a global scale. In German urban areas, pellet boilers (small scale), larger wood chip heating plants with some 100 kW_{th} up to several MW_{th} (medium scale) and to some extent wood chip heat and power plants with up to 40 MW_{th} and 20 MW_{el} offer the greatest heat contribution. Single-room combustion units are more likely to be installed for reasons of comfort. The co-combustion of biomass in coal firing plants plays only a subordinate role and the number of bio-methane CHP and the proportion of bio-methane in gas heating systems are also marginal. In contrast, urban areas in emerging countries (such as Chile) still have simple wood stoves and heating furnaces in operation, which then also contribute significantly to air pollution.

With regard to development structures and climatic conditions, the authors assume that, due to the requirements defined above, a fundamental adaptation of bio-heat technologies will occur in urban areas in the course of this century. Nevertheless, heat from biomass will play an essential role in the future energy system, because: (1) People like their own and secure heating system, even if it is only a stove; (2) biomass is storable and therefore easily available in the winter; and (3) combined heat and power systems with biofuels can provide residual load with high efficiency in the winter and during long periods of cloudy weather with low wind speeds.

In addition to the technology, there is also the central issue of biogenic fuels. Apart from bio-methane, which can be transported from the surrounding areas into the urban areas with the (possibly) existing natural gas grid, solutions are needed for fuel deployment from waste disposal. Biogenic residues and waste in urban areas could be aligned and integrated to the following categorisation: A) relatively wet, perishable components (food scraps and much waste from food production) and B) drier, lignin-containing residues and waste material (landscaping materials, such as, e.g., leaves and grass, waste wood). Depending on these two categories, two basic formatting options can be distinguished: Anaerobic digestion into biogas or bio-methane, or preparation and conditioning into storable and transportable solid fuels. For the latter approach, two basic pathways can be considered: B1) conversion to wood chips and brikletts (small flowable briquettes within robust conveyor systems) with a minimal effort, and used in robustly constructed conversion plants with high emission control (possibly: sorting, screening, washing, and drying). B2) production of high-end compacts like pellets, pelletinos (pellets with diameters of less than 6 mm) or upgraded firewood substitute briquettes, which can pass through various processing steps such as washing, sorting, screening, mixing, torrefaction, hydrothermal treatment, etc., prior to compacting. The fuels thus obtained are suitable for small-scale and micro cogeneration units because they meet the relevant higher quality standards. This market segment also offers higher prices needed to cover the pre-processing. By means of customized treatment paths, either cost-effective, low-quality fuels can be produced for large-scale plants with adequate conversion technologies, or high-end fuels for small plants of the future can be obtained.

5.1 Heating Technologies Based on Solid Biomass

Automatic stoves with water pockets (small scale). In the future, houses for one or two families and small municipal and commercial properties with a high insulation standard (not passive houses) will be heated with a heat pump, maybe in combination with a solar thermal system. To address the double winter slump (electricity shortages and lower returns of solar heat and lower performance factor of the heat pump), there is, today, already the promising option for an optimized renewable heat supply that involves integrating a stove with water pockets and connecting it to the central heating system. For low emissions and optimum integration into the heat supply, the stoves will have an electrical control of the combustion process (air control) and integrated catalysts and downstream precipitators in the future. The systems will have either thermoelectric generators, to ensure the power self-supply (for control units, ventilators, regulators and ignition), or they can be heated in emergency cases without electricity. The thermal output could be between 2 and 10 kW. In terms of urban supply, high-end fuels (B2) will be used. Currently, stoves are not very flexible. Once they have been started, they have to burn a full batch of fuel in most of the installations without adjustment of the load. In the future, there may be pre-gasification systems with automatic starting systems and automatic load control.

Micro gasifiers with fuel cells in the home area (small scale). In 2050, in addition to fuel cells powered by bio-methane, micro gasifier concepts for extracting gases from solid biomass, in conjunction with fuel cells, will be available for houses for one or two families, as well as for well-insulated apartment houses and smaller municipal and commercial properties with and without solar thermal plants. At the moment, only gasifiers for solid biomass that have more than 30 kW electrical output are available. But, in the future, systems with less than 5 kW will be on the market. In order to be able to realize the gasification in cost-effective gasifiers and to make sure that the gas can be used without costly gas purification, high-quality fuels with reliable product quality are needed. For this, there will be high-end pellets made of urban wood, natural wood waste, washed leaves, grass, etc. In order to improve the gas generation, "difficult" fuels must be subjected to a consistent torrefaction process or a hydrothermal treatment in advance (B2). With these high-quality fuels, a more flexible and less expensive operation of the gasifier is possible than with thermally untreated fuels. On a laboratory scale, a very flexible gasification of charcoal has already been demonstrated (Krüger and Ortwein 2014).

Medium size gasifiers with high-temperature fuel cells and additional ORC technology (medium scale). In the power range between 20 kW_{el} and 1000 kW_{el}, flexible plant operation is paramount for stabilizing the local power grid. In addition to the supply of a heating network, the integrated heat storage makes it possible to provide a portion of heat necessary for restarting the gasifier and for maintaining the temperature of the high-temperature fuel cell. Gasifiers, fuel cells and ORC modules are available. The integration of the system and the controller are currently under investigation and development. The gasifiers will probably run with wood and some specially prepared agricultural fuels: B1 fuels (pre-sorted, size-graded, and optionally washed and dried). Integrated gas storages permit less flexibility of the gasifier while running the fuel cell with high flexibility. In this fashion, the power output can be well adjusted to the needs of the power grid. With thermal storages, the produced heat can be used exactly at those times when it is needed. An integrated heat-and-power controller manages the running times for more or less full use of the heat and an optimized impact on grid stability.

Boilers and heating plants (large scale). Even if a high-value energy usage of biomass requires a combined heat and power generation for maximum efficiency, there may be premises that contain biomass combustion plants (boilers) providing only heat. The technology is well developed. Major improvements may occur for emission reduction and increasing the flexibility of the furnaces; e.g., integration of pre-gasification into the reaction process (Windhager 2016). The size of the installations can vary from some kW of thermal output to some MW. Typically, they only use inferior solid biofuels (B1), in particular, fuels with limited suitability for biomass gasifiers. Typical areas of implementation might be premises that need comparatively high heat output at very short notice, so that a power grid-oriented combined heat and power generation would only be possible with buffers of an unjustifiably large size. The furnaces have to guarantee, in these cases, and despite of very unfavorable fuel properties, maximum of flexibility of operation. The worse the fuels are, the lower, typically, the flexibility of the boilers is. For the future, also for low quality fuels, the flexibility of the boilers has to be improved significantly in order to fill any heat gaps.

Industrial high-temperature heat appliances (large scale). In cases where steam is required with high temperature and/or high pressure for industrial applications, biomass is a favorable renewable option. Depending on the applications, the size of the boilers can range from some hundreds of kW to some MW of thermal output. The fuel is mainly B1 fuel, sorted and screened for use. The plants can operate flexibly to the extent required for the production process. Due to the high temperatures and, sometimes, varying needs, the boilers' flexibility could be increased in the future.

5.2 Solid Biomass for Production Purposes

Use of biogenic fuels in thermal production processes (cement production, steel production). In certain production processes, extremely high temperatures are necessary, e.g., 1450 °C for cement production. Therefore, in this case, the fuel is often mixed with the raw materials to be processed in a (melting) furnace. It should be noted that, in addition to very high temperatures (little water content), certain trace gases of the fuel could often have a positive influence on the final product, or, they should have. Accordingly, the fuel has to be selected and processed (B1 to B2). Because the processes are in a closed system, there are no direct requirements (such as flexibility) from external users. Excess heat may be used for external applications.

5.3 Heating Technologies Based on Gaseous Biofuels

Combined heat and power (CHP) from biogas (anaerobic digestion). Anaerobic processes produce a methane-rich biogas from wet, non-woody biomass, including various biological residues. The biogas can be directly burned in combustion engines, to provide electric power and heat via a power generator and heat exchangers. The advantage of these biogas CHP units is their waste reduction potential (including related emission reduction and waste valorization), stand-alone capability, high efficiency, flexibility, and the non-necessity for an upgrading system for bio-methane. The capacities range from tens of kW_{el} to a few MW_{el} output. By installing several small CHP units via a micro gas grid, even smaller units may be installed to serve single buildings. With the standard layout, apartment building blocks, smaller neighborhoods and industrial areas can be supplied with heat and power. In particular, sectors of the food processing industry are an interesting option for biogas CHP plants, because they could process the biological residues to provide (process) heat and power to industrial sites close by, thus reducing the transport of waste and fuel.

Because biogas CHP units can be operated according to demand (heat or power), as they are not based on fluctuating sources (such as wind and solar radiation), they are suitable for complementing these fluctuating renewable energies by providing the residual load needed to match the demand for both heat and power. The challenge lies in a technological, as well as in a managerial solution, that can match the differing demand curves of both energy forms. Available solutions for biogas CHP units involve flexible engines (capable of start/stop operation, quick load changes, and high partial load efficiency) in combination with storage for biogas and heat (in the case of operation driven by power demand) or, possibly, with power storage. Current research is related to improving the as yet unattained flexibility of the digestion process itself, in order to achieve a variable gas output, to decrease or eradicate the biogas storage; this should be an established technology by 2050. The urban heat distribution grid may also render heat storage obsolete if its base load absorbs the provided heat completely. Biogas CHP units may also provide peak loads, not only in the range of the installed conversion capacity but also by providing biogas to a reserve gas burner or using its power output to drive a power-to-heat system. A connection to the gas grid is conceivable for periods of depleted gas storage, to obtain bio-methane for continuous operation at peak load (emergency backup). Provided that market conditions in 2050 are similar to current ones, the power output, based on its flexibility, can be sold as a high-value product in the day-ahead or intraday market, as balancing power, services for the local distribution grid (e.g., redispatch or output reduction in case of over-supply) or via direct contracts (OTC – over the counter) with customers who have their individual demand curves. Additional benefits, such as treatment of urban biogenic wastes and lower grid expenditures might need to be monetarized, in order to be reflected in the economic feasibility of the biogas CHP plants.

Anaerobic digesters can be constructed to match the aesthetics of industrial buildings and thus be integrated seamlessly into the city layout. The storage of substrates and digestates, as well as the digester itself, need to be well controlled to avoid odor nuisances and methane emissions. The engine itself produces a low frequency noise that is contained by proper sound insulation – already a standard today. These measures are necessary for social acceptance.

For the future, besides combustion engines, fuel cells for small- and medium-scale applications based on bio-methane may also enter the market. Positive effects could be reduced emissions, lower noise levels, and an even more flexible operation:

Bio-methane mini fuel cells (small scale). For single-family houses that are not optimally insulated, as well as for insulated apartment buildings or smaller municipal or commercial properties with existing gas connections, the use of bio-methane mini fuel cells is conceivable. Their optimal field of application, from today's perspective, is for closing the gaps in the power and heat supply, after consideration of all existing renewable energy and storage options (residual load supply). The system operates in a serving manner for the power grid, i.e., the system runs whenever heat demand is foreseeable and the power grid needs positive balancing power. The electricity and heat output is in the range from 1 to 20 kW, depending on the size of the object to be supplied. To offer balancing power, these small-scale systems need to be part of a pool, to fulfil the minimum capacity requirements of this market. Methane that was derived from municipal waste fermentation plants, sewage gas plants, or via thermochemical processes for municipal solid waste or in biomass conversion plants in the surrounding rural region, can be considered as fuel. Because the bio-methane is stored within the grid, the flexibility of the system only depends on the characteristics of the fuel cell.

High-temperature fuel cells with bio-methane and additional ORC technology (medium scale). Bio-methane, both from the urban as well as from the surrounding areas, can also be used in CHP plants offering residual load and balancing power. Here, once again, a very high electrical efficiency combined with a high load flexibility is desired. The actual technical requirements depend on the market strategy of the plant and the pool operators. Typical sizes could be in the range of some hundred kW to some MW of electrical output. The generated heat is distributed via a heating network or is stored in large buffers until the heat demand occurs. The stored heat is also used to keep the fuel cell in a stand-by state at the required operating temperature, ready for a short-term start. When needed, the ORC technology is used to convert the high temperature excess heat of the fuel cell into electricity, in order to generate total electrical efficiencies of up to 60% with respect to the fuel heating value and to reduce the amount of heat that has to be stored.

A comparison of the listed technologies is shown in Table 4.

All mentioned technologies have a good chance of providing favorable future solutions for the varied heating requirements in urban areas. The future potential of the different technologies depends on many aspects, such as price development of biomass, technical improvements of the technologies, regulations of energy markets, price development for system services, social acceptance, and so on. Further research is needed not only with respect to the technologies but also according to market options and the legislative framework.

Table 4 Comparison of promising supply systems for heat from biomass in urban areas for the future.

Criteria	Automatic stoves with water pockets	Micro gasifiers with fuel cells in the home area	Medium gasifiers with high temperature fuel cell + ORC	Boiler and heating plants/ Industrial high-temperature heat applications	Use of biogenic fuels in thermal production processes	Combined heat and power (CHP) from biogas (anaerobic digestion)		
						Combustion engine	Mini fuel cell	High temperature fuel cell + ORC
Operating flexibility – load change rate 30–100% and 100–30%	10–15 min	1–3 min	5–10 min	15–30 min	Depending on the production process	Below 1 min, duration depending on gas storage, partial load usually not below 50%; biogas production 12–24 h	1–3 min	5–10 min
Operational flexibility – cold start and stop time	15–30 min/20–30 min	15–30 min/5–10 min	60–120 min/30–60 min	6–12 h/1–2 h	Depending on the production process	3–5 min/1–2 min; depends on engines used, some are not designed for start/stop operation	5–10 min/ 3–5 min	60–120 min/30–60 min
Specific investment costs (roughly estimated 2015)	200–500 €/kW_{th}	7000–10,000 €/ kW_{el}	3000–5000 €/ kW_{el}	300–800 €/kW_{th}	Depending on production process; especially costs for fuel feeding	5000–7000 €/ kW_{el}	6000–8000 €/ kW_{el}	5500–7500 €/kW_{el}
Starting material of fuels	Wood residues	Wood residues	Leaves, grass etc.	Any kind of solid biogenic residues and wastes	Wood as well as biogenic wastes	Manure, wet biogenic wastes and residues, grass, sewage sludge		

Fuels (fuel processing expenses)	Medium (sorting, homogenization and pelletization)	High (sorting, torrefication, homogenization and pelletization)	Very high (sorting, washing, torrefication, homogenisation and pelletization)	Low (cutting, homogenization, drying depending on humidity)	Very high (sorting, homogenization, possibly charcoal production)	Precondition depending on fuel source – Anaerobic digestion has integrated fuel handling; for some fuels, special preconditioning (temperature, pressure) necessary		
Fuels (storability)	High – in sheltered storage	High – in sheltered storage	High – in sheltered storage	Medium	High – in sheltered storage	Depending on fuel, low to medium		
Near-zero emission technology	Electronic combustion control, catalyst and precipitator	Yes	Yes	Combustion control, precipitators, catalytic emission reduction	Depending on production process	Catalytic emission reduction after engine; avoiding leakages in biogas process	Avoiding leakages in biogas process; avoiding leakages in bio-methane conversion	
Local fuel source	Wood from landscaping	Wood from landscaping and wood processing industry	Leaves, grass, etc., from landscaping	Any kind of solid wastes and production residues	Mainly wood residues or even forest wood	Wet biogenic waste from households, restaurants, food and fodder production, trading, water cleaning, etc.		
Noise emissions	Low	Very low	Low	Medium to high	Depending on production process	Medium to high	Very low	Low
Odour emissions	Very low	No	No	Possibly from fuel	Depending on production process	Only as long as primary fuels (wastes) are handled and in some cases during handling of process residues; no odour from conversion		

(continued)

Table 4 (continued)

Criteria	Automatic stoves with water pockets	Micro gasifiers with fuel cells in the home area	Medium gasifiers with high temperature fuel cell + ORC	Boiler and heating plants/ Industrial high-temperature heat applications	Use of biogenic fuels in thermal production processes	Combined heat and power (CHP) from biogas (anaerobic digestion)		
						Combustion engine	Mini fuel cell	High temperature fuel cell + ORC
Aesthetics/ optics	No problem	No problem	No problem	View of an "industrial" building; exhaust gas chimney	Depending on production process	Digester, depending on size, looks like "industrial" buildings; conversion technology integrated in small buildings or existing ones, thus, no additional problem		
Acceptance of technology	Still high	New technology	New technology	Transport frequency and fear about emissions	Depending on production process	Some resistance against anaerobic digestion	Some resistance against anaerobic digestion; new conversion technology	
Space requirements	1 m^2 stove, 1–2 m^2 fuel storage	2 m^2 fuel cell and gasifier, 2–3 m^3 for fuel	10m^2 gasifier and fuel cell, 5–10 m^2 storage	10,000–30,000 m^2 for building and fuel storage	Depending on production process	10,000–50,000 m^2 depending on size, fuels, and storage capacity of fuels		

Source: Lenz and Thrän (2015)

6 Change Management – Integrating People and Processes

So far, the focus was on renewable technologies for generating heat. In the following, the issue of how the transformation of the heat sector can be realized in strongly agglomerated areas is investigated. The "how" should be understood in terms of "Change Management" that can build a bridge between technology and local actors.

Change management involves the monitoring and handling of processes and activities, as well as the assistance and coordination of stakeholders, the identification of the crucial points. Overall, the change management has to secure a long-term coherent support of the transformation process and coherent conditions (UBA 2015). It should be tried to transfer positive experiences with the implementation of (collective) renewable energy heating projects in rural areas to urban regions.

Besides the communication pathway – discussed in the following – the German government has set up some regulations for new buildings as well as for the reconstruction of existing buildings: e.g., an energy saving ordinance ("Energieeinsparverordnung") with regulations about the insulation and the efficiency of heat generation, and the renewable heating act ("Erneuerbare Wärme Gesetz") with regulations on the use of renewable energies in new buildings. Additionally, numerous funding programs for new, renewable heating systems exist in Germany, e.g., the market incentive program ("Marktanreizprogramm").

Challenges in the city. In addition to the general challenges of the energy transition, which include security of supply, affordability, and acceptance, there are additional aspects that need to be considered in agglomerated areas. The reasons for this are very different conditions in terms of existing buildings, remediation conditions, and forms of heat supply. This leads to a large heterogeneity in town segments such as districts and blocks or even within a single building (see above).

Moreover, there is also a wide variety and heterogeneity among the stakeholders involved, e.g., tenants and landlords. The latter can be divided into small private providers, commercial landlords, municipal housing companies, and housing associations. The ownership rate differs widely between regions and rural and also urban areas. For example, the proportion of owners in owner-occupied dwellings, compared to all inhabited dwellings in residential buildings in Berlin, was almost 16% in 2011 (Statistische Ämter des Bundes und der Länder 2014). This results in a bilateral dependence of tenants and landlords concerning the allocation of the costs of energy-saving renovation, which is also called the "landlord-tenant dilemma": The landlord may transfer the costs of energy-saving renovation only to a limited extent to the tenants, although they benefit, in the long-term, from cheaper energy costs (Neitzel et al. 2011). This can lead to a lack of incentives for the energy modernization and a stalemate concerning the allocation of costs.

If the relationship between tenant and landlord is at the level of buildings or residential units, other stakeholders who are involved in comprehensive considerations about using renewable energy resources emerge. These include local/regional energy providers, urban management and urban development agencies, or the local/regional policy. Excluding the area of residential buildings, the number of stakeholders is also spread over the commercial or even industrial sector, education, health care, etc.

In addition to the challenges described in the context of existing buildings and of the participants, there are very often overall legal requirements about building standards, including insulations requirements and demands about the use of renewable heat, as well as possible emission limits. Besides these "strong" regulations, every city can have individual (urban policy) goals and regulations that need to be considered during the implementation of the transformation of the heat sector. This concerns, for example, the creation or continuation of social housing, in order to counteract spatial segregation and gentrification.

Therefore, it is clear that the urban transformation of the heat sector takes place at different levels that can be located at the level of living units, buildings, building blocks or districts/neighborhoods, depending on the specific issue. In addition, there is a great structural heterogeneity and the requirement to involve various groups of stakeholders and interests. Unlike power generation, the heat supply and the demand for heat must always be brought together locally. For this purpose, management on a variety of levels is necessary.

Change Management. Grid-connected solutions for the supply of heat are usually dominated by influential actors such as public utilities, whilst the respective owners make decisions about the object supply.

Urban planning plays a central role in the basic control of the heat supply (WBGU 2011). Formal instruments, such as land-use planning and the implementation of urban rehabilitation measures, but also increasingly informal instruments and measures, can supply basic choices for the transformation of the urban energy supply. Integrated overall concepts, such as master planning or city district development plans, play an important role in the development of cross-property solutions such as heat power networks, which can profit from experiences in other cities. The necessary information can be put together in a heat register, where energy needs are recorded at the building level. If such a register is integrated in a Geographic Information System (GIS), analysis can easily be carried out on different spatial scales and can be blended with further information (such as building renovation). Unlike the electricity supply, the type of heat supply cannot be changed very easily if heating systems are installed decentrally in the building (i.e., one system for each apartment) and are fixed by different investment cycles. Additionally, landlords and tenants follow different interests in investing in modern, fossil-free heating systems with lower fuel costs and lower greenhouse gas emissions (see landlord-tenant dilemma and see also Bedtke and Gawel, "Linking Transition Theories with Theories of Institutions – Implications for Sustainable Urban Infrastructures Between Flexibility and Stability", in this volume). Novel approaches are required to allocate investment costs and benefits of a renewable heat supply more attractively.

Integration by means of information, communication, organizational and financial participation (e.g., financing of a new heating system by the tenants and the landlord, involving professional contractors for efficient heat provision) provide the opportunity to create new synergies. In urban areas, the population density is much

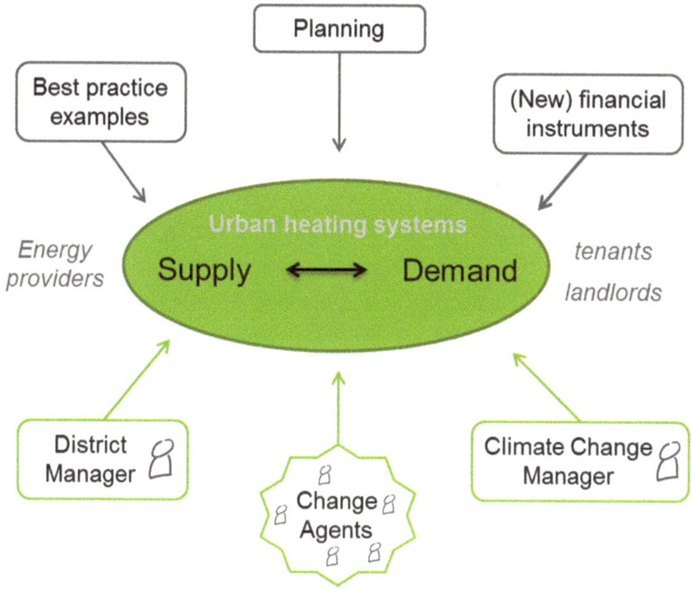

Fig. 3 Tools and actors for change management in urban heating systems

higher than in rural areas. Therefore, informal communication could spread more rapidly and people could be more interested in activities occurring in their part of the urban area.

Change agents are of particular importance in the transformation process: These are individual actors who convey the change of institutions and mentalities in social transformation processes (WBGU 2011). These actors may, for example, be connected to the Local Agenda 21 or belong to the Transition Town movement (Doerr and Carr 2014). However, a committed district management can also help to transform the idea of local cycles into concrete projects and to search for technological solutions for object-dependent urban heat solutions (Fig. 3).

The challenges and chances of a smart heat supply must be communicated to different audiences. Local change agents play a key role in this process. Furthermore, good examples are needed that can be communicated and spread.

7 The Path Ahead for Biogenic Urban Heat Supply

In the past, the development of large cities was only possible if energy carriers with high specific energy density, to be used in industry, transport and, last but not least, heating were explored. Thus, the decarbonization of the heating – and also cooling – sector is one of the central challenges for growing urban areas. Heat from biomass is the most relevant renewable energy carrier today and also, at least, for the mid-term. Additionally, there are huge potentials for biogenic residues and wastes in urban areas, as well as in the surrounding rural regions that can be used for energy provision, e.g., the three regions analyzed in Germany could produce up to 12% of their future heat demand by using all the available biogenic residues and wastes. Hence, in the long term, whilst increasing the resource efficiency of urban infrastructures, a sectoral transformation towards sustainability is needed at the same time.

A full change of heat supply to renewable energies ("Wärmewende") is only possible if solar-thermal systems, heat-pumps, and geothermal systems, as well as heat from surplus renewable electricity (power-to-heat) and heat recovery gain significantly in relevance. Together with better insulation and other efficiency measurements, the role of heat provision from biomass needs to change; the opportunity is to use biomass in combination with other renewables to optimize a cost-effective energy transformation, taking the wide range of buildings, infrastructures, and stakeholders into account, as well as the growing interlinkage between the heating sector, with secure electricity provision, and the mobility sector.

There is a wide range of technical options for transforming biomass into heat and also into other energy forms. An assessment of these options, according to various criteria shows that there are promising options, ranging from small-scale to large-scale applications that fit into the decarbonization of the heat supply. Either they can be realized in the urban areas as stand-alone solutions or in combination with the surrounding rural region. An especially interesting technological option might be the conversion of urban biogenic wastes and residues into high-quality energy carriers, such as bio-methane or premium quality biomass pellets (washed, thermo-chemically treated, compressed), that can be used in highly flexible small- to medium-scale combined heat and power units, thus closing renewable heat gaps and stabilizing the local power grid at the same time (SmartBiomassHeat) (Lenz 2017).

Due to the requirements concerning air quality, urban emission limits will be much stricter than in rural regions. In addition to the primary reduction measures, secondary measures (precipitators, catalysts) will also gain in importance. The resulting sales boost is expected to reduce prices in this product segment.

Developing dedicated processes for heat transformation are a key, including the definition of scenarios and transformation pathways, the adopted planning, broad stakeholder and inhabitant involvement – including local change agents – and information about best practice and successful developments as well. With this approach, adopted solutions for all the kinds of buildings (new and old) and infrastructural standards can be developed and the change of the heat supply can be arranged, over time, in the most effective way for complete urban areas.

In order to find collective solutions that also take into account other objectives of urban development, higher-level central managers (district manager, climate change manager, etc.) are required, as well as a general understanding by all actors involved in city planning and development of the technical possibilities of the renewable heat supply and of efficient management of resources.

References

AG Energiebilanzen e.V (2016) Energy balances 2014. http://www.ag-energiebilanzen.de/7-1-Energy-Balance-2000-to-2014.html. Accessed 2 Aug 2016

BMBF – Bundesministerium für Bildung und Forschung (2010) Megacities – shaping of a sustainable future world. http://future-megacities.org/fileadmin/documents/konferenzen/Megacities_in_Balance_2010/Megacities_Shaping_of_a_sustainable_future.pdf. Accessed 12 Feb 2016

BMJV – Bundesministerium der Justiz und für Verbraucherschutz (2009) Verordnung über Deponien und Langzeitlager (Deponieverordnung – DepV). http://www.gesetze-im-internet.de/bundesrecht/depv_2009/gesamt.pdf. Accessed 2 May 2016

BMVI – Bundesministerium für Verkehr, Bau und Stadtentwicklung (2013) Maßnahmen zur Umsetzung der Ziele des Energiekonzepts im Gebäudebereich – Zielerreichungsszenario. BMVBS-Online-Publikation, Nr. 03/2013. http://www.bbsr.bund.de/BBSR/DE/Veroeffentlichungen/BMVBS/Online/2013/DL_ON032013.pdf?__blob=publicationFile&v=5. Accessed 8 Aug 2016

BMWi – Bundesministerium für Wirtschaft und Energie (2014) Bundesrepublik Deutschland. Nationaler Aktionsplan für erneuerbare Energie gemäß der Richtlinie 2009/28/EG zur Förderung der Nutzung von Energie aus erneuerbaren Quellen. http://www.erneuerbare-energien.de/EE/Redaktion/DE/Downloads/Broschuere/nationaler_aktionsplan.pdf?__blob=publicationFile&v=4. Accessed 3 May 2016

BMWi – Bundesministerium für Wirtschaft und Energie (ed) (2015) Zeitreihen zur Entwicklung der erneuerbaren Energien in Deutschland unter Verwendung von Daten der Arbeitsgruppe Erneuerbare Energien-Statistik (AGEE-Stat). http://www.erneuerbare-energien.de/EE/Navigation/DE/Service/Erneuerbare_Energien_in_Zahlen/Zeitreihen/zeitreihen.html. Accessed 10 Dec 2015

Doerr J-T, Carr C (2014) Dreißig Jahre Transformation und trotzdem noch ganz am Anfang? Der Wandel in Beckerich von der Agenda 21 zur Transition Town. PND online, 3/2014. http://www.planung-neu-denken.de/images/stories/pnd/dokumente/3_2014/doerrcarr.pdf. Accessed 20 Jan 2016

Erhorn H, Erhorn-Kluttig H, Staudt A (2010) Auswirkungen der novellierten Energiesparordnung (EnEV) und des Erneuerbare-Energien-Wärme-Gesetzes (EEWärmeG) 2009 auf die Festlegungen im NEH-Beschluss der Landeshauptstadt Stuttgart. IBP-Bericht WB 148/2010. http://www.domino1.stuttgart.de/web/ksd/ksdredsystem.nsf/dc5e48bde54b0b2941256a6f0036f408/6871d99fafdf6ee6c125770e004787dd/$FILE/Anlage%201%20zur%20GRDrs%20165%20Bericht%20IBP.pdf. Accessed 3 May 2016

Fischedick M, Schüwer D, Venjakob J, Merten F, Mitze D, Nast M et al (2007) Potenziale von Nah- und Fernwärmenetzen für den Klimaschutz bis zum Jahr 2020. Forschungsbericht. https://www.umweltbundesamt.de/sites/default/files/medien/publikation/long/3501.pdf. Accessed 15 Jan 2016

JIHK – Deutsche Industrie- und Handelskammer in Japan (2014) Stromerzeugung: Biomasse hängt Erdwärme ab http://www.japanmarkt.de/2014/07/09/trends/energie/stromerzeugung-biomasse-haengt-erdwaerme-ab/. Accessed 14 July 2014

Krüger D, Ortwein A (2014) Erprobung einer neu entwickelten Mikro-KWK-Anlage zur Nutzung von Holzkohle. In: Nelles M (ed) Bioenergie. Vielseitig, sicher, wirtschaftlich, sauber?! – DBFZ-Jahrestagung am 01./02. Oktober 2014. DBFZ, Leipzig, p 243

Lenz V (2017) SmartBiomassHeat – heat from solid biofuels as an integral part of a future energy system based on renewables. Chem Eng Technol 40(2):313–322

Lenz V, Büchner D, Wurdinger K (2017) Evaluation of combining an air-to-water heat pump with a wood stove with water jacket for residential heating. In: Proceedings of 12th IEA Heat Pump Conference -ISBN 978-90-9030412-0 - Rotterdam, June 2017

Lenz V, Thrän D (2015) Flexible heat provision from biomass. In: Thrän D (ed) Smart bioenergy, technologies and concepts for a more flexible bioenergy provision in future energy systems. Springer, Heidelberg

Merkel A (2015) Pressekonferenz von Bundeskanzlerin Merkel zum Abschluss des G7-Gipfels in Elmau. Die Bundesregierung, Aktuelles. http://www.bundesregierung.de/Content/DE/Mitschrift/Pressekonferenzen/2015/06/2015-06-08-pk-merkel-g7.html. Accessed 02 May 2016

Neitzel M, Dylewski C, Pelz C (2011) Wege aus dem Vermieter-Mieter-Dilemma. Konzeptstudie. http://www.jura.uni-bielefeld.de/lehrstuehle/artz/docs/inwis.pdf. Accessed 7 Nov 2015

Ponitka J, Kretschmar J, Thrän D (2010) Can biomass production on urban brownfields be a viable concept? In: Spitzer J (ed) From research to industry and markets: 18th European biomass conference and exhibition; proceedings of the international conference held in Lyon, France, 3–7 May 2010. ETA, Florence, pp 135–138

Raatz A, Wangelin M, Bodmann J, Witzel A, et al (2014) Regionales energie- und CO2-monitoring. Systematische Datenerfassung und orientierende Bilanzierung mit dem KEEA-Tool. http://www.energiewende-frankfurtrheinmain.de/fileadmin/user_upload/content/pdf/KEEA_Bausteine_Regionales_Energiekonzept_Daten.pdf. Accessed 27 Apr 2016

Rat der Stadt Essen (2009) Integriertes Energie- und Klimakonzept. https://www.essen.de/leben/umwelt/klima/Integriertes_Energie_und_Klimakonzept.de.html. Accessed 02 May 2016

Reusswig F, Hirschl B, Lass W, Becker C, Bölling L, Clausen W et al (2014) Machbarkeitsstudie Klimaneutrales Berlin 2050. http://www.stadtentwicklung.berlin.de/umwelt/klimaschutz/studie_klimaneutrales_berlin/download/Machbarkeitsstudie_Berlin2050_Hauptbericht.pdf. Accessed 8 Jan 2016

Scheftelowitz M, Thrän D, Hennig C, Krautz A, Lenz V, Liebetrau J et al (2014) Entwicklung der Förderung der Stromerzeugung aus Biomasse im Rahmen des EEG. Report 21. https://www.dbfz.de/fileadmin/user_upload/DBFZ_Reports/DBFZ_Report21.pdf. Accessed 27 Apr 2016

Schumacher P, Stroh K, Schurig M, Ellerbrok C, Ramonat A, Link S (2015) Generalkonzept im Rahmen des Masterplans „100% Klimaschutz "der Stadt Frankfurt am Main. http://www.masterplan100.de/fileadmin/user_upload/content/pdf/generalkonzept/Masterplan_Klimaschutz_Generalkonzept_Langfassung.pdf. Accessed 12 Mar 2016

Statistische Ämter des Bundes und der Länder (2014) Zensus 2011. Gebäude- und Wohnungsbestand in Deutschland. http://www.statistik-portal.de/Statistik-Portal/Zensus_2011_GWZ.pdf. Accessed 12 Nov 2015

Stryi-Hipp G, Baur F, Borggrefe F, Gerhardt N, Hauer A, Horst J et al (2015) Erneuerbare Energien im Wärmesektor – Aufgaben, Empfehlungen und Perspektiven. Positionspapier des ForschungsVerbunds Erneuerbare Energien. http://www.fvee.de/fileadmin/publikationen/Politische_Papiere_FVEE/15.EEWaerme/15_FVEE-Positionspapier_EE-Waerme.pdf. Accessed 3 May 2016

Thormann L, Pfeiffer D, Bloche-Daub K, Thrän D, Kaltschmitt M (2016) Biomasse im energiesystem. In: Kaltschmitt M, Hartmann H, Hofbauer H (eds) Energie aus Biomasse. Grundlagen, Techniken und Verfahren. Springer, Heidelberg, pp 9–63

Thrän D (2009) Nebenprodukte, Rückstände und Abfälle. In: Kaltschmitt M, Hartmann H, Hofbauer H (eds) Energie aus Biomasse. Grundlagen, Techniken und Verfahren. Springer, Heidelberg, pp 135–170

UBA – Umweltbundesamt (2015) Transformationsstrategien und Models of Change für nachhaltigen gesellschaftlichen Wandel. Wie Transformationen und gesellschaftliche Innovationen gelingen können UFOPLAN-Vorhaben – FKZ 371211103. https://www.umweltbundesamt.de/sites/default/files/medien/376/publikationen/wie_transformationen_und_gesellschaftliche_innovationen_gelingen_koennen.pdf. Accessed 23 Nov 2015

WBGU – German Advisory Council on Global Change (2011) World in Transition: A Social Contract for Sustainability http://www.wbgu.de/fileadmin/templates/dateien/veroeffentlichungen/hauptgutachten/jg2011/wbgu_jg2011_en.pdf. Accessed 12 Jan 2016

Wien Energie GmbH (2016) Spittelau waste incineration plant. The Spittelau waste incineration plant, designed by Friedensreich Hundertwasser, will be completely revamped by 2015. https://www.wienenergie.at/eportal3/ep/channelView.do?pageTypeId=72164&channelId=-51715. Accessed 8 Aug 2016

Windhager Zentralheizung GmbH Deutschland (2016) PuroWin. Die Revolution der Hackgutheizung. http://www.windhager.com/de/produkte/hackgut/purowin/. Accessed 2 May 2016

From Shrinkage to Regrowth: The Nexus Between Urban Dynamics, Land Use Change and Ecosystem Service Provision

Annegret Haase, Manuel Wolff, and Dieter Rink

1 Introduction

Urban shrinkage, commonly understood as population loss of an urban area, and post-shrinkage regrowth have become prominent pathways of urban development across Europe. While many cities, mostly in Eastern Europe, are still shrinking today, other cities that shrunk in the past now see new growth. Numerous cities have undergone this transformation from shrinkage towards new growth within only one or two decades – a relatively short time. Both shrinkage and regrowth have considerable impacts on urban land use, be it on densities or types of use. Both offer a variety of potentials and risks for sustainable use of urban land as well as for the provision of green spaces and urban ecosystem services (UES). On the one hand, there is the risk that new growth after shrinkage puts pressure on the qualities that emerged and/or were created in the time of shrinkage such as less density, more green areas, or spaces for experimentation and innovation, will be questioned in their existence and benefit. On the other hand, regrowth after shrinkage offers the great opportunity to make use of those qualities in order to build more sustainable regrowing cities and to ensure good provision of environmental qualities and ecosystem services for large segments of the population. Here, many trade-offs emerge, and many new challenges have to be addressed. Often, there is a complex setting of actors and interests that make it complicated to negotiate solutions.

Set against this background, this paper deals with the impacts of urban dynamics related to urban shrinkage and post-shrinkage regrowth on changes in urban land use and the provision of UES. The focus will be on the analysis of existing challenges, synergies, and trade-offs. Based on this analysis, the article discusses how

A. Haase (✉) • M. Wolff • D. Rink
Department of Urban and Environmental Sociology, Helmholtz Centre for Environmental Research-UFZ, Leipzig, Germany
e-mail: annegret.haase@ufz.de; dieter.rink@ufz.de; manuel.wolff@ufz.de

S. Kabisch et al. (eds.), *Urban Transformations*, Future City 10,
DOI 10.1007/978-3-319-59324-1_11

urban actors and existing governance arrangements cope with the challenges. The arguments will be illustrated by examples from several shrinking and regrowing cities across Europe.

This contribution addresses the following key questions:

- Which impacts do urban shrinkage and (post-shrinkage) regrowth have for urban land use changes and the provision of UES?
- Which trade-offs and challenges can be identified?
- How do different actors, policies, and modes of local governance respond to them?

As a result, this paper will argue that urban shrinkage produces specific problems for land use management but also chances for the (or: a better) provision of ecosystem services. Regrowing cities face the challenge of retaining many of these chances while successfully remediating the problems. The final outcome will depend mainly on negotiations between various actors and interests as well as on specific local conditions that determine how the interplay between urban dynamics, land use, and provision of environmental services can be shaped and managed.

This paper draws on long-term research by the authors on urban dynamics such as shrinkage and regrowth as well as on related challenges for land use change and ecosystem service provision; it includes empirical evidence from several projects or analyses. Evidence comes from a variety of urban contexts across the European realm, among them the urban region Leipzig-Halle. With respect to the overall focus of the volume, the paper focuses on the aspects of resource efficiency, related to (urban) land use, and quality of life, in terms of opportunities for urban dwellers to benefit from urban green and UES.

2 Urban Shrinkage and Regrowth

During recent decades, shrinkage and regrowth have come to represent two urban dynamics that are characteristic for European urban development. In particular, post-industrial transitions affected cities in Western Europe, for example in France or the UK, since the 1970s and has led to *urban shrinkage*. Furthermore, the "pole" of shrinkage moved to post-socialist Eastern Europe after 1990, due to rapid changes in political, economic, and urban systems. In the early 2000s, approximately 40% of all large cities (>200,000 inhabitants) across Europe were shrinking (Turok and Mykhnenko 2007).

Because urban shrinkage is, in most cases, conceptualized through and measured by population decline, it must be emphasized that also the contexts of deindustrialization and postindustrial change have led to shrinkage in terms of economic decline and abandonment in these cities. As a result, large-scale brownfield sites (post-industrial, post-railway, post-military, post-mining, etc.) emerged that are not found, in terms of their numbers and spatial extent, in non-shrinking cities.

Although the scholarly community dealing with shrinkage remained quite small, its research received increasing attention, not least in the light of post-socialist transformation and the recent economic and financial crises, during which shrinkage and related problems (e.g., abandonment, vacancies) emerged as problems. Thus, we find a large number of studies and an extensive body of literature on shrinkage, among them some comparative studies (Haase et al. 2016; IJURR symposium 2012; Rink et al. 2012) and related European research projects such as Shrink Smart – The governance of urban shrinkage within a European context (EU 7FP Project Shrink Smart 2009–2012) or CIRES – Cities Regrowing Smaller (EU COST Action CIRES 2009–2013). Basically, these investigations point out that the duration, speed, and scope of shrinkage were variegated. Furthermore, several studies underline that this trajectory represents the non-linear urban development that cities, most of all, are facing (Kabisch and Haase 2011; Turok and Mykhnenko 2007) – to put it differently: that growing cites might shrink and shrinking cities may return to new growth is a natural urban phenomenon.

Since the 1990s, many formerly shrinking cities have seen new growth that is referred to as *post-shrinkage regrowth.* This new, emerging trend can be, most prominently, observed in Eastern Germany and the UK. Moreover, the fact that examples from Western, Southern and even post-socialist Europe also exist (Haase 2015) allows the labelling of this regrowth as a 'European phenomenon'. Although one might argue that regrowing cities demonstrate many characteristics that can also be found in continuously growing cities, there is a crucial, qualitative distinction: Regrowth follows a (longer) phase of shrinkage and thus builds on the legacies of shrinkage, i.e., housing vacancies, abandoned land, large-scale brownfield sites, underused infrastructure, tight (public) budgets, low interest in investment, etc. At the same time, the "positive legacies" of shrinkage, e.g., more green and open spaces, less density through abandonment and demolition, offer many generous opportunities for a more sustainable shaping of the new growth than in continuously growing cities, which usually do not have such prerequisites at their disposal.

Similar to shrinkage, the speed and scope of regrowth vary. Whereas some cities have seen continuous regrowth for more than 20 years (e.g., Antwerp, Manchester), others constantly oscillate between growth and new losses (e.g., Brno, Timisoara, Genoa). In addition, the regrowth of cities can differ in extent, leading to a variety of regional contexts. In some cases, the entire urban region sees new growth, whereas in others, population increase occurs in the core city while the hinterland is (still) shrinking. It is factors at different spatial scales, from local to global, that impact on regrowth. Some studies have identified national contextual factors (Kabisch et al. 2012) and compared different cities in terms of the causes and consequences of their turn from shrinkage to regrowth (Rink et al. 2012). Among the various factors influencing urban development, especially external ones cannot be steered by the cities themselves and contribute to the non-linearity of cities' trajectories. Thus, regrowth has only recently also become a subject of research, but it will probably continue to do so and, thus, it will become increasingly important for urban research.

Because the focus of this chapter is on the interlinkages between shrinkage and regrowth and their impacts on land use (change) and the provision of UES we will not elaborate on the various causes and effects of shrinkage and regrowth (Haase et al. 2016; Rink et al. 2012). Instead, this contribution will explicitly link the following *two strands of research*: a) urban dynamics and their impacts on urban development, including land use and land use change, de- and redensification and vacancies/abandonment (Haase and Rink 2015; Haase et al. 2016; Rink et al. 2012; Rink and Wolff 2015); and b) the nexus between urban shrinkage/regrowth and land use policies/practices, together with the provision of UES (Haase et al. 2014; Rink and Arndt 2016). The following section details briefly the relationships between urban dynamics of shrinkage and regrowth, urban land uses changes, and the provision of UES/environmental qualities/open and green space.

2.1 The Nexus Between Urban Dynamics and Land Use Changes and the Provision of UES

The connection between different dynamics of urbanisation beyond just growth (shrinkage, regrowth after shrinkage) and its impacts on land use and UES provision is still relatively rarely described in the literature. There is a number of papers directly addressing the relationship between urban shrinkage and its impacts on land use (Haase et al. 2012; Kroll et al. 2012) or on specific land use settings in shrinking cities, such as green interim uses (Lorance Rall and Haase 2011) or urban forests (Rink and Arndt 2016). Haase et al. (2014) analysed the nexus between urban shrinkage, land use changes conditioned by it, and consequences for the provision of UES in a more conceptual fashion; this chapter builds on this research but goes beyond it in a twofold way: First, it looks at the coupled dynamics of shrinkage and regrowth, a case that is increasingly typical for many European cities (Rink et al. 2012; Haase 2015; Wolff et al. 2016). Second, it directly addresses the relationship between urban dynamics, land use (changes) and UES provision for this coupled context.

Figure 1 shows the basic nexus between *urban shrinkage* and the provision of UES (Haase et al. 2014). Shrinkage leads to impacts on the physical infrastructure and, more precisely, to decreasing demand and underuse. This results in abandonment, e.g., housing vacancies and brownfields, with corresponding effects on the urban land uses. Moreover, changes in land use, such as de sealing or demolition, also have impacts on the provision of UES – especially on fresh air, biodiversity, carbon dioxide storage, cooling effects, and leisure functions. Broadly speaking, the potential provision of these UES is supported by land use changes that are quite typical for shrinking cities; these include the creation or enlargement of green areas and parks after abandonment or demolition of built structures. A better or more efficient UES provision emerges from direct connections with processes of dedensification and the need for an, at least temporary, reuse of abandoned urban land. In sum, urban shrinkage involves a risk for land use and its efficient usage but, at the same time, it creates a window of opportunity for more green areas and ecologically sustainable uses.

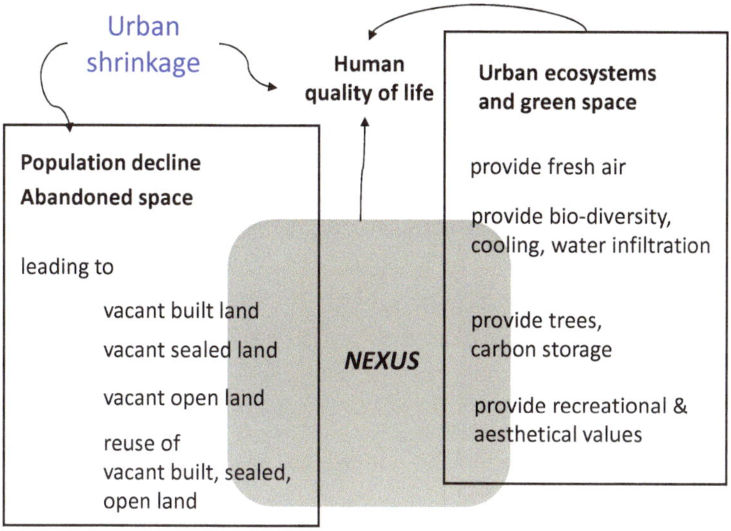

Fig. 1 Nexus between urban shrinkage and the provision of UES (Source: Haase et al. 2014)

The framework conditions for urban development essentially change when a city experiences new *population growth after a period of shrinkage* (Haase et al. 2015). Due to this new growth, an increasing pressure to (re)use urban land for new housing and infrastructure is observed. This includes redensification, infill, and construction on formerly open or green space, including brownfield sites. Redensification is becoming a general trend in regrowing cities also because high-rise buildings are prioritised, compared to low buildings with low densities (such as town houses, which had been built in many shrinking cities). However, this is accompanied by new challenges for the provision of UES, because abandoned or green areas are exposed to new, developing pressure and constantly disappearing. The extent to which green uses and related environmental services that were created during the period of shrinkage can be maintained depends on many factors (new priorities of urban strategic development, involved actors, etc.). On the one hand, regrowth has also been achieved in many cities through substantial investments in green and open spaces, as well as an improvement of residential environments: less dense, greener. On the other hand, the new pressure of reusing urban land to meet the increasing demand often threatens these qualities – redensification and reuse of open space for new construction receive highest priority. Moreover, the redensification of a city also contributes to increasing compactness which, in turn, has positive effects on UES; these include shadow effects of street areas due to high buildings or less (traffic) noise for buildings in the second row (rear buildings). In sum, new opportunities for the provision of UES emerge during regrowth but also new risks for them as well as for the quality of urban life (Fig. 2).

These rather generalized reflections on the nexus between the provision of UES and the two basic trajectories of shrinkage and regrowth are now further underpinned by providing specific examples from several European contexts. Based on these, a

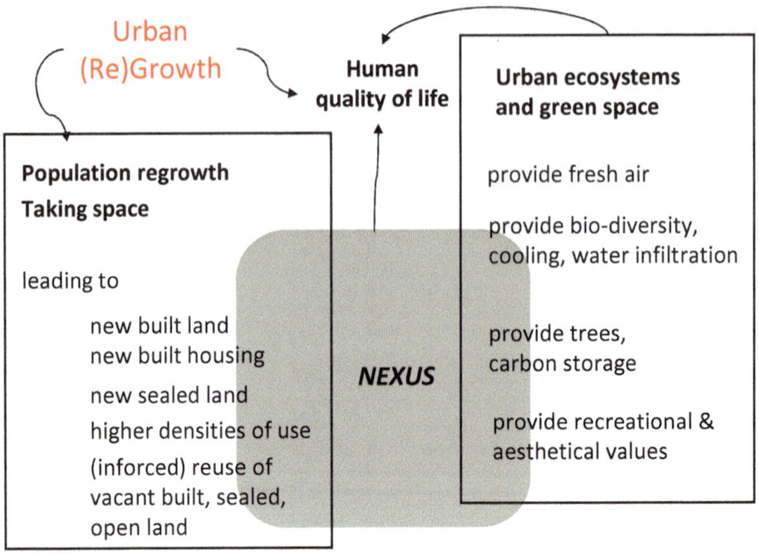

Fig. 2 Nexus between urban regrowth and the provision of UES (Source: Haase et al. 2015)

discussion will mirror the challenges, synergies, and trade-offs together with a consideration of how urban stakeholders and other actors deal with those challenges.

2.2 *Examples for the Interplay of Urban Dynamics, Land Use Changes, and the Provision of UES*

The following paragraph mirrors the previously mentioned interlinkages by illustrating examples from various cities that have been investigated in an explorative way. All of these examples of shrinkage and regrowth need to be considered in relation to each other.

2.2.1 Shrinkage

Shrinking cities usually undergo dedensification and dissolution of the existing built-up structures. For instance, rear buildings were demolished and, instead, green areas or at least partial greening was implemented. As a consequence of demolition or underuse, other forms of uses were required. In the Eastern German city of Leipzig, which used to be a prominent example of a shrinking city, temporary uses were implemented, using a planning instrument (Haase and Lorance 2010a, b). With this, niches came into existence, e.g., for urban gardening, which, in turn, fostered social communication and solidarity ("places of encounter") or the cultural scene, etc. Moreover, green areas in general and urban gardens in particular also contribute to health via healthy food,

Fig. 3 Urban garden project, Leipzig, Germany (Source: Annegret Haase)

leisure, social solidarity, etc. (Fig. 3). The relatively small scale of the created green areas (because, basically, only one building was demolished) opened up the opportunity to also establish green areas in disadvantaged and former high-density districts.

Urban shrinkage even allows for projects that have a *pilot or experimental character*. In the cities of Leipzig and Halle, projects on urban forests were implemented (Fig. 4). These aimed at the reforestation of urban areas that were not supposed to be reused for new construction in the future. The specific character of these areas is that, from a planning perspective, they lose their legal status for building and are, instead, converted to the status of ordinary forests. These projects or instruments save money because extensively designed, high maintenance parks with a minimum infrastructure provision are not needed (lighting, rubbish bins, and benches are not provided). Rather, these areas substantially strengthen the biodiversity and therefore contribute to an enriched provision of UES. What is more, they also serve to use the areas sustainably and as a counter strategy against litter pollution and dilapidation through their use (cycling, walking, walking dogs along predetermined ways).

In addition, large scale demolition allows the *creation of new parks and green areas* such as in Liverpool (Everton Park) or Leipzig (Lene Voigt Park, Rabet Park). These are successful examples that show the interrelation of the requirement for demolition and the creation of new quality of life in the form of urban parks within the central/inner city.

Moreover, even entire districts, e.g., in Manchester (Castlefield district) are converted and newly designed. In this case, a former highly industrialised district was upgraded by residential and leisure functions while considering and maintaining its

Fig. 4 Urban forest, Leipzig, Germany (Source: Annegret Haase)

urban nature, such as green and blue (i.e. water) areas, as urban amenities. Several other projects have followed this principle, commonly known as *water- or river-front developments* (in Bristol, Newcastle, Amsterdam, Malmö, etc.) (Fig. 5).

In several shrinking cities, however, *new construction and investments in infrastructure* can be observed. In the inner cities of Liverpool, Leipzig, or Pittsburgh (US), several types of town houses had been constructed, representing a new, less dense form of dwellings and allowing for more green and open areas (Fig. 6). However, to what degree these developments contribute to an improved provision of UES is still under discussion (Haase et al. 2014). In Liverpool, Genoa, or the Upper Silesian Industrial District, new urban qualities have been created by the decontamination and renovation of former industrial infrastructure such as channels, docks, or industrial buildings that had been converted to residential use.

Table 1 summarizes examples of the described land uses or use strategies with respect to their impacts on land use change and UES provision.

2.2.2 Regrowth

When a city is regrowing, we usually find redensification and re-compaction processes, especially in the inner city, which thus drive the development of the entire city, as in Amsterdam, Manchester, or Leipzig, for example. Most regrowing cities follow an inward development strategy, i.e., they deliberately concentrate new construction

Fig. 5 Riverfront development in Newcastle, UK (Source: Annegret Haase)

Fig. 6 Low-density housing in Liverpool's inner city (Source: Annegret Haase)

Table 1 Land uses and land use strategies in shrinking cities and their impacts on land use change and UES provision (examples)

Land uses and use strategies	Description	Impacts on land use change	Impacts on UES provision	Examples
Urban gardens	small-scale community gardens in residential areas serving local food production, communal activities and communication	non-residential land use, and large unsealed areas, in contrast to the built environment	gardens provide space for encounter, outdoor activities and communication, healthy food and recreation; support rainwater runoff and fresh air filtration	Annalinde, Leipzig, Germany; Greening of Detroit, US South Liverpool Urban Garden (SLUG), UK
New parks	de-sealing and greening of formerly built-up lots; decrease in density of the built environment	new land use on formerly built-up lots, increase in non-sealed, green land use	new opportunities for recreation, leisure, sport and encounter; provision of aesthetic values and healthy environments	Lene-Voigt Park Leipzig, Germany Everton Park, Liverpool, UK Southside Park, Pittsburgh, US
Urban forests	reuse of formerly built-up lots for afforestation and steered succession	new land use on formerly built-up lots, increase in non-sealed, green land use	new opportunities for walking, cycling and walking dogs; provision of aesthetic values and healthy environments	Urban forest pilot projects in Leipzig and Halle, Germany Tree farms in Detroit, US
Interim uses	various use options, e.g., green interim uses for gardens or parks, location for trailer parks, parking lots	generally lower-density land use, non-residential in a narrower sense, partial de-sealing and greening	degree and quality of UES provision depends on the type of interim use; it is high in the case of gardens or parks and lower in the case of parking lots	Urban gardening projects in Leipzig, Detroit and other cities Millennium field (urban farming as an art project) in Leipzig, Germany
Low-density housing	reuse of urban land for low-density residential purposes, mostly no change in type of land use but in density	lower density, more green per square km, lower degree of sealed land than before	provision of more UES than in high-density housing, e.g., fresh air, rainwater runoff, green (depends on equipment and shape of housing)	Town houses in Leipzig, Germany Town houses in Liverpool, UK Suburban-type housing in inner Pittsburgh, US

Source: authors' compilation

and infrastructure development activities in the inner parts of the city or the core city. Whereas, during shrinkage, less dense buildings, such as town houses, had been preferred, under conditions of regrowth more and more *multi-level buildings* (Fig. 7 left photo) are required, due to strong immigration and the demand for flats. However, this does not automatically led to less green in the surrounding areas because, due to large scale refurbishment, the accompanying green areas did not need to be touched and, if new constructions are needed, they are usually equipped with greening of the facades, the inner courtyards, balconies, or the roof itself; this has occurred in many European cities (Green Surge WP3 report 2015). The cooling effects of such re-densification are still being discussed because, on the one hand, denser and larger buildings provide more shade but, on the other hand, they may block fresh air corridors. Inward development and re-densification means, on the one hand, less provision of UES such as fresh air, rainwater runoff, green/biodiversity through concentration of use and densities, but, on the other hand, better access to urban green spaces through shorter distances, as well as opportunities for high-quality green spaces in the urban hinterland. Moreover, *new housing* is being built, in many cases, *close to high-quality green areas*, such as parks (Fig. 7 right photo). This surely provides benefits in terms of UES (cooling, air filtration, biodiversity, recreational opportunities) for those who live in the housing, but it might lead to exclusionary displacement of lower income groups because such types of new housing are usually rather expensive.

Regrowing cities, in general, *reuse their brownfields*, albeit with different foci. In Ostrava (Karolina area), Liverpool (docklands), Leipzig (Plagwitz) or Manchester (Castlefield), many brownfields have been converted into new commercial, residential, and even nightlife areas. This often results in a sealing of formerly non-sealed or even green areas. In a few cases, the existing green and blue infrastructure was incorporated into the new built-up structure (as in Manchester, Liverpool, or Birmingham). Other prominent examples of such developments can be found in the former harbour areas of Liverpool or in large, previously highly industrialized areas in Manchester, Ostrava, or Leipzig. This emphasises that the degree to which green plays a role for the provision of UES depends highly on the new uses of the former brownfields (Fig. 8).

However, *green and blue infrastructure* is increasingly being used as a *strategic upgrading component* (living in a green environment, living next to water- or riverfronts, etc.) with the associated increasing prices for rents, flats, and land prices. In many cases, these policies have led to selective upgrading and a displacement of lower income groups. This phenomenon has been discussed as "eco-gentrification "or "green gentrification" in an evolving debate since the mid-2000s (Dooling 2009; Quastel 2009; Banzhaf and McCormick 2007), although proof for a real causality is still lacking and the role and impact of environmental improvement on socio-spatial segregation is a complex and multi-faceted one. More green areas that are created when new buildings are constructed have, in sum, a positive effect on the urban population (filtering effects, mitigated heat effects, etc.) but a closer look reveals that especially those residents who can afford the higher prices actually benefit most from these green areas (see: the Phoenix project in Dortmund, which provides luxury housing and leisure areas on a former industrial ground; http://www.phoenix-dortmund.de/de/home/index.html), thus reinforcing socio-spatial differentiation.

Fig. 7 (**a**) Redensification in Liverpool and (**b**) new, upmarket housing close to a park in Leipzig (Source: Annegret Haase)

The *increasing settlement pressure* due to very high rates of immigration also has *trade-offs for UES provision*. The niches, mentioned above, which emerged during shrinkage, are subjected to further pressure, especially those with temporary green uses (e.g., community gardens, cultural projects, caravan parks, or informal uses) and more and more such niches are disappearing. Increasing rates of sealed land cover, growing infrastructure systems, and increasing traffic can even lead to new environmental stress because the remaining green areas are over-used, resulting in lower quality of the green areas. Increased densities of built environment, more traffic, and more use of infrastructures and public areas also impact on the capabilities of the urban ecosystems to provide UES such as fresh air, carbon dioxide storage capacity, filtering of pollutants, cooling effects, or rainwater runoff. Here, the capacities of UES to balance environmental pollution and burdens (for the urban population) are often exhausted and/or exceeded. Several studies on various European cities have confirmed that more urban green is not balancing out the pollution produced by the constantly increasing traffic in a regrowing city (Kain et al. 2016; Larondelle et al. 2016; Baro et al. 2015; Lauf et al. 2016).

Table 2 summarizes examples of the described land uses or use strategies with respect to their impacts on land use change and UES provision.

a

Fig. 8 Reuse of former large-scale brownfield sites: (**a**) Princes Dock, Liverpool and Czermaks Garden, Leipzig, (**b-1**) as interim use area in 2004 and (**b-2**) under reconstruction in 2016 (Source: (a) Annegret Haase, (b-1) J. Volz (2004), (b-2) Dieter Rink (2016))

Fig. 8 (continued)

Table 2 Examples of land uses and land use strategies and their impacts for land use change and UES provision in regrowing cities

Land uses and use strategies	Description	Impacts on land use change	Impacts on UES provision	Examples of cities
Redensification; infill	newly built high-density housing in gaps	increase in densities, partly new sealing, change in land use from non-residential to residential	less provision of UES such as fresh air, rainwater runoff, green/biodiversity, opportunities for rooftop and vertical greening	Leipzig, Liverpool, Manchester, Pittsburgh
Redensification: multi-storey new housing	newly built high-density housing on former non-built lots	increase in densities, partly new sealing, change in land use from non-residential to residential	less provision of UES such as fresh air, rainwater runoff, green/biodiversity, opportunities for rooftop and vertical greening	Leipzig, Liverpool, Manchester, Pittsburgh, Baltimore
New neighbourhoods	new construction of housing and infrastructure on a larger scale	increase in densities, partly new sealing, change in land use from non-residential to residential	less provision of UES such as fresh air, rainwater runoff, green/biodiversity, opportunities for roof top and vertical greening and high-quality green areas between buildings	Leipzig, Liverpool, Manchester, Birmingham
New commercial/ infrastructure areas	new construction of buildings on former brownfield sites or non-built lots	increase in density, partly new sealing, partly change in land use type	less provision of UES such as fresh air, rainwater runoff, green/biodiversity, (limited) opportunities for rooftop and vertical greening	Liverpool, Timisoara
Inward development	concentration of new construction and infrastructure development activities on the inner or core city	increase in density, partly new sealing, partly change in land use type towards land for building, commerce, or traffic	less provision of UES such as fresh air, rainwater runoff, green/biodiversity through concentration of use and densities, better access to urban green spaces through shorter distances, opportunities for high-quality green spaces in the urban hinterland	Leipzig, Liverpool, Pittsburgh (partly)

Source: authors' compilation.

3 Which Challenges Can Be Identified from the Interplay of Urban Dynamics, Land Use Change, and UES Provision?

The following section reflects in a systematic way on the challenges that emerge for land uses and the corresponding provision of UES in phases of shrinkage and regrowth. Existing synergies/benefits and trade-offs are of particular interest.

Urban shrinkage is widely perceived as a problem with many negative consequences for urban society and urban space, including the sustainable use of urban land and the provision of a good quality of life. One of the main challenges thereby is it to deal with vacancies, abandonment, and demolition and to balance these with greening of areas and the quality of life for the remaining population. Most shrinking cities experience difficulties in tackling these challenges, due to decreasing human capital and budgets.

However, shrinkage does not just represent a problem for sustainable land use and UES provision in cities. It also offers opportunities and potentials for more sustainable land use and better provision of UES, both factors that may lead to an increased quality of life for the (remaining) inhabitants. For example, town houses and new housing next to green and blue infrastructure provide residential qualities and retain population in these districts. Although new construction means additional soil sealing, these measures often contribute to an improvement of the housing conditions in terms of leisure, recreation, and even health aspects. Newly created parks and urban forests or interim/temporary uses such as urban gardening provide more space for green and thus for biodiversity and movement (sports). This represents an essential link between land use changes and the provision of UES (biodiversity, recreation, fresh air, etc.; see Haase et al. 2014; Lorance and Haase 2010; Rink and Arndt 2016).

In fact, in many cities that have seen a period of shrinkage, these new opportunities and created qualities have helped to encourage residents to stay and to attract new residents. In doing so, low-density land use, more green spaces, opportunities for alternative uses and niches, as well as higher environmental qualities contributed, in many cases, to a halt of shrinkage and to new growth. To put it differently: the trade-offs of the shrinkage context were translated into benefits for the urban environment (land use, UES) and quality of life. Even if this was, in most cases, only possible with large-scale public financial support and in conjunction with the creation of new job opportunities, it makes it obvious that only the context of shrinkage provided the framework for new developments and qualities that made the cities attractive for their inhabitants to stay there and for new in-migration.

When, after a longer period of shrinkage, a city is regrowing, views on land management, reuse of abandoned areas, greening and interim uses change. Although regrowth is generally a favourable situation for a city because it means new people, increasing attractiveness, and often also new investment, it is not just a benefit when it comes to land use (change) and potentials for UES provision. New challenges resulting from emerging or changed requirements and interests have to be balanced.

On the one hand, new demographic and settlement pressure requires a constant redensification and construction of new houses. On the other hand, the conservation of hitherto created green and open areas, as a pull element, is necessary. The corridor of how sustainable urban development, including land use and UES provision, may be shaped ranges between these two poles. What the real world result looks like depends, however, on several factors. Among these are the overall strategy and leitmotifs of a city, the constellation of decisive actors, ownership issues, existing institutional and legal conditions, and, not least, the available budget. Regrowth must not automatically lead to new segregation, pressure on green spaces and displacement but, in most cases, documented in the literature, new growth brought about exactly these processes, although in different degrees of depth and speed and focused on specific parts of the cities. In these parts, usually the most attractive areas – green and blue – become part of upgrading strategies, leading, in the worst case, to eco- or green gentrification (see above).

By and large, there are many more ambivalences around the nexus between urbanisation dynamics and its impacts on land use (change) and the potentials and constraints for UES provision. Whilst urban shrinkage, despite all its problems, offers the space for new ways of thinking, less density, more green and experimentation, urban regrowth sets new challenges for the maintenance of green qualities that were created or established within the period of shrinkage. Potentially, the context of regrowth offers the chance to steer new growth in a more sustainable direction by building exactly on this "green legacy" from the period of shrinkage. In fact, however, this remains a theoretical option in many cases. In most cases for which the authors have evidence, new growth brought an immediate return to market-oriented and pro-growth strategies and a delimitation of space for experimentation, innovation, niches, as well as for affordable or non-market solutions. Whereas, during shrinkage, the market was dominated by the supply side, it is highly demand-driven during regrowth, allowing the market to react by creating new and high priced housing. Therefore, especially in the context of post-shrinkage regrowth, the creation of a balance between the new settlement pressure and the maintenance of green areas remains a continuing challenge. In shrinking cities, the mentioned potentials for more green, less density, and experimentation are often restricted by tight budgets and an insistence on pro-growth strategies, instead of using the opportunities of abandonment and emptying for green solutions and an enhancement of the quality of life.

Thus, it is necessary to look more closely at all trade-offs between UES, land use change, and the changing urban context: Which positive aspects emerge from re-densification, not just for residential quality and provisioning with green areas, but also for socio-spatial configurations, physical improvements, or even displacement. We emphasise here that an integrated perspective is essential in order to capture all effects and to bridge debates that deal with the topic from different perspectives, e.g., the debate on land use change under conditions of non-linear urban development, UES and ecological sustainability, as well as the debate on connections and trade-offs between green/blue-based upgrading and socio-environmental justice, or socio-spatial segregation/displacement. To date, only a few investigations have addressed these connections and trade-offs explicitly (Berbes-Blazquez et al. 2016; Kabisch et al. 2016, for example). Here, more mutual consideration and interest is needed to achieve a truly integrated view.

4 How Do Actors, Policies, and Governance Respond to These Challenges?

The synergies listed above and trade-offs between shrinkage or regrowth with respect to land use changes and the provision of UES, as well as the resulting opportunities and challenges show that how actors cope with the respective situation given is crucial. Therefore, the following section will expand on the responses of different actors, their motivations, and when or how shrinkage and regrowth have been treated as a problem or chance.

In shrinking cities, the management of, or planning for, shrinkage represented a new challenge for local actors. There were no blueprints about how to respond to a situation that was opposite to the overall growth paradigm and in which a decline of the population, economic base, infrastructure and, in general, demand, had to be governed. Within this context, space also emerged for experimentation, new coalitions of actors and new models for land use related to growth/market expectations/mechanisms beyond (pure) market orientation.

In the case of regrowth, the framing conditions, in contrast to shrinkage, change. Empirically, most shrinking cities suddenly turn to (re)growth rhetoric and planning as soon as some insipient signs of stabilization appear. Policies are re-adjusted to growth orientation; the management of shrinkage is no longer seen as necessary. Investment (also private) often assumes a larger role and, in many cases, the cities are (still) dependent on it. Generally, they see a recovery of their real estate market and the local economy. Simultaneously, the former support for green, less dense, and less market-oriented uses decreases.

Two meaningful examples of this situation are the cities of Leipzig and Halle; whilst Leipzig officially labeled its post-2000 stabilization immediately as new growth, Halle declined any deliberate planning for shrinkage (which was also backed by a written strategy) immediately after it started to tentatively regrow in the years after 2010 (Wolff and Haase 2015). Under the condition of regrowth, the management of shrinkage is seen as no longer necessary; instead, the new growth was to be stabilized and increased. Whereas, during the period of shrinkage, local actors typically are comparatively weak and limited in their capacities to act and depend on external investment and decisions (Rink et al. 2014), the situation changes (at least partly) when a city turns to new growth. In the period of regrowth, investment in the city increases, as does the demand for urban space. The existing building stock is mainly renovated in attractive inner-city locations, and new housing is built. Re-densification takes place through renovation of existing, vacant housing, infill and new multi-storey or high-rise construction. Generally, land and the real estate market recover and increase in value (Couch and Cocks 2013; Haase and Rink 2015). Given this context, non-market interim uses that were typical for the shrinkage period, tend to decline in importance or face increasing pressure. They do not have a strategic importance for urban planning anymore. Interim uses are increasingly displaced by market-oriented uses, as described, e.g., by gentrification research.

Existing governance structures, including interim uses and users, are subjected to increasing pressure. The much more purpose-driven coalitions of actors from the context of shrinkage that oriented to stabilization, maintenance, and attenuation of decline, usually break or disappear in the context of regrowth and market and benefit-oriented logic returns. New conflicts of interest between actors who formerly cooperated emerge, e.g. when interim users want to continue to use their spaces or land and defend them against market-oriented interests. Differences in interests between actors of different origin become larger and always more difficult to balance since municipal actors and entrepreneurs prefer (again) the classical pro-growth strategies. Purpose-driven coalitions from the period of shrinkage that created win-win situations for private owners, housing companies and entrepreneurs are replaced by pro-growth interests and logic. The speed of this change depends on the scope and type of regrowth. Whether sustainable solutions are subsequently pursued and "green qualities" maintained depends on how they fit into the new growth concepts and exploitation interests. This might be the case when a city adopts smart or sustainable growth strategies or when creative or innovative interim uses are later useful for marketing purposes, to keep the city attractive for new (better-off) residents. While et al. (2004) introduced the term of "sustainability fix" for the extent to which green or sustainable developments are employed as strategic priorities for a city's future ("city of gardens", "innovative city", "eco-city"). Of course, the maintenance of "green solutions" also depends on how interim users may counteract market interests, and how they can, e.g., mobilise the public in favor of their interests.

By and large, green qualities and sustainable solutions created in the period of shrinkage can be only maintained when they are situated on public land or in public ownership. Examples of this are the new parks or urban forests that were created in shrinking cities in former railway areas or other (industrial and residential) brownfield sites, such as Everton Park in Liverpool, Lene Voigt Park in Leipzig, Emscher Landscape Park in the Ruhr area, or the "forest city" Halle-Silberhöhe. Other examples are urban gardens that are situated on public land. It is crucial to guarantee public ownership before market-oriented actors can buy the land because then, in most cases, interim and/or green uses are displaced stepwise or just become niches. This can be observed in some regrowing cities. It shows that the steering of regrowth cannot be left to the market because, in this case, benefit-oriented uses will win – and they do not need to be sustainable, green, or inclusive uses. The municipalities have a decisive role in this process because they have to plan for the long term and have to pay attention to the maintenance of sustainable qualities. Of course, different interests and use developments have to be negotiated, e.g., between redensification strategies on former brownfield sites that are now open/green spaces and new land consumption. From a scientific standpoint, the new conflicts also lead to the question of how shrinkage and regrowth may be assessed from a sustainable or UES standpoint, and it becomes obvious that there are benefits and trade-offs: Despite all its problematic aspects, shrinkage offers space and opportunities for sustainable and green solutions supporting livability and well-being; regrowth may set new limits to such solutions or even "sacrifice" them to market- or benefit-oriented interests that are less or not sustainable/green.

It is obvious that the consideration of green qualities and the provision of UES are not easy tasks for governance to fulfill under the conditions of regrowth. Complications arise from differing or even opposing interests and demands. The crucial question is whether the long-term guarantee of UES provision can be defended against (more short-term) market-oriented interests and how these two issues can be brought together in a way that is also socially sustainable. The only actor who can defend sustainable developments is the municipality. It can form an alliance with other actors (civic society, green economy) to strengthen its position. Because municipalities are also interested in supporting regrowth, their task consists of constant processes of negotiation and balancing. To date, relatively few examples of how these various interests and priorities can be successfully combined exist. There is only little evidence for which interests will finally prevail or how market and sustainability interests might be coupled and/or balanced in a reasonable way. Here, we see a new "avenue of research" for interdisciplinary social-environmental sciences but also for the transdisciplinary dialogue between science and urban practice.

Generally, a bundle of factors decides to what extent shrinkage and regrowth represent a problem or a chance for cities, in terms of sustainable land use and provision of UES. These factors include structures of ownership and types of leadership, availability/sources of funding, strategic goals of development, political and institutional settings, as well as the level of inclusion of different actors into decision-making. When considering the political framework, it is important to note that, in this contribution, the examples used describe cities in a welfare state context with an elaborated environmental policy and sustainability goals (Germany, UK); the situation would change considerably if we considered Central Eastern European cities with a much stronger neoliberal orientation or also cities in crisis countries such as Greece, Italy, Portugal, or Spain, where consequences of crises but also of response policies (e.g. of privatization, see Haase et al., "On the Connection between Urban Sustainability Transformations and Multiple Societal Crises", in this volume) are much less favourable for discussions on sustainability or for transformation towards more sustainability.

5 Conclusions

Coming back to our original research questions raised, we conclude that there is a complex and by no means clear-cut or one-dimensional relationship between contexts of shrinkage and regrowth and their impacts on land use change and UES provision. Whilst shrinkage represents a problem for cities, due to abandonment, underuse of urban land, decline of the economic base, and tight budgets, it also offers new opportunities for creating solutions for reuse of areas or re-shaping brownfield sites in a green, sustainable way. This generally (and excluding differences and unevenness here) contributes to the quality of life, as well as the health and well-being of the remaining residents. This development might contribute to an increasing attractiveness of a city and act a source to make people stay or even to attract new residents. The main challenge in the context of regrowth is it to balance different interests (more or less benefit-oriented) and to steer new growth in a way that maintains at least some of the green, UES provisioning uses, and solutions that were created in the period of shrinkage. Because many of those uses are not market-oriented, the probability that they will continue is, given today's primacy of profit maximization as a guiding principle in many cities, relatively small. This is also due to the dependency of many regrowing cities on private capital. At the same time, regrowth also represents a chance for compact city development in a resource-efficient way (short distances, high-density housing). This can contribute to decreasing environmental burdens (less traffic, cooling effects of high buildings, energy-efficiency of high-density structures) and increasing accessibility, e.g., of green areas such as public parks. Here, the role of municipal policy, policy priorities, and public ownership of green areas seem to be crucial for maintaining green (re)uses in formerly shrinking cities.

Here, we see the emergence of an interesting new avenue for future research. At this point, the idea of the above-mentioned "sustainability fix" of While et al. (2004) might serve as a conceptual framework to detect which role land use and green quality play when it comes to strategies for urban development and to what extent these strategies overcome one-dimensional thinking. This would, indeed, help to reveal the relationships between different developments "below the surface", e.g., impacts of land use change on socio-spatial patterns and, even more, to shed light on intended and unintended developments accompanying such changes. The emerging debate about the relationship between UES and power relations, which leads to uneven distribution of UES (winners and losers), can be enriched by a reinforced cooperation between social and environmental science. In this context, many new questions arise, such as: How long can the qualities created during shrinkage that support the turn towards regrowth be maintained? How important are green areas and the recognition of benefits through UES (see Hansjürgens et al., "The TEEB Approach Towards Sustainable Urban Transformations: Demonstrating and Capturing Ecosystem Service Values", in this volume) for housing and relocation choice in a context of rising housing costs and tighter housing markets, and, what effects does redensification have on both residential and environmental quality of life? How should redensification be shaped in order to avoid segregation and allow for affordable solutions and niches for experimentation/innovation?

Factors such as land use, UES provision and the political, societal, or governance framework can hardly be examined individually. The contexts of urban shrinkage and regrowth are – although meaningful – just examples to support this general argument. Thus, we strongly recommend an integrated perspective that considers all three elements and their integrative relationships. In research, as well as in policy practice, this integrated view is rarely followed; instead, individual problems and challenges are addressed and the side effects of causes and problem solutions are ignored. These debates have to be linked very closely for what this paper would like to contribute and they represent core elements for a fruitful exchange.

When we return, finally, to the overall context of this volume and the relationship of our subject to the general context of urban transformation, it should be highlighted that urbanisation has to be considered as a non-linear process. Because, for the future, shorter periods of shrinkage and regrowth are probable, the impacts of non-linear development have to receive more consideration in the discussion on transformation. The nexus between urban land and provision of UES is in the centre of urban transformation towards more sustainability; as the chapter has attempted to show, this nexus is a complex one and there are many factors that determine whether shrinkage or regrowth are a problem or benefit. A better and more informed understanding of emerging and existing trade-offs and ambivalences is crucial for a full grasp of the challenges and potentials that shrinkage and regrowth offer for sustainable land use and UES provision.

References

Banzhaf HS, McCormick E (2007) Moving beyond cleanup: identifying the crucibles of environmental gentrification. Andrew Young School of Policy Studies, research paper series, working paper 07/29. http://aysps.gsu.edu/publications/2007/index.htm. Accessed 9 June 2016

Baro F, Haase D, Gomez-Baggethun E, Frantzeskaki N (2015) Mismatches between ecosystem services supply and demand in urban areas: A quantitative assessment in five European cities. Ecol Indic 55:146–158. doi:10.1016/j.ecolind.2015.03.013

Berbes-Blazquez M, Gonzales JA, Pasqual U (2016) Towards an ecosystem services approach that addresses social power relations. Curr Opin Environ Sustain 19:134–143. doi:10.1016/j.cosust.2016.02.003

Couch C, Cocks M (2013) Housing vacancy and the shrinking city: trends and policies in the UK and the city of liverpool. Hous Stud 28(3):499–519

Dooling S (2009) Ecological gentrification: a research agenda exploring justice in the city. Int J Urban Reg Res 33(3):621–639

EU 7FP Project Shrink Smart – The governance of urban shrinkage within a European context. (2009–2012). www.shrinksmart.eu. Accessed 9 June 2016

EU COST Action CIRES (2009–2013) Cities regrowing smaller – fostering knowledge on regeneration strategies in shrinking cities across Europe (TUD COST Action TU0803). http://www.cost.eu/COST_Actions/tud/TU0803. Accessed 9 June 2016

Green Surge, EU-Project (2013–2017). WP3 report 2015. http://greensurge.eu/. Accessed 5 Nov 2016

Haase A (2015) European cities between shrinkage and regrowth: current trends and future challenges. In: Tomaschek N, Fritz J (eds) Die Stadt der Zukunft. Waxmann, Münster/New York, pp 89–120

Haase D, Lorance ED (2010a) Gestattungsvereinbarungen: Zwischennutzungsform urbaner Brachen. Stadt Leipzig: Statistische Quartalsberichte 1/2010, 44–46

Haase D, Lorance ED (2010b) Gestattungsvereinbarungen: Zwischennutzungsform urbaner Brachen (Teil 2). Stadt Leipzig: Statistische Quartalsberichte 2/2010, 40–42

Haase A, Rink D (2015) Inner-city transformation between reurbanization and gentrification: Leipzig, eastern Germany. Aust Geogr 15(2):226–250

Haase D, Haase A, Kabisch N, Kabisch S, Rink D (2012) Actors and factors in land-use simulation: the challenge of urban shrinkage. Environ Model Softw 35:92–103. doi:10.1016/j.envsoft.2012.02.012
Haase D, Haase A, Rink D (2014) Conceptualizing the nexus between urban shrinkage and ecosystem services. Landsc Urban Plan 132:159–169. doi:10.1016/j.landurbplan.2014.09.003
Haase D, Quantz J, Rink D, Haase A (2015) Trade-offs von Landnutzungsdynamiken auf ÖSD im Kontext von städtischer Schrumpfung und erneutem Wachstum. Presentation at the Congress for Geography in Berlin (unpublished)
Haase A, Bernt M, Grossmann K, Mykhnenko V, Rink D (2016) Varieties of shrinkage in European cities. Eur Urban Reg Stud 23(1):86–102
IJURR Symposium on shrinking cities (2012) Int J Urban Reg Res 36(2):213–414
Kabisch N, Haase D (2011) Diversifying European agglomerations: evidence of urban population trends for the twenty-first century. Popul Space Place 17(3):236–253
Kabisch N, Haase D, Haase A (2012) Urban population development in Europe 1991–2008 – the examples of Poland and the UK. Int J Urban Reg Res 36(6):1326–1348
Kabisch N, Frantzeskaki N, Pauleit S, Naumann S, Davis M, Artmann M et al (2016) Nature-based solutions to climate change mitigation and adaptation in urban areas: perspectives on indicators, knowledge gaps, barriers, and opportunities for action. Ecol Soc 21(2). doi:10.5751/ES-08373-210239
Kain J-H, Larondelle N, Haase D, Kaczrowska A (2016) Exploring local consequences of two land-use alternatives for the supply of urban ecosystem services in Stockholm year 2050. Ecol Indic 70:615–629. doi:10.1016/j.ecolind.2016.02.062
Kroll F, Müller F, Haase D, Fohrer N (2012) Rural–urban gradient analysis of ecosystem services supply and demand dynamics. Land Use Policy 29(3):521–535
Larondelle N, Frantzeskaki N, Haase D (2016) Mapping transition potential with stakeholder- and policy-driven scenarios in Rotterdam City. Ecol Indic 70:630–643. doi:10.1016/j.ecolind.2016.02.028
Lauf S, Haase D, Kleinschmidt B (2016) The effects of growth, shrinkage, population aging and preferenceshifts on urban development—A spatial scenario analysis of Berlin, Germany, Land. Use Policy 52:240–254. doi:10.1016/j.landusepol.2015.12.017
Lorance Rall E, Haase D (2011) Creative intervention in a dynamic city: a sustainability assessment of an interim use strategy for brownfields in Leipzig, Germany. Landsc Urban Plan 100(3):189–201
Quastel N (2009) The political ecologies of gentrification. Urban Geogr 30(7):694–725
Rink D, Arndt T (2016) Investigating perception of green structure configuration for afforestation in urban brownfield development by visual methods – a case study in Leipzig, Germany. Urban Forestry & Urban Greening 15:65–74. doi:10.1016/j.ufug.2015.11.010
Rink D, Wolff M (2015) Wohnungsleerstand in Deutschland. Zur Konzeptualisierung der Leerstandsquote als Schlüsselindikator der Wohnungsmarktbeobachtung anhand der GWZ 2011. Raumforsch Raumordn 73(5):311–325
Rink D, Haase A, Grossmann K, Couch C, Cocks M (2012) From long-term shrinkage to re-growth? A comparative study of urban development trajectories of Liverpool and Leipzig. Built Environ 38(2):162–178
Rink D, Couch C, Haase A, Krzysztofik R, Nadolu B, Rumpel P (2014) The governance of urban shrinkage in cities of post-socialist Europe: policies, strategies and actors. Urban Res Pract 7(3):258–277
Turok I, Mykhnenko V (2007) The trajectories of European cities, 1960–2005. Cities 24(3):165–182
While A, Jonas AEG, Gibbs D (2004) The Environment and the entrepreneurial city: searching for the urban 'Sustainability fix' in manchester and leeds. Int J Urb Reg Res 28(3):549–569
Wolff M, Haase A (2015) Stadtregion Leipzig-Halle jenseits der Schrumpfung: neues Wachstum und Stabilisierung. Statistischer Quartalsbericht I/2015, Amt für Statistik und Wahlen, Stadt Leipzig, 36–42
Wolff M, Haase A, Haase D, Kabisch N (2016) The impact of urban regrowth on the built environment. Urban Stud. doi:10.1177/0042098016658231

Potentials of Urban Brownfields for Improving the Quality of Urban Space

Ellen Banzhaf, Thomas Arndt, and Jan Ladiges

1 Urban Brownfields as Land Resources

Urban brownfields have been an international phenomenon for decades and are part of most industrialised countries worldwide (de Sousa 2008). Brownfields are understood as previously used land that has been abandoned and may suffer from contamination. They occur in shrinking cities as well as in growing cities. In particular, urban locations that changed from agglomerations with heavy industry to service centres transformed these urban spaces into brownfields. Several international institutions and expert networks deal with concerted actions for brownfields, to develop such valuable urban areas (CABERNET; US Environmental Protection Agency (EPA), European Environmental Agency (EEA), etc.). Despite substantial efforts, many of these previously used industrial sites could not be assigned to new functions and, therefore, no further use was made available. Re-using brownfields can help to reduce the amount of newly sealed land (e.g., Rink and Banzhaf 2011). Land recycling, in particular, is considered a solution because it offers potential sites for residential and commercial constructions (EEA 2015). For the process of reurbanisation, brownfields play a particularly important role in recycling former industrial

E. Banzhaf (✉)
Department of Urban and Environmental Sociology, Helmholtz Centre for Environmental Research - UFZ, Leipzig, Germany
e-mail: ellen.banzhaf@ufz.de

T. Arndt
Federal Agency for Nature Conversation, Division II 4.1 Landscape Planning, Spatial Planning and Urban Areas, Leipzig, Germany
e-mail: thomas.arndt@bfn.de

J. Ladiges
Institute for Geosciences and Geography, Martin-Luther-University Halle-Wittenberg, Halle, Germany
e-mail: janladiges@web.de

S. Kabisch et al. (eds.), *Urban Transformations*, Future City 10,
DOI 10.1007/978-3-319-59324-1_12

sites into land that meets the needs of today's cities. In urban areas, land is subjected to strong dynamics and is a tenuous and competing resource, in view of ongoing global urbanisation. Therefore, the development of commercial brownfields is an important municipal task for sustainable land management. Their potential redevelopment is vital with regard to resource-efficient land management, on the one hand, and the newly developed land-use type as an option for increased urban resilience against natural hazards on the other. Consequently, redevelopment can have marked effects on the quality of life and sustainable urban development (Martinat et al. 2016; Norrman et al. 2016; Bartke et al. 2016).

Urban brownfields can, therefore, have positive implications as land resources for new urban development, despite their negative images as contaminated sites or as symbols of urban decay and blight. When considering the redevelopment of formerly abandoned sites, there are two major options for dealing with urban brownfields. The aspect of revitalisation is concerned with the re-use of these brownfields for commercial purposes and has been identified as an important tool to prevent further urban sprawl. The second option – renaturation – treats brownfields as important green spaces for humans, animals, and plants. Although this aspect has high social and ecological values, it has, so far, only played a minor role when revaluating derelict land (BBSR 2013, p. 93). In this regard, redeveloping brownfields is important in terms of urban transformations towards sustainability and it shows how different development options may help to fulfil two substantial, yet divergent objectives of urban sustainability: reducing urban sprawl or maintaining urban ecosystem services.

Understanding the potentials of urban brownfields has been discussed comprehensively in the literature (e.g., Keil 2005; Lafortezza et al. 2008; Budinger and Gruehn 2012). For a city, their specific potentials depend on their respective size and allocation, their overall area, spatial distribution, and connectivity.

Location-based development incentives have been used for decades. A challenge for urban planning is to find a balance between brownfields' actual contribution to real estate investment decisions that stimulate infill growth in urban areas and their value for improving environmental quality, based on their degree of vegetation cover and succession stages.

2 The Potential Re-Use of Urban Brownfields

Brownfields have diverse functions for urban development and are in "conflict between conservation and structural re-use" (Hansen et al. 2012, p. 126). On one hand, they provide important land-use options for reducing the consumption of land. This aspect supports the goals of the National Strategy for Sustainability (Presse- und Informationsamt der Bundesregierung 2016). On the other hand, brownfields are important sites for urban biodiversity and are part of the urban green infrastructure (Die Bundesregierung [i.e. Federal Government of Germany] 2002). Thus, these aspects support the goals of the National Strategy for Biodiversity (exemplified by BMU 2007). These conflicting objectives, in particular with regard to further greening of cities, are a challenge that has to be met on a local scale and within an integrated urban planning approach (Arndt and Werner 2015). As a consequence, local planning authorities face the challenge of either revitalising existing brownfields, or conserving the site to enhance green infrastructure and/or to safeguard its biodiversity (BMU 2007). In many cases, the priority is set to economical reactivation, especially in the inner city (Mathey and Rink 2013). In any case, re-using abandoned land still is a complex process, from incipient remediation through demolition and new design to the rehabilitation of a brownfield site into the structure of the city.

2.1 Revitalisation

Municipalities are mainly interested in an economic recovery of unused land because it generates tax revenue, revives the economy, and – in the case of commercial sites – creates jobs. Therefore, potential recovery areas need to be well connected, logistically, and have to be offered on acceptable terms. In addition to economic development, a municipal aim is also to minimize the consumption of undeveloped land, in order to restrict urban sprawl and to manage urban land in a more resource-efficient manner. For this purpose, the intention is to re-use existing space potentials by the complete recovery of derelict land.

Because they are named as "prime candidates for urban revitalisation efforts" (De Sousa 2008, p. 251), it is obvious how precious these sites are and what kinds of urban benefits can emerge from their refurbishment. As mentioned above, different international organisations in North America and Europe, as well as local and regional governments, have initiated incentives to make the recycling of brownfields attractive to land owners. Such innovative strategies include environmental and economic policies and programs that provide relief from the costs and risks immediately connected with brownfield remediation and redevelopment. Liabilities for contaminated sites keep owners from selling their land, so that valuable space in central urban areas lies abandoned, rather than being returned to the development cycle of land as a resource. Because they are unsure about regulations governing who should cover the costs of cleaning the lot, owners refrain from selling. As a consequence, local governments tend to exempt owners from liability to clean-up the premise from pollution. Only by gaining the empowerment to reclaim the real estate can local authorities stimulate new investments in the premises. Redevelopment of brownfields exists for private and public users and makes the land economically viable, depending on the demand for real estate in urban centres. The support for brownfield redevelopment must be shared by various divisions of local governments, such as divisions for urban planning, urban greening, environmental protection, and business development, to make a concerted action profitable for the social, environmental and economic benefit of a neighbourhood or of a local district.

2.2 Renaturation

With respect to a resource-efficient city that takes brownfields as an opportunity to rethink urban planning, preservation and development of open spaces has become of great importance, but stands behind the economic profit of re-using brownfields (EEA 2015, p. 44). Nevertheless, it has been recognized that urban open spaces can be used not only for recreational purposes; they also serve essential functions for the urban ecosystem, such as climate regulation (local solutions against heat hazards) and provide habitats and refuge for many rare and endangered plants and animals. Here, brownfields may, in particular, represent an important complement to existing open spaces if they are understood as part of the urban green infrastructure (EC 2013). Recultivated land complements the urban ecological inventory and also supports economic and socio-spatial revitalisation. Taking nature conservation on brownfields into consideration, and making use of them in terms of their local proximity to nature, can only be implemented with the involvement of socio-demographic demands for design and usability in densely populated urban areas. Beyond this, urban brownfields fail to be perceived as providers of cultural ecosystem services, due to their lack of aesthetics (different succession stages, and untended vegetation).

The controlled development of urban brownfields into intensive or extensive landscapes has often been cited as a proactive strategy for dealing with brownfields (Rößler 2010). In addition, however, there are also unused areas that have no social interference. A few years after an area has become a brownfield, some kind of natural greening usually occurs. In the early 1990s, Kowarik (1993) introduced the concept of a specific *urban-industrial nature* to which urban brownfields belong. The natural succession areas in a city have, in particular, high nature conservational and ecological value, because they become important stepping stones for animals and plants. A problematic aspect here is that green fallow succession sites do not have administrative protection, because these potentials are not recognized in the urban planning process, with respect to their qualitative characteristics. Brownfields are recorded in the municipal registers, according to their original dedication, either as commercial or residential sites. Although awareness of the ecological function of urban brownfields exists at the local urban planners' level, permitting succession is, currently, definitely not a proactive greening strategy (Ahern 2007).

2.3 Multifunctional Land Use Options

Studies on the renaturation of brownfields show that legal planning instruments exist that are relevant when considering nature conservation and the respective planning aspects. The economic focus of city planning, however, makes such an implementation somewhat problematic (Rößler 2010; Kühnau et al. 2016). Decisions about how a specific urban premise can be used are often made on an individual

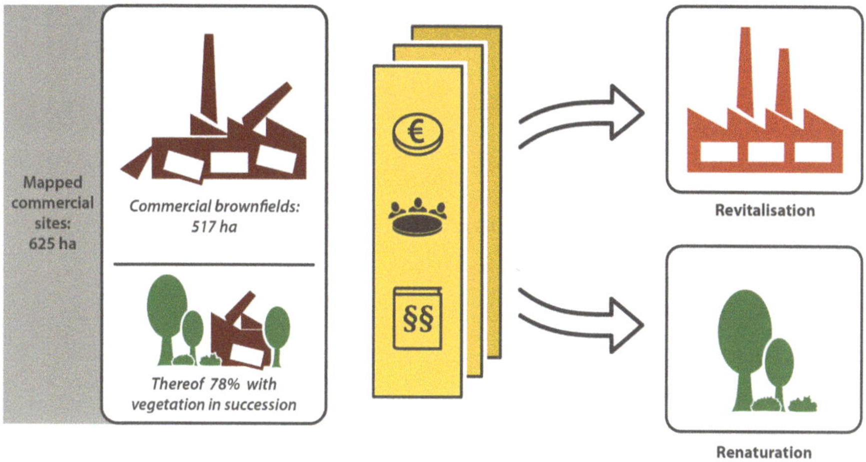

Fig. 1 Multiple factors influence priority setting for redevelopment of commercial brownfields; illustrated by figures from our own survey for the City of Leipzig (*left side* of graphics)

basis. Hence, appropriate guidelines that could help to decide if an area is more suitable for revitalisation or for renaturation are still not available. A good example is a German study, funded by the Research and Development Projects of the Federal Office for Nature Conservation (BfN), on inner urban development (i.e. in contrast to suburban development) and urban planning that deals with this topic (Kühnau et al. 2016). The objective was to develop a catalogue to support decision-making for ecologically important or nature conservation sites that would exclude structural re-use or the need to be adjusted accordingly. Thereby, specific multifunctional parameters were considered for each site with respect to spatial coverage, condition of the vegetation, accessibility, etc. The project results demonstrate that sustainable urban development has to be taken seriously and should not be biased towards economic prosperity. Social and, especially, ecological aspects have to be involved as well, to help in providing ecosystem services and, therefore, to improve the quality of life in cities (see Kabisch et al., "Local Residential Quality from an Interdisciplinary Perspective: Combining Individual Perception and Micrometeorological Factors", in this volume concerning local residential quality and also Fig. 1 for the factors that set priorities for redevelopment). To make well-balanced decisions, it is particularly important to have detailed qualitative information on brownfields, in addition to quantitatively obtained figures.

3 Potential Spaces for Regrowing Cities – The Case of Leipzig, Germany

In recent years, the City of Leipzig has undergone a drastic change and has developed from a shrinking city into a regrowing city (see Haase et al., "From Shrinkage to Regrowth: The Nexus Between Urban Dynamics, Land Use Change and Ecosystem Service Provision", in this volume dealing with urban dynamics). Although brownfields had existed for a long time, the amount of derelict land increased rapidly after reunification in Germany, and currently represents a large stock of land reserves. The majority of brownfields is located in industrial sites, along railway tracks, in former military zones, and in vacant residential housing. As in many formerly industrialised areas worldwide, it was common in all eastern German cities that many of their brownfields were contaminated.

Since the mid-1990s, the City of Leipzig has implemented an integrated urban development strategy (STEP). The problem of brownfields became an important area of action during the debate on urban redevelopment ("Stadtumbau") at the end of the 1990s. Tackling brownfields became inevitable; their monitoring and new instruments (e.g., interim use, permanent green spaces) were established. In the early phase of this urban redevelopment, Leipzig started to improve the quality and quantity of its public spaces by setting up its "Guidelines for Urban Renewal". These expressed the central objective of "more green, less density," which, amongst other issues, requires a higher quantity and quality of public parks, as well as alternative usage of open spaces (Stadt Leipzig 2009). Such alternative uses included interim use with short- to medium-term afforestation of brownfields. Around the turn of the millennium, the city started to support the temporary use of brownfields by citizens. In 2005, about 14 ha of new green and open spaces could be developed that covered a wide spectrum (e.g., green areas, pocket parks, urban gardens; Muschak et al. 2009). Existing brownfield sites were initially developed, primarily to establish gardens, and thus pursuing the target of improving the quality of life in the affected neighbourhoods. As a result, permanent green spaces were established and, additionally, a variety of green spaces with temporary uses were created. Although this new way of dealing with brownfields has also been applied in other cities (Burkhardt et al. 2008, p. 105ff), the case of Leipzig has been acknowledged as a model for this pattern of re-used brownfields nationwide (Rink and Arndt 2016).

In 1999, the City of Leipzig first adopted a development plan for commercial construction areas, with an update in the urban development concept a decade later, in 2009. Setting the priority on inner urban development, namely its densification process, and against external development which favours further building activities in the outskirts, has led to an upgrading of brownfields as potential spaces for redevelopment (Ferber et al. 2010). Therefore, the aim of managing and redeveloping commercial and industrial brownfields focussed on revitalising the industrial sector. Furthermore, a differentiated supply of space for various economic policy objectives was created. One of these objectives was to focus, in particular, on a demand-driven development of re-usable land. The minimum size of an individual commercial site was defined to be at least three hectares, because this size was considered to be the smallest that could have an economic influence on the development of adjacent sites.

About a decade ago, the process of shrinkage in Leipzig came to a halt and, ever since, the city has reversed towards reurbanisation, thus making Leipzig the fastest growing city in Germany, at present. This trend raises new challenges for the use of limited space, resulting in decreasing amounts of brownfields and open spaces. Due to its extremely fast growth and its high expectations for sustainable environmental development, the City of Leipzig faces the challenge of balancing revitalisation and renaturation of brownfields. Both types offer various advantages for the further growth of the city. As mentioned above, economic reasons drive many municipalities to revitalise brownfields into commercial sites. Leipzig represents an example of where this option has been made feasible through a city development plan (Stadtentwicklungsplan der Stadt Leipzig – STEP [urban development plan of the City of Leipzig]). Because not all potential sites are suitable, information about the sites' conditions is mandatory and brownfield monitoring assures its updating (Stadt Leipzig 2009; Corebro Project 2016). On the higher hierarchy level, this development plan is part of the urban development concept (the so-called Stadtentwicklungskonzept – SEKo), which also attributes economic aims to certain districts.

In this regard, the guidelines of Leipzig's Urban Development Plan list a "strategic area precaution". Accordingly, commercial sites are allocated as required by enterprises or provided flexibly and prepared for any future use. Urban planning needs a quantitatively and qualitatively differentiated set of tools, in order to realize the greatest possible range of demands (Stadt Leipzig 2012). For economic development it is important to gather information on the quantitative and qualitative structure of land reserves. Therefore, the City of Leipzig has monitored brownfields within its administrative boundaries since 1998 to guarantee their sustainable development. Additionally, a project group that deals with the recording and analysis of "**co**mmercial and **re**sidential **bro**wnfields" (Corebro Project 2016) was established at the Helmholtz Centre for Environmental Research – UFZ GmbH. During the last field mapping in 2012, 1014 commercial sites with a total area of 6.25 km^2 have been evaluated (Fig. 1). Of these sites 783 (5.17 km^2), have been considered as brownfields, and 191 sites were redeveloped between 2007 and 2012. Few remained unknown, due to their inaccessibility during field work.

Apart from its location and degree of integration into the public infrastructure, the size of a brownfield is a significant factor for its subsequent use. In this respect, Leipzig faces a major challenge, because most sites do not meet the prerequisites of three hectares as a threshold for direct redevelopment. Instead, a large share (80%) of these sites covers an area of less than one hectare, which is simply too small for revitalisation, as such. Figure 2 demonstrates the large number of smaller commercial brownfields in the urban area for which it has become a prerequisite, for urban planners, to consolidate neighbouring sites into a larger connected area before they can be offered for further use. Such a step initiates a sequence of procedures that forces planners to negotiate with the owners of adjacent sites to cooperate, to ensure that they fulfil all the clean-up requirements and meet the request for future development of the newly combined area that then has the appropriate shape and size.

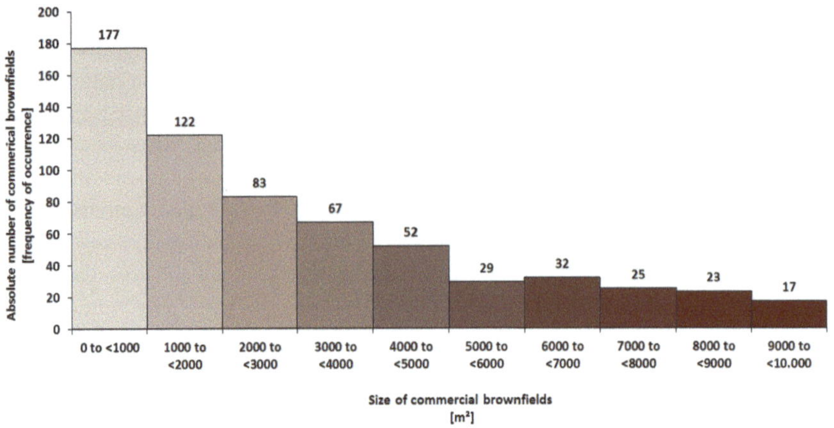

Fig. 2 Absolute number and size of commercial brownfields in Leipzig

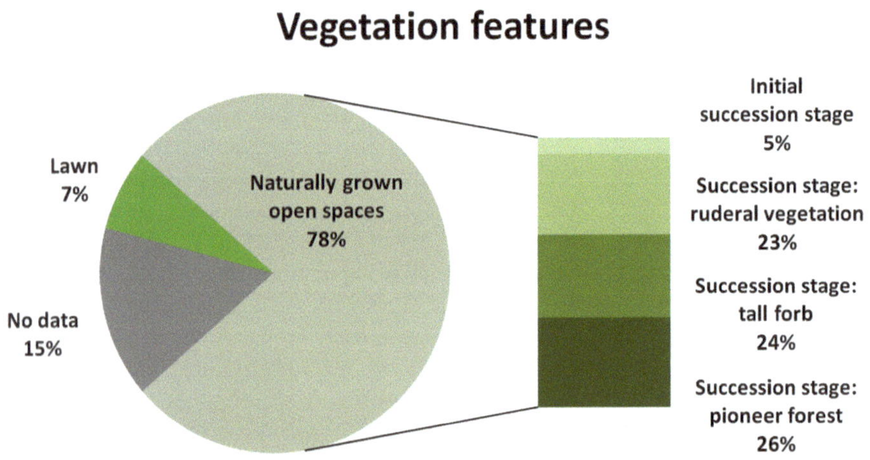

Fig. 3 Composition of vegetation features in urban brownfields in Leipzig

Leipzig has a long history of brownfields with extensive vegetation cover; thereof, 78% are naturally grown open spaces with vegetation in succession, especially in advanced succession stages. Half of the sites have a vegetation cover of at least 75%, classified as tall forb (24%) and pioneer forest (26%) (Fig. 3).

Therefore renaturation is an option to retain the existing vegetation in these succession stages as a basis for new green spaces. Fostering a green network in the city implies counteracting climate change issues and providing positive environmental well-being for citizens. Especially those sites that show advanced succession stages are also of high ecological value. In addition, certain linear brownfields might even be appropriate for traffic use by cyclists and pedestrians, which also improves the quality of life for citizens. To jump at the opportunity and profit from their share of vegetation and biodiversity, the quality of living in neighbourhoods can thus be enhanced (Lucius et al. 2011) and resilience against natural hazards such as summer flash floods or heat waves can be reinforced.

Positive aspects of brownfield renaturation are improved ecological and social conditions in neighbourhoods, and these are targeted in projects such as afforestation of an urban brownfield site (Rink and Arndt 2016). Concerning the experience with this project, one must keep in mind that residents might see a brownfield site as wasteland and, thus, problematic because of its uselessness and non-aesthetic appearance, especially because advances in succession stages and afforestation take time.

4 Leeway in Decision-Making

The City of Leipzig stands for highly dynamic urban processes, previous shrinkage and current, intensive regrowth. Communal brownfield development strategically aims at a qualitative inner urban development to use these existing spatial potentials for integrated urban development with respect to ecological, social and economic aspects. "Urban development is a joint venture that is partly governed by the public authorities. Civic society – e.g. members of associations, private foundations, dedicated individuals, entrepreneurs, property owners and their tenants – play an equally important role. We also work with neighbourhood management offices and civic forums, various study and action groups, district centres and organizers of communication platforms, all of whom are needed to complete a common task." (Stadt Leipzig 2012, p. 10). Brownfields are no longer merely understood as an environmental burden but rather as a valuable and non-renewable land resource. The current pressure on land and its potential supply for such a growing city must be governed cautiously to preserve sustainable land use. The offset of the newly available land to be re-used thus provides rewards for the obstacles to be overcome: its location within the city, accessibility, public infrastructure. Densification processes and land recycling are considered to be solutions for a resource-efficient land management that is a prerequisite for sustainable urban development (Stadt Leipzig 2016a; EEA 2011). In this context, the city council takes the question "how can a dual inner development be successful?" as a guideline for its planning strategies to fulfil its sustainability goals.

Revitalisation of brownfield sites has become an important instrument for intensifying building activities in such neighbourhoods and to thus improve previously deprived locations (Stadt Leipzig 2016b; EEA 2015, p. 44). When rethinking urban planning strategies for a city to improve urban design, opportunities for community initiatives for innovative redevelopment must also be provided. Whether a brownfield is revitalised or restored has to be decided individually at the local level, and this also depends on the site's location in focal planning spaces. In this respect, setting priorities to restore some derelict land produces social benefits such as recreation and leisure, balancing out the various needs of citizens, and supports precautious climate and environmental policies (Stadt Leipzig 2016b). With successful land-use management, newly created sites can serve to regenerate the central urban area and foster a resource-efficient handling of brownfields, from an urban planning perspective, in the face of a rapidly regrowing city that aims at diminishing land consumption towards sustainable urban development. Therefore, the City of Leipzig is exemplary for concerned and responsible strategic urban planning; other city councils should also keep both options in mind, if they intend to support smart and sustainable growth of cities.

Acknowledgement We very much appreciate the intensive collaboration with the City Council of Leipzig, Department for Urban Development and Construction, City Planning Office, Division for Urban Development Planning. Under their guidance, the mapping of commercial brownfields was carried out and funded under the project "Nachhaltiges Flächenmanagement, Baustein Flächenmonitoring". Being much obliged to Jan Richert, who refined our understanding of transdisciplinary aspects and continuously discussed various priorities between researchers and planners with us, we want to express our gratitude to him and for his engagement.

References

Ahern J (2007) Green infrastructure for cities: the spatial dimension. In: Novotny V, Brown P (eds) Cities of the future towards integrated sustainable water and landscape management. IWA Publishing, London, pp 267–283

Arndt T, Werner P (2015) Naturschutz und Landschaftspflege in der integrierten Stadtentwicklung. Argumente, Positionen, Hintergründe. Bundesamt für Naturschutz (BfN). https://www.bfn.de/fileadmin/BfN/siedlung/Dokumente/NuL_in_der_integrierten_Stadtentwicklung_11_2015.pdf. Accessed 28 July 2016

Bartke S, Martinát S, Klusácek P, Pizzol L, Alexandrescu F, Frantál B et al (2016) Targeted selection of brownfields from portfolios for sustainable regeneration: user experiences from fice cases testing the Timbre Brownfield prioritization tool. J Environ Manag 184:94–107. doi:10.1016/j.jenvman.2016.07.037

BBSR–BundesinstitutfürBau-,Stadt-undRaumforschung(ed)(2013)Innenentwicklungspotenziale in Deutschland – Ergebnisse einer bundesweiten Umfrage und Möglichkeiten einer automatisierten Abschätzung. BBSR, Bonn

BMU – Bundesministerium für Umwelt, Naturschutz und Reaktorsicherheit (2007) Nationale Strategie zur biologischen Vielfalt. BMU, Berlin

Budinger A, Gruehn D (2012) Bedeutung von Freiräumen und Grünflächen für den Wert von Grundstücken und Immobilien in der Stadt Bonn. LLP, Dortmund

Burkhardt I, Dietrich R, Hoffmann H, Leschner J, Lohmann K, Schoder F, Schulz A (2008) Urbane Wälder, Abschlussbericht zur Voruntersuchung für das Erprobungs- und Entwicklungsvorhaben "Ökologische Stadterneuerung durch Anlage urbaner Waldflächen auf innerstädtischen Flächen im Nutzungswandel – ein Beitrag zur Stadtentwicklung". Bundesamt für Naturschutz, Bonn-Bad Godesberg

Corebro Project (2016) Monitoring of commercial and residential brownfields in the City of Leipzig (corebro). Potentials to renaturalise and to revitalize urban brownfields. http://www.ufz.de/index.php?en=40466. Accessed 11 Aug 2016

De Sousa CA (2008) Brownfields redevelopment and the quest for sustainability. Elsevier, Amsterdam

Die Bundesregierung (2002) Perspektiven für Deutschland. Unsere Strategie für eine nachhaltige Entwicklung. Berlin, Bundesregierung

EC – European Commission (2013) Grüne Infrastruktur (GI): Aufwertung des europäischen Naturkapitals. KOM, Brussels, p 249
EEA – European Environmental Agency (2011) Green infrastructure and territorial cohesion. The concept of green infrastructure and its integration into policies using monitoring systems. Technical report, No. 18/2011. EEA, Copenhagen
EEA – European Environmental Agency (2015) Urban sustainability issues – What is a resource-efficient city?. Technical report, No. 23/2015. EEA, Copenhagen
Ferber U, Grimski D, Glöckner S, Dosch F (2010) Stadtbrachenpotenziale: Von Leuchttürmen und Patchwork. Informationen zur Raumentwicklung 1:1–11
Hansen R, Heidebach M, Kuchler F, Pauleit S (2012) Brachflächen im Spannungsfeld zwischen Naturschutz und (baulicher) Wiedernutzung. Bundesamt für Naturschutz, Bonn-Bad Godesberg
Keil A (2005) Use and perception of post-industrial urban landscapes in the Ruhr. In: Kowarik I, Körner S (eds) Wild urban woodlands: new perspectives for urban forestry. Springer, Berlin, pp 117–130
Kowarik I (1993) Stadtbrachen als Niemandsland, Naturschutzgebiete oder Gartenkunstwerke der Zukunft? In: Wittig R, Zucchi H (eds) Städtische Brachflächen und ihre Bedeutung aus der Sicht von Ökologie, Umwelterziehung und Planung. Natur & Wissenschaft Hieronimus & Schmidt, Solingen, pp 3–24
Kühnau C, Böhme C, Bunzel A, Böhm J, Reinke M (2016) Von der Theorie zur Umsetzung: Stadtnatur und doppelte Innenentwicklung. Empirische Ergebnisse einer Befragung deutscher Großstädte und Handlungsempfehlungen. Natur und Landschaft 91(7):329–335
Lafortezza RC, Sanesi G, Brown RD (2008) Visual preference and ecological assessments for designed alternative brownfield rehabilitations. J Environ Manag 89(3):257–269
Lucius I, Dan R, Caratas D, Mey F, Steinert J, Torkler P (2011) Green infrastructure. Sustainable investments for the benefit of both people and nature. http://www.surf-nature.eu/uploads/media/Thematic_Booklet_Green_Infrastructure.pdf. Accessed 25 July 2016
Martinat S, Dvorak P, Frantal B, Klusacek P, Kunc J, Navratil J, Osman R, Tureckova K, Reed M (2016) Sustainable urban development in a city affected by heavy industry and mining? Case study of brownfields in Karvina, Czech Republic. J Clean Prod 118:78–87
Mathey J, Rink D (2013) Urban redevelopment and quality of open spaces. In: Loftness V, Haase D (eds) Sustainable built environments. Springer, New York, pp 719–732
Muschak C, Weiland U, Banzhaf E (2009) Brachflächen in Stadtentwicklung und kommunalen Planungen am Beispiel der Städte Leipzig und Stuttgart. In: UFZ-Bericht 02/2009. UFZ, Leipzig
Norrman J, Volchko Y, Hooimeijer F, Maring L, Kain J-H, Bardos P et al (2016) Integration of the subsurface and the surface sectors for a more holistic approach for sustainable redevelopment of urban brownfields. Sci Total Environ 563-564:879–889. doi:10.1016/j.scitotenv.2016.02.097
Presse- und Informationsamt der Bundesregierung (2016) Deutsche Nachhaltigkeitsstrategie. Neuauflage 2016. Entwurf https://www.bundesregierung.de/Webs/Breg/DE/Themen/Nachhaltigkeitsstrategie/1-die-nationale-nachhaltigkeitsstrategie/nachhaltigkeitsstrategie/node.html. Accessed 25 July 2016
Rink D, Arndt T (2016) Investigating perception of green structure configuration for afforestation in urban brownfield development by visual methods – a case study in Leipzig, Germany. Urban Forestry & Urban Greening 15:65–74. doi:10.1016/j.ufug.2015.11.010
Rink D, Banzhaf E (2011) Flächeninanspruchnahme als Umweltproblem. In: Gross M (ed) Handbuch Umweltsoziologie. Springer Fachmedien, Wiesbaden, pp 445–463
Rößler S (2010) Freiräume in schrumpfenden Städten: Chancen und Grenzen der Freiraumplanung im Stadtumbau. Rhombos-Verlag, Berlin
Stadt Leipzig (2009) Stadtentwicklungsplan Zentren 2009, Blaue Reihe – Beiträge zur Stadtentwicklung 49. City of Leipzig, Leipzig
Stadt Leipzig (2012) Leipzig – Integrierte Stadtentwicklung. 5 Jahre Leipzig-Charta/Leipzig – Integrated Urban Development. 5 Years Leipzig Charter, Blaue Reihe – Beiträge zur Stadtentwicklung 54. In German and English. City of Leipzig, Leipzig

Stadt Leipzig (2016a) Integriertes Stadtentwicklungskonzept „Leipzig 2030". Stadt Leipzig, Bauen und Wohnen. http://www.leipzig.de/bauen-und-wohnen/stadtentwicklung/stadtentwicklungskonzept-insek/. Accessed 11 Aug 2016

Stadt Leipzig (2016b) Der Fachteil Brachen. Stadt Leipzig, Bauen und Wohnen. http://www.leipzig.de/bauen-und-wohnen/stadtentwicklung/stadtentwicklungskonzept-insek/stadtentwicklungskonzept-seko/die-fachkonzepte-des-stadtentwicklungskonzepts/fachteil-brachen/. Accessed 11 Aug 2016

Part III
Quality of Life and Ecosystem Services

Outline

Sonja Knapp and Florian Koch

Sustaining the urban quality of life is basic to human wellbeing and, thus, a political goal. Whilst the quality of life has increased in many areas and for many social groups, environmental burdens remain and social inequalities also persist. Therefore, quality of life, from a social-environmental perspective, needs to be considered in urban transformations towards sustainability, because it can help to legitimate transformation processes.

This part of the book focusses on the quality of life from an environmental perspective. Its main topics are the availability of and access to green infrastructure and urban ecosystem services, the connection among ecosystem services and biodiversity, and inequalities in the spatial distribution of environmental burdens. **Kabisch et al.** zoom into a local residential area in the City of Leipzig, Germany. They assess how the citizens of a large housing estate perceive their local residential quality and which factors (such as protection from heat and the availability of green infrastructure) affect these perceptions. By combining data from household surveys, environmental data, and simulations of micrometeorological conditions, they have developed a new, interdisciplinary, methodological approach for understanding the complex relationships of human-environmental interactions. **Kindler et al.** emphasize that environmental justice is still lacking with respect to airborne outdoor exposure of urban dwellers to air pollution. Across the City of Berlin, Germany, they show that exposure is higher in inner urban areas with lower social status. This helps to prioritize actions for increasing the quality of life in a large and highly heterogeneous city. Similarly, **Banzhaf et al.** focus on the configuration of urban green infrastructure – and thus residential access to those ecosystem services that green infrastructure can provide. Their area of research is the growing Latin American urban area of Santiago the Chile, for which they consider targeted spatial analysis on multiple scales. By identifying large inequalities in the amount, distribution, and vegetation cover of green infrastructure in three extremely different municipalities, they emphasize the importance of public green spaces for the quality of life, especially for those urban dwellers with lower economic status. The spatial characteristics of green infrastructure are underpinned by its perceptions by various socio-economic groups, thus providing differentiated answers to the question of what matters in planning green infrastructure towards sustainable urban transformations. In parallel, **Knapp et al.** argue that urban planning should target the multiple functions of green infrastructure and not only selected functions, when designing and managing green spaces. In particular, they show that various goals of sustainable urban transformations – the protection of biodiversity and the delivery of ecosystem services – do not always go hand-in-hand. Here, again, a "one size fits all" solution does not apply, and case-by-case approaches are needed.

Results from these studies should also be taken into account when aiming at an increased resilience and resource efficiency of urban landscapes: resilience with respect to, e.g., floods (Part IV in this volume) might increase safety but might, at the same time, affect another aspect of quality of life, such as access to rivers. Generally speaking, land is a restricted resource and different land-use options compete for available urban areas (see also Part II in this volume).

The implementation of urban transformations through political strategies should reflect these results. Rather than being grounded on general assumptions about the relationships between ecosystem services and the quality of life, urban transformations should be based on place-specific approaches that also consider the existence of possible trade-offs and limiting factors.

Local Residential Quality from an Interdisciplinary Perspective: Combining Individual Perception and Micrometeorological Factors

Sigrun Kabisch, Maximilian Ueberham, Uwe Schlink, Daniel Hertel, and Abdelrhman Mohamdeen

1 Introduction

Local residential quality (LRQ) is one particular expression of quality of life. The quality of life concept reflects the extent to which important needs and values of individuals are satisfied under certain conditions (Steg and Gifford 2005). With our contribution, we specify quality of life as one of the core concepts of urban transformations (see Kabisch et al., "Introduction: Urban Transformations – Sustainable Urban Development Through Resource Efficiency, Quality of Life, and Resilience", in this volume, Introduction). We focus our attention on the local level by applying an interdisciplinary methodological approach for assessing LRQ. This innovative approach enriches the scientific debate about residential satisfaction (e.g., Dekker et al. 2011) and place attachment (e.g., Scannell and Gifford 2010). It pays particular attention to the interrelation between subjective perception and environmental measurements of residential characteristics as well as to simulations of the local environment.

LRQ is determined by both objective features of the surroundings close to the housing location and their subjective perception by residents. The objective features encompass the daily required housing conditions, infrastructure facilities, and environmental conditions, as well as neighborhood networks. The subjective perception includes the relationship between residential expectations and the residential conditions, manifested in, e.g., housing satisfaction, place attachment, and well-being. The local scale refers to the apartment, the residential building, and the

S. Kabisch (✉) • M. Ueberham • U. Schlink • D. Hertel • A. Mohamdeen
Department of Urban and Environmental Sociology, Helmholtz Centre for Environmental Research-UFZ, Leipzig, Germany
e-mail: sigrun.kabisch@ufz.de; maximilian.ueberham@ufz.de; uwe.schlink@ufz.de; daniel.hertel@ufz.de; mahmoud.mohamdeen@ufz.de

S. Kabisch et al. (eds.), *Urban Transformations*, Future City 10,
DOI 10.1007/978-3-319-59324-1_13

nearby surrounding housing, preferably at the neighborhood scale. The apartment provides a private space that should not be disturbed by noise, odor, or heat stress. The building structure and its state of repair influence well-being in the apartment, e.g. through insulation quality or provision of elevators and balconies. The housing's surroundings foster LRQ in terms of availability of green spaces, fresh air, and quietness. This mixture of a variety of objective features and their subjective perception by the residents is decisive for LRQ.

In this paper, we propose an approach for studying LRQ in relation to environmental conditions that is based on our expertise in urban environmental science. We provide evidence that studying a vast range of features rather than single factors delivers a more complete understanding of LRQ.

We account for this by using results from a household survey, in combination with environmental data and simulations of a micrometeorological model. New options for climate modeling offer novel approaches to combining social and environmental data. We test this in order to obtain additional insights into assessing LRQ. For this purpose, we selected survey results that reflect the individual perceptions of residents about the housing conditions in their local surroundings. This survey was carried out in a large housing estate, Leipzig-Grünau, which has approx. 42,000 inhabitants (Kabisch et al. 2016). We combine these survey results with land-use data such as land-use metrics, vegetation density, building height, and wind intensity to simulate micrometeorological conditions and to explain the heat exposure of the residences. The originality of the presented approach lies in a synopsis of social science and natural science approaches, as well as modeling, to provide evidence about the diversity of factors that influence LRQ. This contribution sheds new light on the complex system of environmental features, housing conditions, and residents' perceptions, with particular attention on the neighborhood scale. Specifically, we pursue the following research hypotheses:

- The LRQ is influenced by a variety of factors such as building structure, social conditions, and environmental features.
- There is strong interrelation between neighborhood-specific housing conditions and perceived residential quality or housing satisfaction.
- A synopsis of data based on surveys, measurements, and modeling can deepen the understanding of the interplay of impacts on LRQ.

In the next sections, the factors that influence LRQ will be elaborated. The social science approach for measuring the housing satisfaction will be explained, as well as atmospheric conditions, human thermal comfort, and the influence of green space provision. Our case study area encompasses the large housing estate Leipzig-Grünau. At the neighborhood scale, we focus on the two districts Grünau-East and Grünau-North, which differ in their building density, as well as green space availability. In the last part, we reflect on the results and benefits of this interdisciplinary approach, as well as its limitations, and present some recommendations for further research.

2 Conceptual Framework for Local Residential Quality (LRQ)

Statistical institutes in Europe have launched a variety of initiatives to measure quality of life in a comprehensive way. All approaches have in common that they pursue a multidimensional concept that includes issues ranging from material living conditions to personal and social well-being, as well as environmental circumstances (Garcia Diez 2015, see Kindler et al., "Socio-Spatial Distribution of Airborne Outdoor Exposures – An Indicator for Environmental Quality, Quality of Life, and Environmental Justice: The Case Study of Berlin", in this volume). Therefore, subjective indicators such as the individual perception of the environment are at least equally important as more classic objective indicators such as poverty or crime rates (Marans 2015). Mainly two approaches are used. The first is through applying a set of indicators from aggregated data, using official sources of statistical data such as censuses. The second is related to the application of sample surveys that measure people's subjective assessments.

In the recent Eurostat-report (Eurostat 2015), nine overarching sets of indicators are proposed, along with several relevant subtopics. The report emphasizes the combination of social, economic, and environmental factors. For our understanding of LRQ, especially relevant indicators in this publication are: housing conditions and satisfaction as well as the natural and living environments. Several investigations also reveal the influence of environmental quality on the quality of life of humans (Banzhaf et al. 2014; Billota and Evans 2012). Especially against the background of climate change, it is therefore crucial to monitor the environment on a local scale, to identify the impact of climate-related stressors that can disturb the LRQ. Factors of major importance are acoustic noise, ambient temperature, and air quality (Rivera et al. 2007; Kjellstrom and McMichael 2013; Weber et al. 2014a).

However, these indicators are often designed to provide aggregate measurements (such as total exposures to pollutants over time) and their usefulness for assessing individual LRQ perception is rather limited. Hence, we need a comprehensive assessment of the immediate exposure of people living in a specific environment, e.g., neighborhoods, considering features of surrounding green areas, buildings, infrastructure, housing features, sociodemographic factors, and distinct environmental stressors (Fig. 1). This includes, besides measurements of quantitative data, the qualitative self-reporting of the subjectively perceived quality of the environment by the residents (Bonnes et al. 2007; Fornara et al. 2010).

In the context of LRQ, it is important to emphasize that, subjective perception, in particular, is a matter of scale and place. Based on the framework of Marans (2015), it is suggested that satisfaction with living conditions can be evaluated on multiple levels of scale. Starting from the city perspective, we want to zoom into an estate and place our focus on two neighborhoods (districts), to look at particular differences.

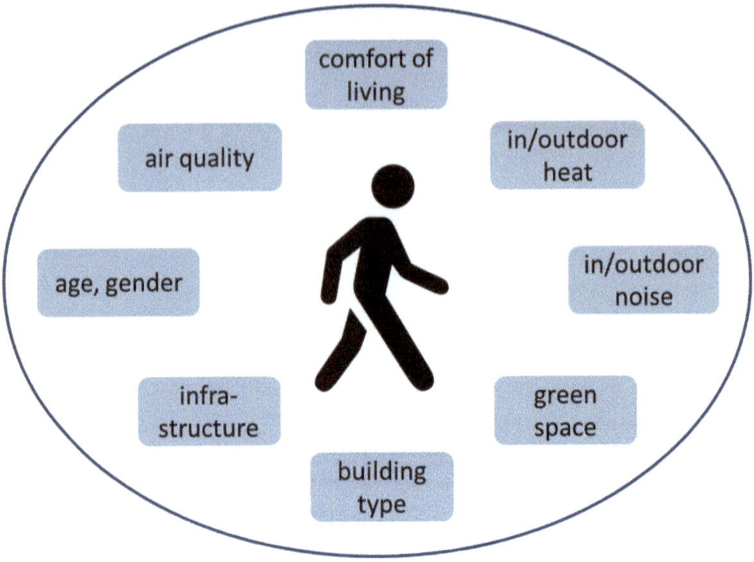

Fig. 1 Factors characterizing the LRQ in an urban neighborhood (Source: Authors' adaptation of Pantisano et al. (2014) and VDI (1998))

2.1 *Social Science Perspective to Grasp the Individual Perception*

To assess LRQ as an expression of urban quality of life, quantitative measurements as well as subjective evaluations based on individual perceptions are required. Many approaches to grasping the quality of life concept deal with income and provision and/or health, including mortality, to quantify the particular values (e.g., Knecht 2010). Other experts in this field prioritize happiness (e.g., Helliwell et al. 2013) or satisfaction with the living conditions, and thus have a more qualitative perspective (e.g., Marans and Stimson 2011).

To obtain a comprehensive picture of the perceived LRQ, both perspectives need to be considered and integrated. At the local scale in a residential environment, physical, social, environmental and financial conditions influence residential quality. Statistical data are used to describe the population number and density in a given district. Furthermore, data concerning housing market offers, including rents and housing prices, social and technical infrastructure provision, green spaces and their uses, as well as leisure time facilities, are available. In addition to the facts represented by these socio-demographic and land-use data, the level of residential satisfaction, place attachment, or well-being in a certain housing district demand separate data acquisition techniques, such as household surveys. In our survey, a restricted number of questions to assess the LRQ was used, e.g., "Do you feel comfortable in the estate?"; "What are the advantages of your district?"; "Do you intend to move out of your apartment during the next two years?". However, these items were not sufficient to reveal deeper insights into the local housing conditions and their appropriateness for both different residents and households with respect to the impact of environmental conditions.

To differentiate between expectations and perceptions, a variety of indicators, measurements, and statistical data are needed. A realistic approach considering advantages and disadvantages in detail on several scales from the perspective of different residential groups allows the identification of cumulative benefits and synergies, as well as of obstacles, conflicts, or trade-offs. These insights enable specific decisions about targeted and prioritized improvements.

The LRQ, as part of the general quality of life, focusses on places where people live and reside, where they have their physical home embedded in a particular social and natural environment, and the feeling of being at home. Thus, subjective dimensions have to be considered. Whereas objective dimensions can be measured in terms of financial values and costs, subjective dimensions reflect people's feelings about objective socio-environmental conditions and people's behaviors. Subjective dimensions are investigated primarily through survey research and interviews, together with corresponding measures of objective conditions.

The use of this research design yields evidence about the access to resources and the distribution of risks among the residents. It reveals and explains socio-spatial patterns of LRQ. Existing socio-spatial differences in cities, between districts, as well as between neighborhoods, can be discovered.

In our contribution, we highlight the importance of environmental issues for LRQ in relation to social conditions and individual perceptions of the immediate housing conditions. The urban environment influences human physical, social, and mental well-being (Lawrence 2011). For the measurement of local residential quality, indicators such as air quality, pollution, thermal comfort, distance, access to green spaces, recreation areas, public transport, shopping facilities, etc., are indispensable (Banzhaf et al. 2014). By using survey results, micrometeorological data and data about urban green areas, we can identify housing conditions and individual perceptions in a specific residential environment.

2.2 Micrometeorological Factors Characterizing Local Exposure

Urban atmospheric conditions (including air quality) have a direct impact on individuals and are essential elements of the LRQ. Airborne contaminants, such as particulate matter (e.g., PM10), as well as heat and air humidity, are often associated with adverse health effects. For example, heat stress impairs the quality of life of specific population groups (Großmann et al. 2012). Moreover, even the combination of anthropogenic aerosols with air humidity increases the risk of respiratory symptoms (Leitte et al. 2009).

For a comprehensive evaluation of urban air quality and thermal conditions, two specific factors have been defined (VDI 1998, 3787, Blatt 2). They are, firstly, the thermal factor. It comprises air temperature, air humidity, wind velocity, and short wave and long wave radiation, all of which have a thermo-physiological effect on citizens. The significance for health is associated with the close link between thermoregulation and circulatory regulation.

Secondly, the air quality factor involves all solid, liquid, and gaseous natural and anthropogenic air pollutants. Air pollution conditions depend on the emission sources and the transmission conditions, which are determined by atmospheric stratification (extreme cases: atmospheric inversion and smog formation), wind, precipitation, air humidity, and solar radiation.

The focus of the present study is on the thermal comfort, which is defined on the basis of the heat balance between the human body and the surrounding environment. Thermal comfort can be measured by the predicted mean vote (PMV) (Fanger 1972), the physiologically equivalent temperature (PET) (Höppe 1993), or the universal thermal comfort index (UTCI) (Fiala et al. 2001). Generally, such indices are a mixture of several atmospheric variables and some subjective or generalized assumptions that quantify thermal comfort by terms such as "comfort" and "discomfort". With this differentiation, it is possible to predict human reactions to thermal (weather) stress and physiological strain (Yan 2005). A mapping of the thermal comfort to urban regions is provided by the "Klima-Michel-Modell" (DWD 2016).

A particular urban phenomenon that occurs within large urban areas is the urban heat island effect (UHI) (Oke 1982). Nocturnal air temperature in downtown regions is elevated, in comparison to the rural surroundings of the city. In cities with a complex built-up area, an archipelago of nocturnal urban heat can arise (Schwarz et al. 2012). In the context of social disparities, it is possible to relate urban structure parameters, e.g., building density, with the urban heat island effect: The higher the building density, the higher the temperature (Romero et al. 2010). The temperature excess in cities depends on multiple factors but is predominantly associated with the number of inhabitants (Oke 1973), the extent, the land-use setting, as well as the urban structures (Oke 1987; Mahmood et al. 2014) and regional climate (Zhao et al. 2014).

Whilst high temperatures during the night impair the quality of sleep and recreation, dissatisfaction with the thermal comfort during the day is a stressor that superimposes on the basic level of stress perceived by various social and demographic population groups. This affects, in particular, people at work or on their way to carry out fixed obligations without any opportunity to avoid heat stress (Großmann et al. 2012). The day-time heat stress interferes with the conduct of daily life and demands subjective adaptation strategies and an involvement of adaptation measures in urban planning. For example, the latter can account for the use of vegetation (trees, bushes) or architectural arrangements (shadowing) that provide the much-needed shade (Mayer et al. 2008).

Periods of excessive urban heat are already occurring more frequently (Christidis et al. 2015). They will bring essential modifications and contributions to the complex, multifactorial personal exposure of urban inhabitants. Assessing the personal exposure, the mobility of individuals is the primary factor (Schlink et al. 2011). To measure mobility impacts in a real-world context, persons can carry a variety of environmental sensors and make micro-scale measurements over a certain time (Schlink et al. 2014).

2.3 Influences of Urban Landscape Features on Environmental Stressors

The urban landscape and land-use structure have significant impacts on exposure to environmental stressors. Several studies reveal an influence of urban green, especially on temperature, acoustic noise, and air quality (Irga et al. 2015; Hart and Sailor 2009). However, climate change and dense building structures in cities lead to rising temperatures, and to augmented frequencies and durations of temperature extremes, which adversely affect human well-being and residential quality (Martiello and Giacchi 2010; Franck et al. 2013). For example, in the city of Leipzig, the number of heat days (maximum day-time temperature at 2 m above ground level $\geq$30°C) is expected to double by 2060 (PIK 2010).

Depending on the surface and building material and its ability to accumulate heat, orographic conditions, air exchange, and cold-air corridors, some regions may suffer more from urban heating than others. Urban landscapes are very heterogeneous in terms of density and structure. In addition, indoor temperatures depend on those outdoors and are influenced by the type of urban structure, characterized by different housing densities, building types, and urban green spaces (White-Newsome et al. 2012; Franck et al. 2013).

In this context, urban green space is associated with the concept of urban ecosystem services as a decisive factor for LRQ. More specifically, it is possible to identify different vegetation patterns in residential areas by remote sensing technologies and geoinformatics and their impacts on ecosystem services (Banzhaf and Kollai 2015). For instance, Jaganmohan et al. (2015) found that the size and type of urban green space have an influence on the cooling effect for the adjacent surrounding.

Another approach to analyzing the influence of land-use on micrometeorological conditions is to use landscape metrics. Landscape metrics operationalize specific spatial characteristics of elements; e.g., land-cover, elevation, building types and heights. Numerous land-use and landscape studies have used landscape metrics to assess the impacts of form, patterns, and configurations of built and non-built land covers on ecological processes, bio-physical properties of the earth's surface, biodiversity (Schindler et al. 2013), the quality of residential areas (Cushman et al. 2012), and land-use change (Hassett et al. 2012). Weber et al. (2014a, b) demonstrated the usefulness of landscape metrics for forecasting noise, particle exposure and surface temperature during heat waves for different urban structure types.

In the following section, we introduce a methodological design that combines land-cover, socio-demographic and survey data with micrometeorological factors to assess LRQ for our case study area.

3 Methodological Design

3.1 *The Case Study: Leipzig and Its Large Housing Estate Grünau*

With about 560,000 inhabitants, Leipzig currently belongs to the ten largest German cities. During the last 6 years, the population number grew to around 60,000 and it is expected to continue. The municipality undertakes intensive efforts to maintain the high quality of environmental conditions.

One important instrument here is the recently published City Climate Report (*Stadtklimatische Untersuchungen*, Hoffmann and Behrens 2016). The report deals, in particular, with the heat island effect that indicates a temperature difference of up to 11 °C between the city center and the fringe. Furthermore, a questionnaire survey was carried out among Leipzig's population to reveal the state of knowledge about issues related to heat and climate change, their perception and respective adaptation measures (Stadt Leipzig 2014). The survey results show spatial differences of perceptions within the city districts, and the need to consider different spatial scales.

For our in-depth analysis, we selected Grünau, a large housing estate at the western fringe of the city covering about 10 km^2. The estate is divided into five districts. 42,000 inhabitants are currently living there, which encompasses 8% of the entire population of the city. This estate was erected from 1976 till 1989 by industrial panel construction of blocks with 5–16 floors. This leads to a residential density in some inner districts of up to 12,000 residents per km^2.

The estate is well equipped with public transport facilities, enabling a fast connection with the city center. Transit traffic is concentrated on two main roads; thus, within the estate, there is much car-free space. A large shopping center with a cinema, a public indoor swimming pool, two churches, and a number of cultural offers for different age groups are located in the estate. In addition, 24 schools and 18 kindergartens complete the facilities. Refurbishments and new construction of these facilities are in progress and support the improvement of the appearance of the estate.

Following a period of massive population shrinkage during the 1990s and 2000s, a smooth increase in population is currently occurring. Improved housing conditions and the marked population growth of the city also affects the large housing estate. The growing attractiveness of the estate is characterized not only by considerable investments in the housing stock, but also by the development of large open and green spaces including diverse playgrounds for children and paths that are very suitable for bicycling and walking in general. The proximity of larger parks and a lake with very high water quality is particularly important. However, a closer look into the estate reveals that there are differences in the amount and the type of green infrastructure.

3.2 *The Methods Used: Household Survey, Air Quality Measurements and Micrometeorological Simulations*

From the social science perspective, we used the results of the recent household survey “Living and Housing in Grünau”, carried out in May and June 2015. (Kabisch et al. 2016). The questionnaire contains core questions about the housing perceptions and valuations of the residents over time. A questionnaire was handed over at 945 addresses. Of these, 705 questionnaires could be recollected and analyzed, which corresponds to a response rate of 75%. With regard to the focus of this article on LRQ, we analyzed the following indicators in relation to different spatial scales such as the entire estate, the district and the apartment:

1. perception and valuation of: housing satisfaction and feeling comfortable, place attachment, indoor heat stress, indoor disturbance due to noise, acoustic isolation, comfort of living, outdoor air quality
2. level of consent to statements: green area maintenance, urban forest as a benefit
3. reference to: age groups, building heights, residential districts, apartment size

Our survey data were complemented by measured values of air quality (TEOM measurements of PM10) and meteorological parameters such as temperature (Pt-100), humidity (polymer film capacity) and wind speed (mechanical anemometer) from the municipal monitoring stations (LfULG 2016).

Applying the ENVI-met software, we simulated thermal conditions as a function of the urban and environmental structure. Details about the software can be found in the contribution by Welz et al., “Adapting Built-Up Areas to Climate Change: Assessment of Effects and Feasibility of Adaptation Measures on Heat Hazard”, in this volume. We selected 2 days (05–06 June 2015) during the survey period to compare the simulated micrometeorological data with the survey data. We utilized simulations from the second day at 2 p.m., taking advantage of the GFS (Global Forecast System) reanalysis data that confirm that this was a typical, warm day, without clouds. These typical summer days ensured that maximum temperature contrasts can occur between the two analyzed districts of Grünau-East and Grünau-North.

The proportion of tree cover was calculated in ArcGIS, based on a 3D–vegetation model of Leipzig (Banzhaf and Kollai 2015; EEA 2015; ATKIS 2013). Furthermore, the proportion of green area for each district was calculated by counting the grid cells with vegetation based on the ENVI-met area input file (www.envi-met.com).

4 Empirical Findings and Their Synopsis

4.1 *Survey Findings*

4.1.1 Estate Scale: Perceptions of Local Residential Quality

A comparison of the concentrations of particulate matter registered by the urban monitoring network reveals that the district of Leipzig-Grünau is less exposed to air pollution (PM10) than other sites in the city. Although all stations are located near a main road (and traffic is a major emission source), the surrounding space and green infrastructure are assumed to influence the concentration.

To evoke a general reflection about the residential quality, we use the answers to the question "Do you feel comfortable living in Grünau?" (Fig. 2). About two third of all respondents answer with "yes", 30% with "yes, but with restrictions", and only very few answer with "no". This is a very high value for the positive perspective of the estate expressed by its residents. The major reasons revolve around the accessibility of large and diverse green areas, the extensive retail and shopping facilities, the easily accessible traffic connections to the city center, and the calm housing environment. Reservations are mostly connected with features of the social milieu, linked with some security and cleanliness deficits.

It is obvious that the older the residents are, the higher the share of respondents confirming the positive perception of the estate. Younger residents note more reservations. This difference is statistically significant (Chi-Quadrat-Test, $p < 0.01$). Many of these older persons have lived in the estate for more than 20 years. It is their home, and their place attachment is quite strong. They experienced many investments in the estate over the time that brought about changes and improvements. With reference to developments over the last 5 years, 51% of all respondents,

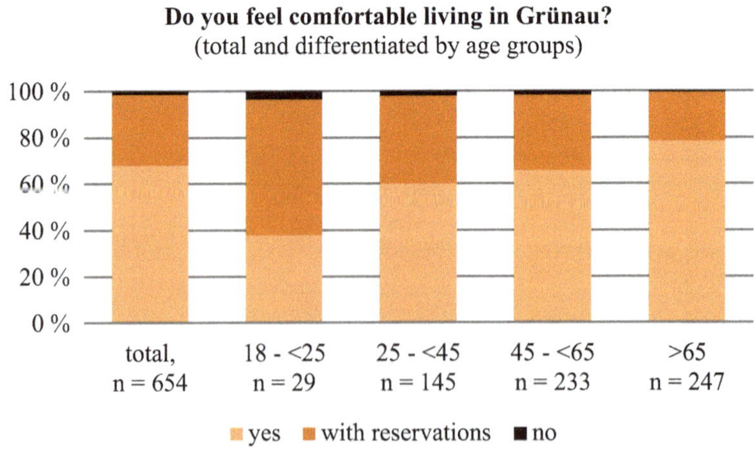

Fig. 2 Feeling comfortable in Grünau; classified in age-groups

recognized an improvement in the estate, only 10% noticed deterioration. 39% observed a stable development. In terms of age groups, the elderly (55%) recognized improvements more clearly than the younger ones (40%). All respondents strongly confirmed the statement "The older residents stick to Grünau" (x = 4.4, 1–5 scale, 58% chose the highest category 5).

To investigate features of environmental stress, we asked the residents how strongly they perceive heat during summer and noise in their apartment. Both mean values are relatively moderate: 3.1 and 3.0, respectively (Scale 1–5). There are some nuances concerning the distribution of the extreme values. Whereas 30% answer that they more or less completely did not notice any heat stress, 40% did not feel disturbed by noise. The latter is an important reason for the acknowledgement of the appreciated quiet housing environment. Whereas all respondents feel very comfortable in their apartment (68% yes, 30% yes with reservations), the satisfaction with particular features of the apartment differs markedly (Fig. 3).

Here, the acoustic isolation is reported to be problematic in comparison to the other features (mean value of 3.6). Approx. 80 respondents mentioned, in their own words, noise disturbance as a major disadvantage of their apartment. Although the apartment is, in general, quite highly ranked at 5.5, those features describing the construction such as heat isolation, state of refurbishment, as well as construction quality as summarizing feature had lower values. Thus, feeling comfortable in the apartment and appreciating the calm housing environment do not exclude dissatisfaction with particular aspects of the housing conditions.

At the estate scale, we realize a general positive perception of the residential quality. The beneficial characteristics prevail. In the following, we focus our attention to the district scale, to check for possible spatial differences.

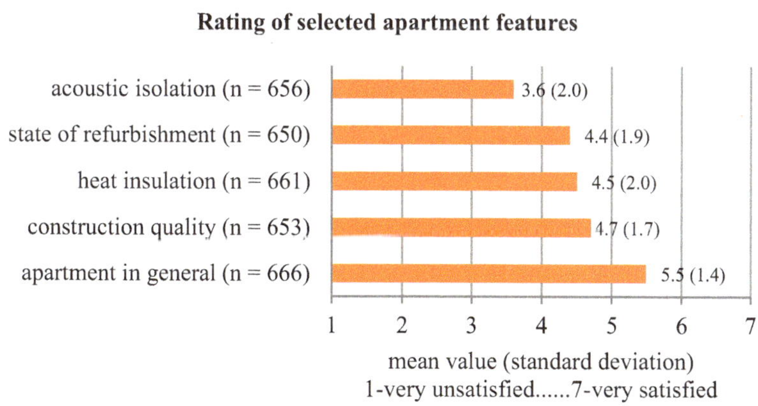

Fig. 3 Rating of selected apartment features

4.1.2 District Scale: Perceptions of Local Residential Quality

Looking at the more detailed district scale, we find an evaluation similar to the entire estate. More or less the same features concerning strengths and weaknesses of the districts are mentioned in general (Fig. 4). For example, of the residents who gave an answer to the open question of strengths and weaknesses, most of them consider supply opportunities as a strength, whilst only very few see this feature as a weakness, because of few offers in their immediate vicinity.

When moving the focus into the districts, we find differences that bring us to a more detailed explanation of LRQ at the neighborhood level. The two selected districts, Grünau-East and Grünau-North, are characterized by a similar area size and population number. The share of residents older than 65 years differs: in Grünau-East, 44% of the respondents are part of this age group; in Grünau-North, 36%. Grünau-East covers 0.95 km^2 and has 7400 residents; Grünau-North covers 1.05 km^2 and has 7400 residents. The differences between these two districts relate to the construction style and the green structure. Grünau-East encompasses the first construction sector, from the mid of the 1970s onwards. The building height is restricted to five stories. The block structure has been built with consideration and preservation of, the old tree population on the location. Furthermore an old park area was included in the district. Thus, this district was highly appreciated from the very beginning.

During the mid-1980s, the construction of the Grünau-North district started. Because of economic weaknesses, the industrial housing construction was changed in favor of a block height of six stories (without elevator) and to 11 stories. Furthermore, a densification of blocks was undertaken, to fulfil the state targets concerning the number of apartments. Green areas and trees had been planted and

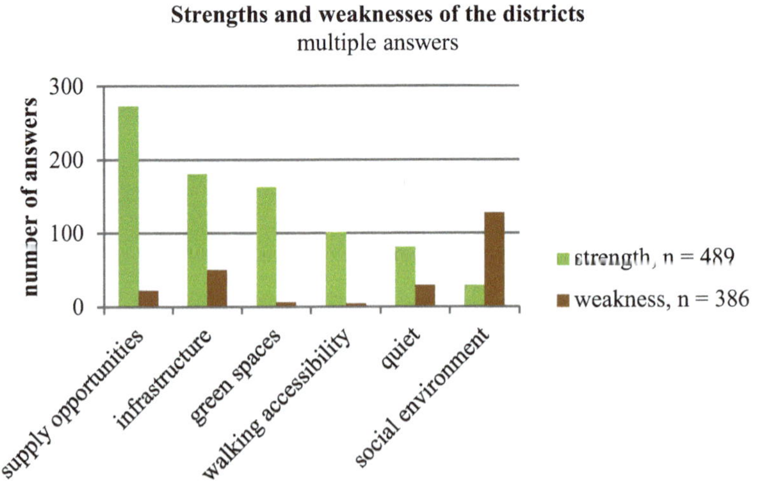

Fig. 4 Condensed answers to the open question: "What are the strengths, and what are the weaknesses of your district?"

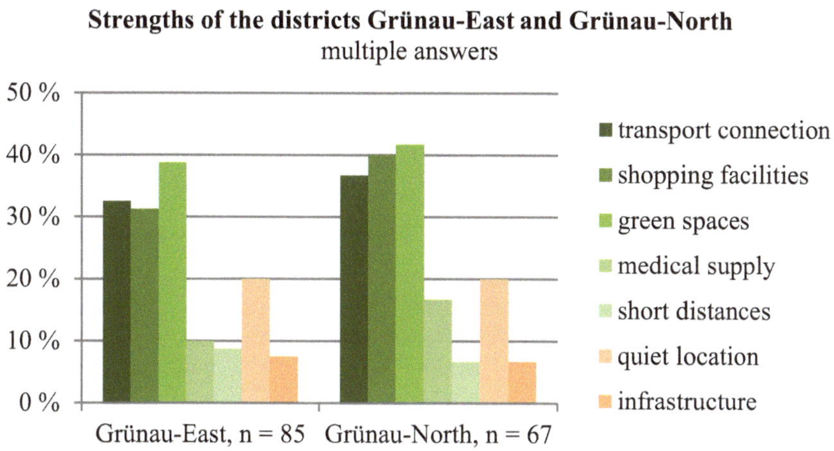

Fig. 5 Strengths of the districts Grünau-East and Grünau-North

developed. After 30 years, there is a lot of green space available, but no old tree population providing shade. In 2013 a new, low-budget project, called "urban forest", was implemented by the municipality. Following the demolition of a large block, an area of 0.055 km^2 was planted with small trees, to enhance the proportion of green space. Thus, in this district, the initial conditions were more difficult, compared to Grünau-East.

Nevertheless, the recent results from the survey show that the respondents from both districts acknowledge the availability of and the access to green spaces as a major strength (Fig. 5). Furthermore, in both districts, the quiet location is appreciated. This result of our study distinguishes both districts from the other three districts within the estate.

These characteristics are part of the aspects that define the residential quality. The results of the indicator "Do you feel comfortable in Grünau" show that 79% of the Grünau-East residents answered "yes", compared to 68% of the Grünau-North residents. This indicates differences in the neighborhood-specific perception. Concerning environmental stressors such as heat, noise, and air quality, we find distinct differences between the two districts in Grünau. The analysis of variance (ANOVA) reveals the most significant ($p < 0.05$) and strongest differences in the residents' perception of heat stress (mean difference 0.52) and noise pollution (mean difference 0.54) (Fig. 6).

The results of the comparison of the two districts demonstrate that residents in Grünau-East perceive less exposure to heat stress and noise pollution indoor and air pollution outdoors, compared to residents of Grünau-North. This mainly depends on the proportion of urban green space. Although the calculation of all green space in relation to the total district areas shows no marked difference, with 52% in Grünau-North and 55% in Grünau-East, it was found that the proportion of trees (over 5 meters high) is much higher in Grünau-East (34.4%) than in Grünau-North (20%). One reason for this is a large park area with a dense tree population and more

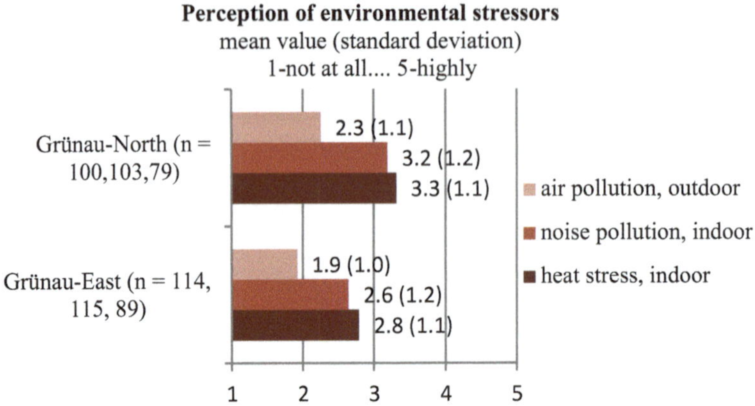

Fig. 6 Perception of environmental stressors according to the number of respondents

tree coverage in between the housing blocks. Our results confirm that trees can provide shadow for a cooler environment and also contribute to better air quality, as others studies show (Bowler et al. 2010, Gromke and Blocken 2015).

To discuss and compare the findings of the household survey in the context of the micrometeorological conditions, we modelled the thermal conditions on a representative summer day, which is described in the next section.

4.2 *ENVI-met Model Findings*

To assess local thermal conditions in the two districts, we extracted temperature maps representing the air temperature on 6 June 2015, 2 p.m., at 1.6 m above ground (Figs. 7 and 8) from ENVI-met simulations. Red and yellow (green and blue) contours indicate warm (cold) areas, respectively. Warmer regions are mainly associated with streets, a high percentage of impervious soils and open places (i.e., spaces with reduced shadowing), such as areas with open grass fields or parking facilities.

In Grünau-North (Fig. 7), the simulation shows the hottest places on the southern border of the district, where a main road and the tram lines are located. A second hotspot is situated on the eastern border, at a large parking area. The coolest parts of this district are mostly in the courtyards of the buildings and next to green spaces, which is caused, on the one hand, by shadowing effects of buildings and trees and, on the other hand by evaporative cooling of the vegetation.

In Grünau-East (Fig. 8), the built-up structures result in thermal patterns similar to Grünau-North. The hottest places are in the center, along the train lines, and at the northwest border. This part is characterized by streets and open places with less shadowing. In contrast to Grünau-North, blue contours indicating relatively low temperatures are more dominant in Grünau-East. This means that all temperature values are lowered to the range of 26.0–28.4 °C (Grünau-North: 26.6–28.9 °C).

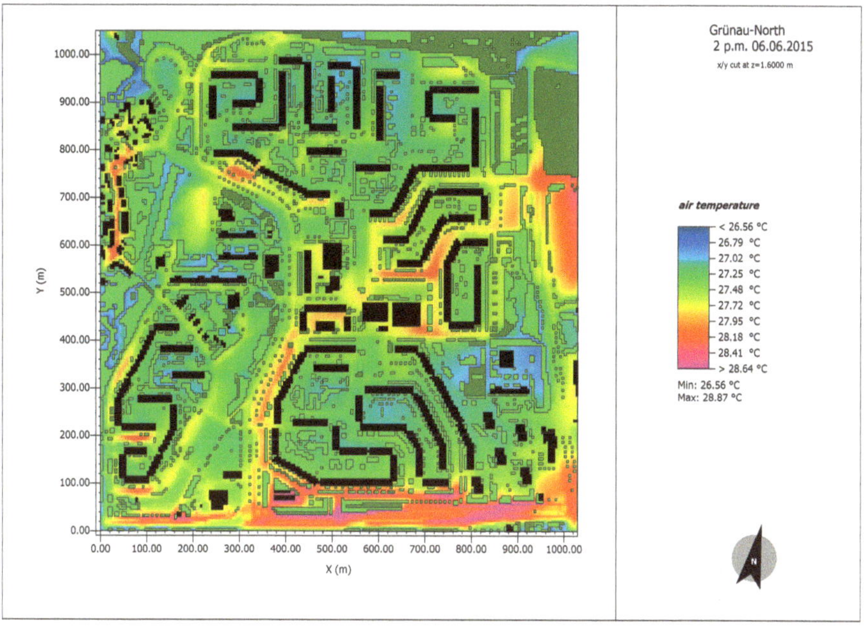

Fig. 7 Simulated air temperature 1.6 m above ground in Grünau-North; *black* represents buildings, *green* represents vegetation, visualized by LEONARDO-2D

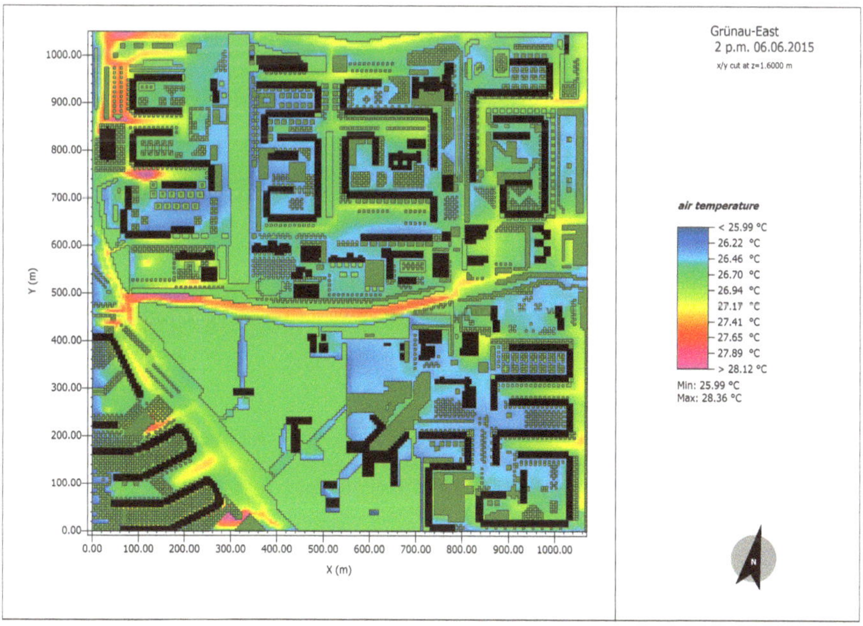

Fig. 8 Same as Fig. 7, but for Grünau-East

One reason for this temperature contrast can be found in differences in the green space distribution between the two districts. Whilst Grünau-North possesses roughly the same proportion of green space as Grünau-East (see Sect. 4.1.2), the percentage of grassy areas is much larger, whereas the proportion of areas with trees is considerably smaller. As a consequence, the evaporative cooling in Grünau-East is more effective than in the other district. Trees can store and evaporate more water than grass, which causes a stronger heat reduction effect (Bowler et al. 2010).

On the basis of the simulated air temperature, we assessed the perception of thermal conditions using three different indicators (see Sect. 2.2): Predicted Mean Vote (PMV), Physiologically Equivalent Temperature (PET), and Universal Thermal Climate Index (UTCI). These indicators represent the average thermal comfort. They depend on the air temperature, humidity, radiant temperature, air velocity, metabolic rates, and clothing insulation. Each individual experiences these sensations slightly differently, according to his or her physiological and mental state.

The thermal comfort indicators calculated by the BioMet module of ENVI-met (for a female person of age 60, weight 73 kg, body height 1.64 m, walking speed 1.25 m/s, summer clothing = 0.5 clo) agree with the simulated air temperatures (Figs. 7 and 8) and, likewise, confirm the results of the survey in the two districts (Table 1). Generally, Grünau-East has lower values for PMV, PET, and UTCI and, therefore is, on average, thermally slightly more comfortable than Grünau-North. This corresponds to the higher proportion of green areas in Grünau-East. Remarkably, this also applies to the indoor perceptions of heat, and even for noise and air pollution perception. All these indicators provide evidence that the environmental quality in Grünau-East is slightly superior to that in Grünau-North.

Table 1 Area-averaged thermal comfort indices and perceptions reported in the survey for Grünau-North and Grünau-East; higher values marked in **bold**

Indicator	Grünau-North	Grünau-East
PMV [range: Very cold (−4), neutral (±0), very hot (+8)]	**3.9**	3.3
PET [neutral = 20 °C]	**36.0 °C**	33.3 °C
UTCI [neutral = 20 °C]	**34.2 °C**	29.4 °C
Proportion of tree area (higher 5 m)	20.2%	**34.4%**
Proportion of all green space	52.0%	**55.0%**
Green area **maintenance satisfaction** [1–5]	3.6	**3.8**
Perception of indoor heat stress [1–**5**]	**3.3**	2.8
Perception of indoor noise pollution [1–**5**]	**3.2**	2.6
Perception of air pollution outdoor [1–**5**]	**2.3**	1.9

4.3 Synopsis of the Results and Discussion

The combination of our household survey results, environmental data and our model simulations yields an improved understanding of LRQ details. We provide evidence about the interlinkages between built housing conditions, the quality of the state of repair of the blocks, the surrounding environmental conditions, and the residential quality perceived by the local residents. Our empirical results suggest differences between Grünau-East and Grünau-North according to their particular local residential conditions.

We demonstrate that our approach can identify hotspots of adverse environmental features and their perception by residents. Obviously, the individual perception of LRQ depends on a number of influencing factors, one of them being the micrometeorological factors. For the two neighborhoods, we demonstrated good agreement between thermal comfort, as perceived by the residents, and model simulations, based on the urban structure.

A result of this extent was not expected, because we put together independent research approaches to test if there are consistencies that generate additional insights into the interpretation of LRQ. We had to accept incompatibilities, e.g., restricted modeling periods and restricted data. Nevertheless, the results of our combined analyses confirm our methodological approach and verify our hypotheses. We found evidence that LRQ is influenced by a variety of factors such as building structure, environmental features, and subjective reflections. The different values for each district lead to differences in LRQ within one estate. Furthermore, we showed a strong interrelation between objective, neighborhood-specific residential conditions and LRQ perceived by the residents. The synopsis of data based on surveys, measurements, and modeling could provide additional details and clear proofs that confirm the different characteristic of LRQ in the two distinct districts.

We would like to stress that our study refers to data gathered at the same locations and during the same time-span. This is a particular advantage of the study and distinguishes it from many health effect studies that use only aggregated data or proxy variables, such as temperature or air quality measured only at more or less representative monitoring sites. Here, we utilized highly resolved data and assessed the exposure at those locations where the people are.

The entire study concept of LRQ is transferable to other geographical and cultural regions. Nevertheless, there are some technical limitations, such as a maximum feasible micrometeorological modeling region of about 1 km × 1 km and restricted computational power.

5 Conclusions

In urban environmental research, interdisciplinary approaches are becoming more important to understand the complex relationships of human-environmental interaction and their impacts on the quality of life. The study at hand, to assess LRQ, demonstrated that an appropriate synoptic consideration of a number of methodical components is necessary. Measurements with technical equipment provide objective data, and survey results inform about subjective perceptions, which influence human behavior and residential satisfaction, as well as place attachment. To obtain a more comprehensive result that includes the interrelatedness of individual quantitative and qualitative data sets, a modeling approach is useful. Furthermore, a multi-scale perspective for assessing LRQ is crucial to reveal the heterogeneity within a given estate and at district level.

However, for a validation of the micrometeorological simulations, additional personal measurements of environmental stressors and perception surveys are indispensable, to sharpen the detailed insights on the specific individual exposure. This includes the residents' daily activity spaces outside but also within their apartments.

Further research on the following points is recommended:

- There is a need for the definition of a clear set of locally specific indicators to capture LRQ; it should take into account the dynamics of urban development, in particular, ongoing housing construction and surface sealing. It should also include specific individual factors (age, behavior priorities, health, etc.).
- We suggest the development of an LRQ index. The application of such an index makes it possible to compare very different environments. This index should be flexible in terms of ranking factors and identification of the most important ones according to the local context. It should include physical variables as well as subjective reflections and extended psychological approaches (Bonaiuto et al. 2003).
- Concerning the research design, an interdisciplinary approach from the very beginning of the research is highly recommended to specify common research aims and appropriate methods. If collaboration starts after the empirical phase, limitations related to common results are possible. Thus, our research should be considered as a pilot phase.
- More advanced, spatially referenced personal mobile measurements of several environmental stressors at one and the same time as well as the same location, could provide deeper insights about the multiple exposures of citizens during their local daily activities and the results would complement our knowledge about LRQ.
- With regard to ongoing and non-linear changes in the urban environment, repeated studies are necessary to assess the LRQ as a dynamic issue, to define corridors and thresholds for acceptable residential conditions, and to provide knowledge and instruments for urban planning.

Summarizing, urban transformations occur under the pressure of climate and demographic change, and inevitably impact on the urban quality of life and, in particular, on LRQ. For the development, implementation, and evaluation of appropriate adaptation measures, the suggested approach is useful and transferable. In this sense, the concept of LRQ and its interdisciplinary methodological design specifies and deepens the understanding of urban human-environment relations at the level of urban districts or neighborhoods.

Acknowledgement D.H. (AZ:20015/373) and M.U. (AZ:20015/411) gratefully acknowledge the financial support given by the Deutsche Bundesstiftung Umwelt (DBU). The work of A.M. was funded by the Alexander-von-Humboldt Foundation (AGY 1158029 IKS).

References

ATKIS – Amtliches Topographisch-Kartographisches Informationssystem (2013) Digitales Basis-Landschaftsmodell. Arbeitsgemeinschaft der Vermessungsverwaltungen der Länder der Bundesrepublik Deutschland, München

Banzhaf E, de la Barrera F, Kindler A, Reyes-Paecke S, Schlink U, Welz J, Kabisch S (2014) A conceptual framework for integrated analysis of environmental quality and quality of life. Ecol Indic 45:664–668. doi:10.1016/j.ecolind.2014.06.002

Banzhaf E, Kollai H (2015) Monitoring the urban tree cover for urban ecosystem services – the case of Leipzig, Germany. The International Archives of the Photogrammetry, Remote Sensing and Spatial Information Sciences, XL-7/W3, 301–305

Billota E, Evans GW (2012) Environmental stress. In: Steg L, Van den Berg AE, De Groot JIM (eds) Environmental psychology: an introduction. Wiley-Blackwell, Oxford, pp 27–36

Bonaiuto M, Fornara F, Bonnes M (2003) Indexes of perceived residential environment quality and neighbourhood attachment in urban environments: a confirmation study on the city of Rome. Landsc Urban Plan 65(1/2):41–52

Bonnes M, Uzzell D, Carrus G, Kelay T (2007) Inhabitants' and experts' assessments of environmental quality for urban sustainability. J Soc Issues 63(1):59–78

Bowler DE, Buyung-Ali L, Knight TM, Pullin AS (2010) Urban greening to cool towns and cities: a systematic review of the empirical evidence. Landsc Urban Plan 97(3):147–155

Christidis N, Jones GS, Stott PA (2015) Dramatically increasing chance of extremely hot summers since the 2003 European heatwave. Nat Clim Chang 5:46–50. doi:10.1038/nclimate2468

Cushman SA, Shirk A, Landguth EL (2012) Separating the effects of habitat area, fragmentation and matrix resistance on genetic differentiation in complex landscapes. Landsc Ecol 27(3):369–380

Dekker K, de Vos S, Musterd S, van Kempen R (2011) Residential satisfaction in housing estates in European cities: a multi-level research approach. Housing Stud 26(4):479–499. doi:10.1080/02673037.2011.559751

DWD – German Weather Service (2016) Klima-Michel Modell http://www.dwd.de/DE/service/lexikon/Functions/glossar.html?lv2=101334&lv3=101438. Accessed 18 June 2016

EEA European Environmental Agency (2015) EUNIS categories. Habitat types key navigation. Category: (G) Woodland, forest and other wooded land. http://eunis.eea.europa.eu/habitats-key.jsp?level=2&idQuestionLink=--%3E&pageCode=G. Accessed 15 Apr 2016

Eurostat (2015) Quality of life. Facts and views. http://ec.europa.eu/eurostat/en/web/products-statistical-books/-/KS-05-14-073. Accessed 20 June 2016

Fanger PO (1972) Thermal comfort, analysis and applications in environmental engineering. McGraw Hill, New York

Fiala D, Lomas KJ, Stohrer M (2001) Computer prediction of human thermoregulatory and temperature responses to a wide range of environmental conditions. Int J Biometeorol 45(2):143–159

Fornara F, Bonaiuto M, Bonnes M (2010) Cross-validation of abbreviated perceived residential environment quality (PREQ) and neighborhood attachment (NA) indicators. Environ Behav 42(2):171–196. doi:10.1177/0013916508330998

Franck U, Krüger M, Schwarz N, Grossmann K, Röder S, Schlink U (2013) Heat stress in urban areas: indoor and outdoor temperatures in different urban structure types and subjectively reported well-being during a heat wave in the city of Leipzig. Meteorol Z 22(2):167–177

Garcia Diez S (2015) Indikatoren zur Lebensqualität. Vorschläge der europäischen Expertengruppe und ausgewählte nationale Initiativen. Statistisches Bundesamt. http://www.destatis.de/DE/Publikationen/WirtschaftStatistik/2015/06/IndikatorenLebensqualitaet_062015.pdf?__blob=publicationFile. Accessed 14 July 2016

Gromke C, Blocken B (2015) Influence of avenue-trees on air quality at the urban neighborhood scale. Part II: traffic pollutant concentrations at pedestrian level. Environ Pollut 196:176–184. doi:10.1016/j.envpol.2014.10.015

Großmann K, Franck U, Krüger M, Schlink U, Schwarz N, Stark K (2012) Soziale Dimensionen von Hitzebelastung in Großstädten. disP – Plann Rev 48(4):56–68

Hart MA, Sailor DJ (2009) Quantifying the influence of land-use and surface characteristics on spatial variability in the urban heat island. Theoret Appl Climatol, 95(3–4), 397-406 doi: 10.1007/s00704-008-0017-5

Hassett EM, Stehman SV, Wickham JD (2012) Estimating landscape pattern metrics from a sample of land cover. Landsc Ecol 27(1):133–149

Helliwell JF, Layard R, Sachs J (eds) (2013) World happiness report 2013. UN Sustainable Development Solutions Network, New York

Hoffmann K, Behrens U (2016) Bericht – Stadtklimatische Untersuchungen in Leipzig. Ergebnisse statistischer Auswertungen langjähriger Klimareihen sowie temporärer Stations- und Profilmessungen. Deutscher Wetterdienst, Potsdam, not published

Höppe P (1993) Heat balance modelling. Experientia 49(9):741–747. doi:10.1007/BF01923542

Irga PJ, Burchett MD, Torpy FR (2015) Does urban forestry have a quantitative effect on ambient air quality in an urban environment? Atmos Environ 120:173–181. doi:http://dx.doi.org/10.1016/j.atmosenv.2015.08.050

Jaganmohan M, Knapp S, Buchmann CM, Schwarz N (2015) The bigger, the better? The influence of urban green space design on cooling effects for residential areas. J Environ Qual 45(1):134–145. doi:10.2134/jeq2015.01.0062

Kabisch S, Ueberham M, Söding M (2016) Grünau 2015. Ergebnisse der Einwohnerbefragung im Rahmen der Intervallstudie „Wohnen und Leben in Leipzig-Grünau". UFZ-Bericht 02/2016. Helmholtz-Zentrum für Umweltforschung – UFZ, Leipzig. http://www.ufz.de/export/data/2/114999_113257_ufz_bericht_022016_.pdf. Accessed 15 May 2016

Kjellstrom T, McMichael AJ (2013) Climate change threats to population health and well-being: the imperative of protective solutions that will last. Glob Health Action 6. doi:10.3402/gha.v6i0.20816

Knecht A (2010) Lebensqualität produzieren, Ressourcentheorie und Machtanalyse des Wohlfahrtsstaats. Springer, Wiesbaden

Lawrence R (2011) Understanding environmental quality through quality of life. Encyclop Environ Health 5:518–525. doi:10.1016/S0169-2046(02)00239-6

Leitte AM, Petrescu C, Franck U, Richter M, Suciu O, Ionovici R, Herbarth O, Schlink U (2009) Respiratory health, effects of ambient air pollution and its modification by air humidity in Drobeta-Turnu Severin, Romania. Sci Total Environ 407(13):4004–4011. doi:10.1016/j.scitotenv.2009.02.042

LfULG - Sächsisches Landesamt für Umwelt, Landwirtschaft und Geologie (2016) Messung der Luftqualität mit Hilfe des sächsischen Luftmessnetzes. http://www.umwelt.sachsen.de/umwelt/luft/3611.htm. Accessed 19 Apr 2016

Mahmood R, Pielke RA, Hubbard S, Niyogi D, Dirmeyer PA, McAlpine C, Carleton AM, Hale R, Gameda S, Beltrán-Przekurat A, Bruce B, McNider R, Legates DR, Shepherd M, Du J, Blanken PD, Frauenfeld OW, Nair US, Fall S (2014) Review: land cover changes and their biogeophysical effects on climate. Int J Climatol 34:929–953. doi:10.1002/joc.3736

Marans RW (2015) Quality of urban life & environmental sustainability studies: future linkage opportunities. Habitat Int 45(1):47–52. doi:10.1016/j.habitatint.2014.06.019

Marans, RW., Stimson, RJ. (2011). Investigating quality of urban life. Theory, methods, and empirical research, Social indicators research series. Springer, Dordrecht

Martiello MA, Giacchi MV (2010) High temperatures and health outcomes: a review of the literature. Scand J Public Health 38(8):826–837. doi:10.1177/1403494810377685

Mayer H, Holst J, Dostal P, Imbery F, Schindler D (2008) Human thermal comfort in summer within an urban street canyon in Central Europe. Meteorol Z 17(3):241–250. doi:10.1127/0941-2948/2008/0285

Oke TR (1973) City size and urban heat island. Atmos Environ 7:769–779. doi:10.1016/0004-6981(73)90140-6

Oke TR (1982) The energetic basis of the urban heat-island. Q J R Meteorol Soc 108(455):1–24. doi:10.1002/qj.49710845502

Oke TR (1987) Boundary layer climates. Routledge/John Wiley & Sons, London/New York

Pantisano, F., Craglia, M., Rosales-Sanchez, C. (2014). New indicators of quality of life: a review of the literature, projects, and applications. European Commission project 1076.

PIK - Potsdam-Institut für Klimafolgenforschung (2010) Projektionen zu "heißen Tagen", STAR II, A1B-Szenario, ECHAM5–1. Lauf, mittlere Realisierung. PIK, Potsdam

Rivera AL, Morse GS, Haase RF, Mccaffrey RJ, Tarbell A (2007) Exposure to an environmental toxin, quality of life and psychological distress. J Environ Psychol 27(1):33–43. doi:10.1016/j.jenvp.2006.12.004

Romero H, Salgado M, Smith P (2010) Climate change and urban climate: relations between thermal zones and the socioeconomic conditions of the population of Santiago, Chile. Revista INVI 70:151–179

Scannell L, Gifford R (2010) Defining place attachment: a tripartite organizing framework. J Environ Psychol 30:1–10. doi:10.1016/j.jenvp.2009.09.006

Schindler S, Wehrden H, Poirazidis K, Wrbka T, Kati V (2013) Multiscale performance of landscape metrics as indicators of species richness of plants, insects and vertebrates. Ecol Indic 31:41–48. doi:10.1016/j.ecolind.2012.04.012

Schlink U, Kindler A, Grossmann K, Schwarz N, Ulrich F (2014) The temperature recorded by simulated mobile receptors is an indicator for the thermal exposure of the urban inhabitants. Ecol Indic 36:607–616. doi:10.1016/j.ecolind.2013.09.017

Schlink U, Ragas AMJ (2011) Truncated levy flights and agenda-based mobility are useful for the assessment of personal human exposure. Environ Pollut 159(8–9):2061–2070. doi:10.1016/j.envpol.2011.02.023

Schwarz N, Schlink U, Franck U, Großmann K (2012) Relationship of land surface and air temperatures and its implications for quantifying urban heat island indicators – an application for the city of Leipzig (Germany). Ecol Indic 18:693–704. doi:10.1016/j.ecolind.2012.01.001

Stadt Leipzig, Amt für Statistik und Wahlen (2014) Befragung zum Klimawandel in Leipzig 2014 – Ergebnisbericht [1/15]. Stadt Leipzig, Amt für Statistik und Wahlen, Leipzig

Steg L, Gifford R (2005) Sustainable transportation and quality of life. J Transp Geogr 13(1):59–69. doi:10.1016/j.jtrangeo.2004.11.003

VDI – Verein Deutscher Ingenieure (1998) Methods for the human biometeorological evaluation of climate and air quality for urban and regional planning at regional level. VDI-Richtlinie 3787, Blatt 2

Weber N, Haase D, Franck U (2014a) Assessing modelled outdoor traffic-induced noise and air pollution around urban structures using the concept of landscape metrics. Landsc Urban Plan 125:105–116. doi:10.1016/j.landurbplan.2014.02.018

Weber N, Haase D, Franck U (2014b) Zooming into temperature conditions in the city of Leipzig: how do urban built and green structures influence earth surface temperatures in the city? Sci Total Environ 496:289–298. doi:10.1016/j.scitotenv.2014.06.144

White-Newsome JL, Sanchez BN, Jolliet O, Zhang Z, Parker EA, Dvonch JT, O'Neill MS (2012) Climate change and health: indoor heat exposure in vulnerable populations. Environ Res 112:20–27. doi:10.1016/j.envres.2011.10.008

Yan YY (2005) Climate comfort indices. In: Oliver JE (ed) Encyclopedia of world climatology. Springer, Dordrecht/New York, p 227

Zhao L, Lee X, Smith RB, Oleson K (2014) Strong contributions of local background climate to urban heat islands. Nature 511:216–219. doi:10.1038/nature13462

Socio-Spatial Distribution of Airborne Outdoor Exposures – An Indicator for Environmental Quality, Quality of Life, and Environmental Justice: The Case Study of Berlin

Annegret Kindler, Heinz-Josef Klimeczek, and Ulrich Franck

1 Introduction

Population growth and rapid urbanization accelerate the scarcity of limited environmental resources such as land, drinking water, energy, and clean air, especially in urban agglomerations (Kabisch and Kuhlicke 2014). Environmental pollution of air, water and soil, noise, and littering adversely affect the environmental quality and quality of life, especially of populations within cities and in their surroundings.

In view of this, essential scientific, technical, management, and governance efforts and advances are needed to achieve the 17 ambitious sustainable development goals of the United Nations "2030 Agenda for Sustainable Development" (UN 2015) (see Koch and Ahmad, "How to Measure Progress Towards an Inclusive, Safe, Resilient and Sustainable City? Reflections on Applying the Indicators of Sustainable Development Goal 11 in Germany and India", in this volume). Amongst others, by 2030, the number of deaths and illnesses from hazardous chemicals and air, water and soil pollution and contamination should be reduced substantially, to ensure healthy life and promote well-being (Agenda goal 3). According to goal 11, cities and human settlements should be made inclusive, safe, resilient, and sustainable.

A. Kindler (✉)
Department of Urban and Environmental Sociology, Helmholtz Centre for Environmental Research-UFZ, Leipzig, Germany
e-mail: annegret.kindler@ufz.de

H.-J. Klimeczek
Senate Department for the Environment, Transport and Climate Protection, Berlin, Germany
e-mail: Heinz-josef.klimeczek@senuvk.de

U. Franck
Department of Environmental Immunology and Core Facility Studies, Helmholtz Centre for Environmental Research-UFZ, Leipzig, Germany
e-mail: ulrich.franck@ufz.de

S. Kabisch et al. (eds.), *Urban Transformations*, Future City 10,
DOI 10.1007/978-3-319-59324-1_14

When aiming at sustainable urban development and at ensuring natural resources and a high quality of life for future generations, the efficient use of limited environmental resources, including the mitigation of environmental pollution, are great challenges requiring new scientific approaches, technological solutions, rethinking, and conscious actions of society and individuals.

2 Effects of Airborne Outdoor Exposure on Human Health and Well-Being

Ambient air pollution is one of the most important environmental problems in urban areas (Schwela 2000; Pascal et al. 2013) and one of the most pressing problems in rapidly industrializing regions (UNEP/UNECE 2016; WBGU 2016). It strongly affects not only the environmental quality but, consequently, also central aspects of urban quality of life. Human health and well-being are both impaired by it. Outdoor air pollution influences the entire urban population. In contrast to waterborne and soil pollutions, there is no effective individual strategy for residents to avoid the intake of outdoor airborne pollutants. In 2012, about 3.7 million premature deaths were attributable to ambient air pollution worldwide, including about 480,000 deaths in Europe (WHO 2014). Heart disease and stroke, with about 40% each, are the most common reasons for premature death attributable to air pollution (WHO 2014). In Europe, air pollution is the single largest environmental health risk (EEA 2015a). Particulate matter (PM), nitrogen dioxide, and ground-level ozone are identified as the three pollutants that most significantly affect human health. According to the EEA, about 90% of city dwellers in Europe are exposed to concentrations of air pollutants that are harmful to health. It was estimated that $PM_{2.5}$ reduces life expectancy in the EU by more than 8 months (EEA 2016b). In Germany, air pollution has been reduced significantly since the early 1990s. Despite reduced emissions, there has not been a decreasing trend in air pollution caused by particulate matter, nitrogen oxide, and ozone since the beginning of the present decade (UBA 2016).

Epidemiological studies have consistently demonstrated that numerous adverse health effects can be caused or exacerbated by exposure to outdoor air pollutants (Chen et al. 2007; Forastiere et al. 2007; Bell et al. 2009; Brunekeef et al. 2009; Anderson and Thundiyil 2012; Faustini et al. 2014; Franck et al. 2015; Crouse et al. 2015; Mills et al. 2015). Solving environmental problems that are linked to clean air requires holistic approaches to analysis and decision-making that also include social and economic aspects (Cooter et al. 2013).

Nitrogen oxides and particulate matter play an important role in many urban agglomerations and also in developed countries, because they are associated with traffic emissions. Many studies show the detrimental effects of the human exposure to inhalable and fine particulate matter (particles with aerodynamic diameters smaller than 10 μm – PM_{10} and 2.5 μm – $PM_{2.5}$). Exposure to $PM_{2.5}$ increases the risks for cardiovascular and respiratory short-term health impacts, as well as long-term effects, including cancer (Brunekreef et al. 2009; Pope et al. 2011). Short-term mortality is increased by 1–7% per 10 μg/m^3 $PM_{2.5}$ (Brunekreef et al. 2009; Franklin et al. 2007; Janssen et al. 2013; Shi et al. 2016). Ambient particulate matter also increases human morbidity. Studies describe rising numbers of hospital admissions on days following days with elevated concentrations of airborne particles (Chen et al. 2007; Franck et al. 2014a, b; Franck et al. 2015; Lall et al. 2011; Slaughter et al. 2005; Vera and Cifuentes 2008) and also due to long-term exposure (Bell et al. 2009; Hruba et al. 2001).

Previous studies also found evidence that an increase of 10 μg/m^3 of the NO_2 concentration in ambient air is associated with a ~ 1–5% increase of mortality due to cardiovascular and respiratory diseases (Cooter et al. 2013; Mills et al. 2015; Perez et al. 2015). The pooled effect on mortality amounts to 4% per 10 μg/m^3 (Faustini et al. 2014). Human exposure to NO_2 also increases the number of emergency hospital admissions (Franck et al. 2014a, b; Franck et al. 2015; Mills et al. 2015; Perez et al. 2015). Summarizing, the European Environment Agency stated, "Despite considerable improvements in past decades, air pollution is still responsible for more than 400,000 premature deaths in Europe each year" (EEA 2015b).

Recent studies suggest that variation in NO_2 concentrations within a city may more strongly influence human health effects than variation between cities (Crouse et al. 2015). This fact underlines the importance of spatially resolved investigations of the concentrations of outdoor air pollutants within cities.

Exposure to ambient air pollutants and effect strength depend on socioeconomic parameters that include unemployment, long-term unemployment, educational level, temporary employment, manual work, residential segregation, and further variables (Barcelo et al. 2009; Brochu et al. 2011; Forastiere et al. 2007; Rice et al. 2014; Yap et al. 2013; Young et al. 2012). Typically, exposure concentrations are higher and health impacts are stronger in socially disadvantaged areas (Prochaska et al. 2014). Hence, from the perspective of environmental justice, more attention has to be paid to the interrelations between environmental quality, health, and the social situation of the population. The unequal distribution of environmental burdens and resources could change the social and ethnic composition of individual neighborhoods or districts, with the possible consequence of social segregation.

3 Airborne Outdoor Exposures Studied from three Perspectives: Environmental Quality, Quality of Life, and Environmental Justice

In this study, airborne outdoor exposure concentrations were investigated from the following three perspectives: environmental quality, quality of life, and environmental justice. Although these three perspectives are based on broad and complex concepts encompassing different dimensions, they are nevertheless strongly related.

Environmental quality can be understood as a set of properties and characteristics of the environment, either generalized or local, as they impinge on human beings and other organisms. It is "a measure of the condition of an environment relative to the requirements of one or more species and/or to any human need or purpose" (Johnson et al. 1997, p. 586). According to the European Environmental Agency, it is a general term that can refer to various characteristics that relate to the natural environment as well as to the built environment, such as air and water purity or pollution, noise, access to open space, and the visual effects of buildings, and the potential effects that such characteristics may have on physical and mental health (EEA 2016a). Environmental quality has a strong influence on the quality of life because there are direct links between primary elements of the environment, such as air, water, and land surface and the quality of life (Brown 2003; Pacione 2003; Banzhaf et al. 2014).

Good local environmental conditions are a prerequisite for human quality of life (WBGU 2016). There are several definitions of quality of life.

Quality of life comprises objective indicators describing the environments within which people live and work, and subjective indicators describing the ways in which people perceive and evaluate conditions around them. The World Health Organization defined quality of life "as individuals' perception of their position in life in the context of the culture and value systems in which they live and in relation to their goals, expectations, standards and concerns. It is a broad ranging concept affected in a complex way by the person's physical health, psychological state, level of independence, social relationships, personal beliefs and their relationship to salient features of their environment" (WHO 1997, p. 1). Noll (2000) specified quality of life as multidimensional concept comprising tangible and intangible, objective and subjective, individual and collective components of welfare. Constanza et al. (2007) developed an integrative definition of quality of life that combines measures of human need with subjective well-being or happiness. "Quality of life is the extent to which objective human needs are fulfilled in relation to personal or group perceptions of subjective well-being" (Constanza et al. 2007, p. 269). According to Marans and Stimson (2011), the concept of quality of life is a multidimensional entity that includes immaterial as well as material, subjective as well as objective, individual as well as collective elements of welfare, satisfaction and happiness. In spite of great heterogeneity in the details, quality of life is consistently understood as a multidimensional concept consisting of physical, psychological, social and ecological aspects and which considers both subjective well-being and objective conditions (WBGU 2016). Eurostat, the statistical office of the European Union, generated a set of key and standard indicators for quality of life that, for the first time, combined objective indicators with subjective evaluations of individuals' situations. Nine thematic dimensions were identified to measure the quality of life: material living conditions; productive or main activity (covering employment); health; education; leisure and social interactions; economic and physical safety; governance and basic rights; natural and living environment; and overall life satisfaction (Garcia Diez 2015; Eurostat 2015a). Our investigation focusses on objective aspects of quality of life: human exposure to air pollutants and social parameters.

Environmental quality is strongly related to *environmental justice*, an approach that focuses on interrelations between environmental quality, health, and the social situation of affected population. The term environmental justice is used in conjunction with environmental campaigns, political debates, research, and policy-making around the world, and covers a wide diversity of environmental risks, benefits, and resources, from the local street level to the global scale (Walker 2012). Due to the different history of their origin, diverse approaches exist in the US and in Europe (Maschewsky 2006; Elvers et al. 2008; Elvers 2011; Hornberg et al. 2011; Laurent 2010; Walker 2012), so that there is no general definition of environmental justice. The debate on environmental justice was initiated by civil rights movements, environmental organizations and activists in the United States in the 1970s and 1980s; it dealt with the socially and spatially unequal distribution of environmental burdens. The US Environmental Protection Agency defined environmental justice as "the fair treatment and meaningful involvement of all people regardless of race, color, national origin, or income with respect to the development, implementation, and enforcement of environmental laws, regulations, and policies. [...] It will be achieved when everyone enjoys the same degree of protection from environmental and health hazards and equal access to the decision-making process to have a healthy environment in which to live, learn, and work" (EPA 2016). In the European context, in the UK and especially in Scotland, environmental justice was established in scientific and political debates in the late 1990s. Compared to the US, the British understanding of environmental justice included the distribution of environmental goods such as healthy living conditions and access to recreational opportunities (Elvers et al. 2008). In accordance with the US approach, Maschewsky introduced the term environmental justice to Germany in 2001, for the first time, and related it to public health and the social city. He understood environmental justice as the social and spatial unequal distribution of environmental burdens, leading to unequal health risks for the affected population (Maschewsky 2001). Up till now, a consistent definition of environmental justice in Germany does not exist (UBA 2015); however, with respect to investigations in urban areas, "environmental justice is understood as a normative concept oriented towards the prevention and reduction of socio-spatial concentration of health-relevant environmental burdens, as well as ensuring socio-spatially just access to environmental resources" (UBA 2015, p. 29; Bolte et al. 2012).

Walker (2012) gives an overview of the different understanding and definition of environmental justice, analyzing academic, activist, and policy literature. In our study, environmental justice is understood as the social and spatial distribution of environmental goods (resources) and bads (harm and risk) (Maschewsky 2004; Köckler 2011; Walker 2012). Environmental justice can be subdivided into three concepts: distributive justice – justice in terms of the distribution or sharing of goods and bads; procedural justice – justice in terms of the ways in which decisions are made, who is involved and has influence, and justice as recognition – justice in terms of who is given respect and who is and is not valued (Walker 2012). With respect to these three concepts, the socio-spatial distribution of air pollution is an indicator of distributive environmental justice.

Air pollution determines the air quality, which is not only one of the key environmental indicators (OECD 2008; UNDP 2011), but also an indicator of quality of life (Eurostat 2015a, b) and environmental justice (Klimeczek 2011; UBA 2015).

4 Aims of the Study

In comparison to the most polluted regions of the world, concentrations of air pollutants in German cities are rather low. For example, the yearly means of PM_{10} in Delhi (India, 2012), Beijing (China, 2013), Berlin and Munich (Germany, 2013) amounted to 229 μg/m^3, 108 μg/m^3, 24 μg/m^3 and 21 μg/m^3, respectively (WHO 2016). Despite the local decline of air pollution during the last two decades particulate matter (PM_{10}, $PM_{2.5}$) and nitrogen oxides still belong to the most important air pollutants with significant health effects in Germany, especially in urban areas characterized by high concentrations of population, traffic, and industry (Minkos et al. 2016). An essential precondition for identifying and solving environmental problems and for developing urban areas in a sustainable way is the monitoring of environmental quality by means of various indicators, e.g., related to air quality, and their impact on the urban quality of life as well.

The study provides a contribution to the transformation field "Quality of Life", as a part of the three fields of urban transformations – resource efficiency, quality of life, and resilience – described by Kabisch et al., "Introduction: Urban Transformations – Sustainable Urban Development Through Resource Efficiency, Quality of Life, and Resilience", in this volume. However, in contrast to urban transformations, defined as radical and fundamental, multi-dimensional and non-linear alterations in the frame of urban resource patterns, population developments, infrastructures, economy, metabolism, and governance regimes, as well as established values, norms and behaviors (Kabisch and Kuhlicke 2014), investigations on urban air quality are assigned to urban transitions that are understood as alterations within an urban system. Transitions describe incremental changes within a given system (Foxon et al. 2009; Kabisch and Kuhlicke 2014; Kabisch et al., "Introduction: Urban Transformations – Sustainable Urban Development Through Resource Efficiency, Quality of Life, and Resilience", in this volume).

In an urban case study, a GIS-based method should be developed to calculate the air pollution by $PM_{2.5}$ and NO_2 for specific spatial units, to find out differences in the spatial distribution of areas with higher or lower air quality, to calculate the number of exposed inhabitants according to their social status, and to identify areas that are both environmentally and socially disadvantaged and in which a need for action exists.

From the perspective of environmental quality, quality of life, and environmental justice, this study aimed to answer the following research questions.

- How high are the annual mean concentrations of $PM_{2.5}$ and NO_2 and how they are spatially distributed within a city?
- How many inhabitants are affected by different concentrations of these pollutants?

- Is there a relationship between the social status of the inhabitants and their exposure to these pollutants?
- Are there areas within the city that are doubly disadvantaged regarding the social status of population and the airborne outdoor exposure?

Together with the socio-spatial distribution of other environmental indicators, the results can be used to develop and adapt measures for mitigating environmental burdens, minimizing health risks, enhancing the quality of life and well-being towards more environmental justice and sustainable urban development.

The investigations are a part of the pilot project "Environmental Justice in Berlin" (Klimeczek 2011, 2012). This project aimed at determining the correlation between the social status of the population and the environmental quality in the planning areas of Berlin. In a first step, the following five key indicators were at the focus of investigation: traffic-related air pollution and traffic noise, as the most important environmental burdens, green and other open spaces as environmental resources, bioclimatic exposure, and social status. Furthermore, additional indicators – e. g. other land-use types, residential locations, and the structure of buildings – completed the indicator set. The novel methodological approach of integrated reporting on environment, health, social situation, and urban development allows the identification of multiple burdened areas. Thus, focal points, comprising areas that are disadvantaged regarding their social and environmental situation can be determined. By combining environmental and social parameters, this monitoring approach provides an important basis for decision-making based on an integrated concept linking urban, traffic, and environmental planning including an adequate consideration of the social dimension.

5 Material and Methods

5.1 *Study Area*

The socio-spatial distribution of outdoor airborne exposure was investigated in the case study of Berlin, the capital of Germany. It is located in an area of low-lying, marshy woodlands with a mainly flat topography on an ice age glacial valley, 34 m–115 m above sea level, and it has a humid continental climate (Franck et al. 2014a). Berlin covers an area of 891km^2. With a population of about 3.61 million inhabitants in 2015 (2009: 3.36 million inhabitants) (Amt für Statistik Berlin-Brandenburg 2016a), it is the largest city in Germany and, currently, the second largest in the European Union. The population density is 4067 inhabitants per km^2 (2009: 3782 inh./km^2). At the beginning of 2015, about 1.369 million motor vehicles (2009: 1.266 million motor vehicles) were registered in the city (Amt für Statistik Berlin-Brandenburg 2016a).

Administratively, Berlin is subdivided into 12 urban districts, 60 prognosis areas, and 138 district regions. In 2006, a new additional spatial unit was introduced, the so-called planning areas (PLAs), for monitoring and planning purposes (Welsch et al. 2011). These planning areas are the spatial base for the analysis and assessment of environmental, social, economic, infrastructural, and further characteristics, including their interrelations from the perspective of environmental quality, quality of life, and environmental justice. The 447 planning areas are characterized by uniform urban structure types and local environment, main roads and traffic arteries, natural barriers (e.g. water bodies) and additionally, by the relatively homogeneous socio-economic situation of the population living there. On average, 8000 inhabitants (2009: 7500 inhabitants) live in each PLA. The area of individual PLAs varies markedly, from 14 ha to 2370 ha, and it increases from the center to the periphery. 217 PLAs (48.5%) are smaller than 100 ha, 216 PLAs (48.3%) have a size between 100 ha and 500 ha, 11 PLAs (2.5%) cover an area between 1000 ha and 2000 ha, and 3 PLAs (0.7%) are larger than 2000 ha.

5.2 Data

For the investigations of the socio-spatial distribution of outdoor airborne exposures, the Senate Department for Urban Development and the Environment of Berlin provided the following data for the year 2009:

- the annual mean concentrations of $PM_{2.5}$ and NO_2 within grid cells measuring 500 m × 500 m and in linear form for segments of the main roads (length and width of about 11,200 road segments) calculated with the Gaussian multi-source dispersion model IMMISnet (IVU Umwelt GmbH 2011) as ArcGIS shape files
- information about the social status of the population, in the form of the status index for each PLA derived from the Social Urban Development Monitoring 2010 (SenStadt 2010)
- the number of inhabitants per PLA on 31 December 2009
- the administrative boundaries of the urban districts and the PLAs (ArcGIS shape file)
- the boundary of the low emission zone (ArcGIS shape file)

The low emission zone was introduced on 1 January 2008 as central measure of the Clean Air Plan and Action Plan 2005–2010 of Berlin (SenStadt 2005) aiming at the reduction of traffic-related air pollution in the densely populated inner urban districts. This zone is located in the city center and limited by the inner ring of the interurban train (S-Bahn). It covers an area of about 87km^2 and about one million inhabitants live there.

5.3 Methods to Determine $PM_{2.5}$ and NO_2 Concentrations as well as Social Status

5.3.1 Calculation of $PM_{2.5}$ and NO_2 Concentrations

In order to determine the annual mean concentrations of $PM_{2.5}$ and NO_2 for each PLA, as the summary of the annual mean of the concentrations derived from the grid cells and the expected additional emissions by traffic on the main roads, the following GIS-based method was applied, using the ArcGIS 10.2 GIS software:

First, the shape file with the grid cells was intersected with the boundaries of the PLAs. The resulting areas contain information about the annual mean concentrations of the two pollutants and the PLA in which they are located. On this basis, the area-weighted annual mean concentrations of $PM_{2.5}$ and NO_2 of the local background were calculated for each PLA.

$$conc_{PLA,local\ background} = \sum \left(A_{Grid,i} \times conc_{grid,i} \right) \div A_{PLA}$$

$conc_{PLA,local\ background}$: annual mean concentration of $PM_{2.5}$ or NO_2 in a PLA

$A_{Grid,i}$: percentage of grid cell i in a PLA

$conc_{grid,i}$: annual mean concentration of $PM_{2.5}$ or NO_2 in grid cell i

A_{PLA}: area of a PLA

In a second step, the segments of the main roads containing the length and width of the roads and the annual mean concentrations of $PM_{2.5}$ and NO_2 were intersected with the boundaries of the PLAs. Using the length and width of the resulting lines, the areas of main road segments were calculated per PLA. For each segment and PLA, the area-weighted annual mean concentrations of $PM_{2.5}$ and NO_2 resulting from traffic were calculated.

$$conc_{PLA,traffic} = \sum \left(A_{traffic,i} \times conc_{traffic,i} \right) \div A_{PLA}$$

$conc_{PLA,traffic}$: annual mean concentration of $PM_{2.5}$ or NO_2 resulting from traffic in the main roads of a PLA

$A_{traffic,i}$: area of a main road segment i in a PLA

$conc_{traffic,i}$: annual mean concentration of $PM_{2.5}$ or NO_2 in the main road segment i

A_{PLA}: area of a PLA

The combination of the area-weighted annual mean concentrations derived from the raster cells and the main roads results in the area-weighted total annual mean of $PM_{2.5}$ and NO_2 in 2009.

$$conc_{PLA,total} = conc_{PLA,local\ background} + conc_{PLA,traffic}$$

$conc_{PLA,total}$: total annual mean concentration of $PM_{2.5}$ or NO_2 in a PLA

Through the addition of the pollutant concentration emitted on the main roads, the air pollution in PLAs with a high share of road traffic areas can be emphasized, to better cope with the aim of identifying PLAs at highest risk for well-being and health.

In order to classify $PM_{2.5}$ and NO_2 exposures, the values were ranked within deciles according to pollutant concentrations. Analogous to the socio-spatial categorization by the status index (SenStadt 2010), the two lowest deciles were classified as low level exposure. The deciles with the highest and second highest pollution were classified as very high and high levels of exposure, respectively. All other deciles were merged as medium level exposure. The combined exposure of both pollutants in a PLA was ranked "very high" if at least one pollutant had a concentration in the 10th decile, "high" if at least one pollutant had a concentration in the 9th decile, "low" if both pollutants had concentrations in the lowest two deciles, and "medium" in other cases. These rankings aimed at optimal differentiation between the PLAs in Berlin but not at European limit values. The procedure pays special attention to the areas with the highest exposure levels in the city because, in the context of environmental quality, quality of life, and environmental justice, mitigation measures should improve the air quality, especially within such areas.

5.3.2 Social Status Index of the Population

To describe the socio-economic situation of the exposed population, we used the standardized status index from the Social Urban Development Monitoring 2010 in Berlin (SenStadt 2010), representing data for the year 2009. The Social Urban Development Monitoring provides detailed information on the socio-structural and socio-spatial development in the PLAs of Berlin. The status index was built on the basis of the following six indicators that describe the social situation of population: unemployment rate in the age group 15–65 years, unemployment rate in the age group 15 to under 25 years (youth unemployment rate), long-term unemployment rate in the age group 15–65 years, rate of non-unemployed basic welfare recipients, rate of non-unemployed basic welfare recipients under the age of 15 years, rate of inhabitants under the age of 18 years with a migration background. The status index was determined using a mathematical procedure of standardization described in detail in SenStadt (2007). The six equally weighted normalized single indicator values were summarized to the standardized status index, which allows the ranking of the PLAs on the basis of deciles. In this ranking, the more favorable positions are associated with lower numerical values and the more disadvantageous positions with higher numerical values. Therefore, the two deciles with the lowest values, indicating best values of the status index, were ranked as "high status". Correspondingly, deciles with highest and second-highest values of the status index, indicating the worst, were ranked as "very low status" and "low status", respectively. The six deciles between the two lowest and the two highest values of the status index were ranked as "medium status".

For 13 PLAs, the status index was not generated because either fewer than 200 inhabitants lived there (8 PLAs) or these PLAs (5 PLAs) were outliers regarding their single indicators, which ranged beyond the span of values of the other PLAs, due to special factors, e.g. large commercial areas or forest (SenStadt 2010). These PLAs were excluded from further analysis of the status index.

6 Empirical Findings

Table 1 gives an overview of the social status index of the 434 PLAs, their area and population in 2009. By definition, 60% of the PLAs have a medium status, 20% have a high status and 10% each have a low or very low status. With about 64%, most of the population lives in PLAs with a medium social status, 11% lives in PLAs of very low or low social status and 14% lives in PLAs of high social status. The different levels of the status index are spatially very heterogeneously distributed. A specific pattern is not visible. Both in the inner-urban areas, within or close to the low emission zone, and in peripheral PLAs (e.g. in Spandau, Reinickendorf, Marzahn-Hellerdorf), a mixture of all status levels exists. By contrast, Steglitz-Zehlendorf, Pankow and Treptow-Köpenick are urban districts in which most of their PLAs have a medium or high status index (see Fig. 1).

The quantitative analysis of the calculated annual mean concentrations of $PM_{2.5}$ and NO_2 for all PLAs yielded the following results:

The concentrations of $PM_{2.5}$ range from 15 μg/m³ to 34 μg/m³ and those of NO_2 from 11 μg/m³ to 47 μg/m³. Compared with the valid annual mean limit values of the Air Quality Directive 2008/50/EC of the European Council (2008), the target value of 25 μg/m³ (which became a limit value in 2015) for $PM_{2.5}$ was exceeded in 10 PLAs and the limit value of 40 μg/m³ for NO_2 was exceeded in four PLAs. With respect to the WHO Air Quality Guidelines (WHO 2006a, b), the annual mean concentration of $PM_{2.5}$ in all PLAs was higher than the guideline of 10 μg/m³. Figures 1 and 2 show the spatially unequal distribution of the annual mean concentrations of $PM_{2.5}$ and NO_2 and the social status index in the PLAs of Berlin in 2009 Fig. 1.

The concentrations of each pollutant vary within the city and decrease from the inner-urban areas to the periphery. Figures 1 and 2 illustrate that the exposure was highest in the inner-urban areas, within or close to the low emission zone, with highest population and traffic densities. More peripherally located PLAs with lower population and traffic densities were exposed to lower concentrations. This fact

Table 1 Social status index, area and population of the analyzed 434 PLAs of Berlin in 2009

Social status index	Number of PLAs	Percentage of PLAs	Percentage of area	Percentage of population
Very low	43	9.9	3.8	11.0
Low	44	10.1	5.6	10.9
Medium	260	59.9	49.7	63.5
High	87	20.1	40.9	14.5

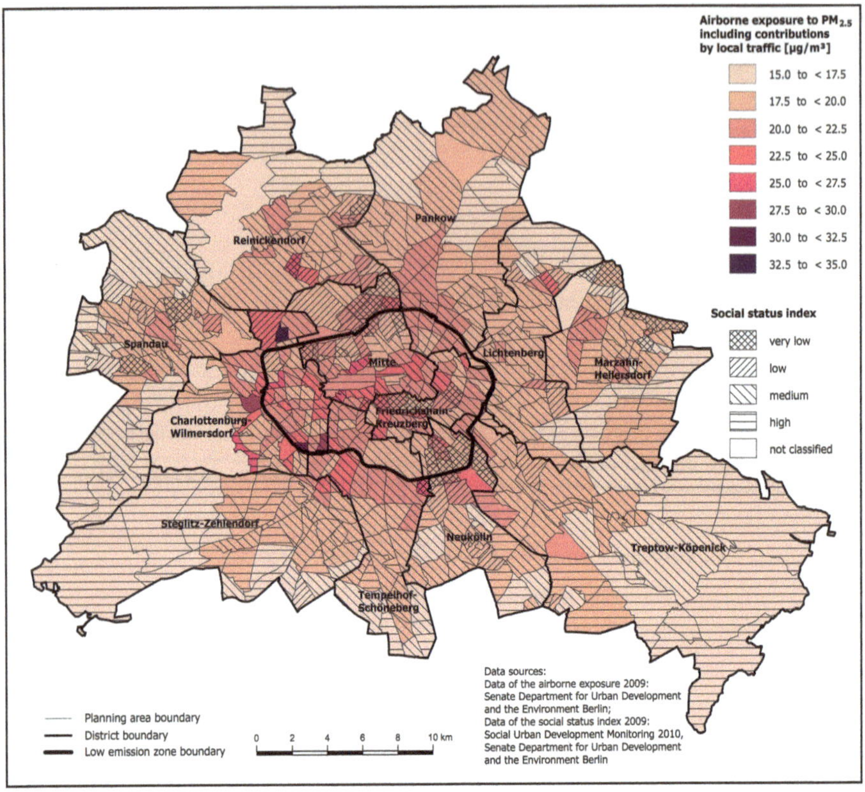

Fig. 1 Socio-spatial distribution of outdoor airborne exposure to $PM_{2.5}$ in the PLAs of Berlin in 2009

underlines the reasonable definition of the boundary of the low emission zone, which was established to protect the health of people within the areas with highest pollutant concentrations and high population density.

With respect to the social status index of the population and exposure to $PM_{2.5}$ or NO_2, the situation is inhomogeneous. PLAs with higher exposure concentrations are often of lower social status. However, there also exist PLAs with a high social status and higher exposure concentrations, especially in the inner-urban area within the low emission zone. This phenomenon could be associated with gentrification processes during recent years and the associated resettlement of young, high-income professionals who may prefer to live in the city center with its greater range of social and cultural facilities and many opportunities for socializing, despite its lower environmental quality, e.g. possible lack of green space, high traffic density, higher noise levels, and higher concentrations of air pollutants.

The findings of classifying the PLAs in terms of their exposure to $PM_{2.5}$ and NO_2 into the four classes "low", "medium", "high" and "very high", including their

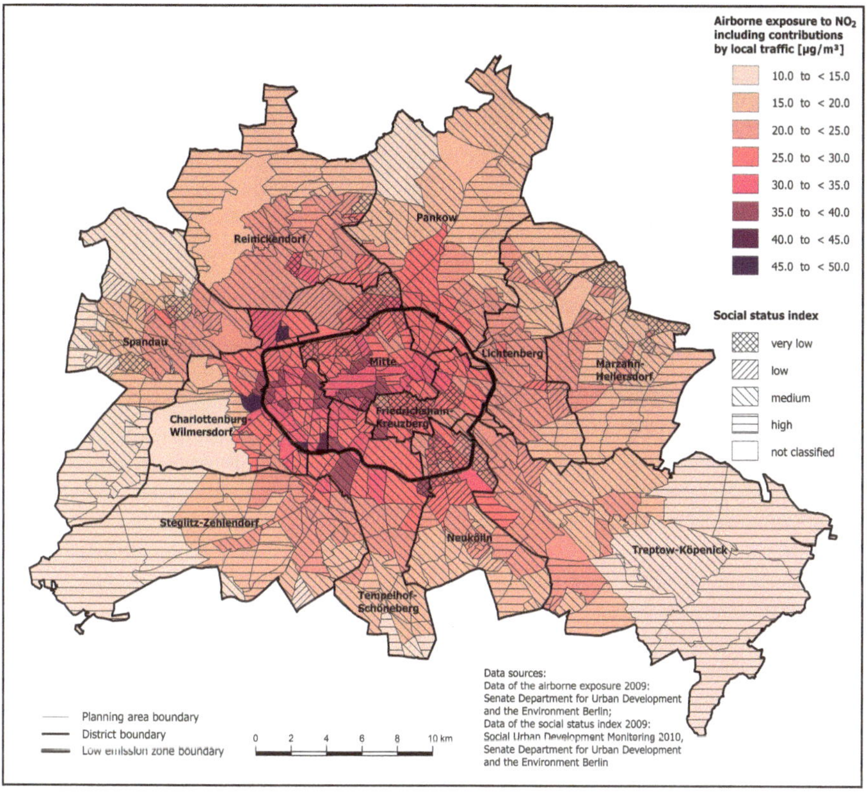

Fig. 2 Socio-spatial distribution of outdoor airborne exposure to NO_2 in the PLAs of Berlin in 2009

social status index, are shown in Fig. 3. The diagram illustrates that, with increasing exposure, the percentage of PLAs with a high social status decreases, and the majority of those with medium, low and very low status increases. Regarding the combined exposure to $PM_{2.5}$ and NO_2 of all 447 PLAs (see Fig. 4), 52 PLAs (11.6%) are "very high exposed", 57 PLAs (12.7%) are "high exposed", 259 PLAs (58%) are "medium exposed", and 79 PLAs (17.7%) are "low exposed". Referring to the area of Berlin and to both pollutants, 4% are "very high exposed", 6% are "high exposed", 44% are "medium exposed", and 46% are "low exposed" Fig. 3.

The unequal socio-spatial distribution of the combined exposure to $PM_{2.5}$ and NO_2 for PLAs in Berlin in 2009 is shown in Fig. 4.

PLAs could be identified that are doubly disadvantaged regarding a high exposure (low air quality) and a low social status of the population. 15 PLAs (3.5%) have a high combined exposure and a very low/low social status. 12 PLAs (2.8%) are characterized by a very high combined exposure and a very low/low social status. In contrast, only two PLAs with a high social status have a high or very high combined

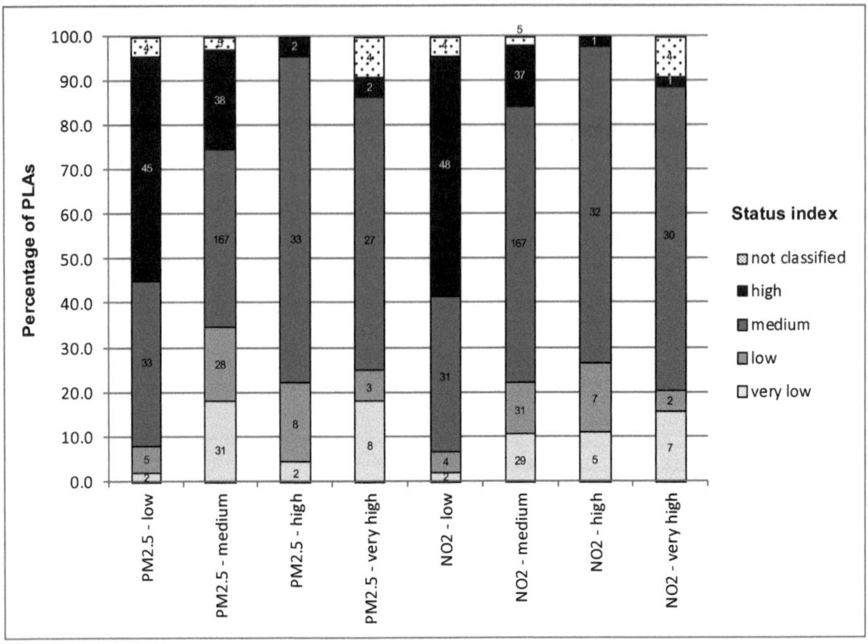

Fig. 3 PLAs in terms of their exposure to $PM_{2.5}$, NO_2, and their social status index. The figures in the vertical bars represent the number of PLAs

exposure. Within PLAs of medium social status, 40 PLAs (8.9%) are "high exposed" and 34 PLAs (7.6%) are "very high exposed" to $PM_{2.5}$ and NO_2.

The findings of the quantitative analysis of the exposed inhabitants (combined exposure) are shown in Fig. 5. About 10% of the population lives in PLAs with very high combined exposure to $PM_{2.5}$ and NO_2, followed by 15% in PLAs with high exposure. 60% of the inhabitants live in PLAs with medium exposure and only 15% in PLAs with low exposure.

Our analysis demonstrates that the combined exposure to $PM_{2.5}$ and NO_2 is unequally distributed in the PLAs of Berlin. Only 46% of Berlin's area, with 79 PLAs and 15% of the population, are "low exposed". In contrast 10% of the area, with 109 PLAs and 25% of the population, are "high" or "very high" exposed. The remaining 44% of the area are "medium exposed", which corresponds to 259 PLAs and 60% of the population. These results show small-scale differences of the outdoor airborne exposure within the city. 27 PLAs are doubly disadvantaged regarding the low air quality (high/very high exposure) and a very low/low social status (see Fig. 6). These PLAs cover an area of 20.5 km^2 (2.3%) in which about 228,000 inhabitants (6.8%) live. These doubly disadvantaged PLAs are not spatially concentrated only in the low emission zone. 11 of these PLAs are located within the low emission zone, in the districts of Friedrichshain-Kreuzberg (4 PLAs), Mitte (4 PLAs), Neukölln (2 PLAs) and Tempelhof-Schöneberg (1 PLA). 9 PLAs are located north and south of the low emission zone. Only 7 PLAs are located more peripherally, in

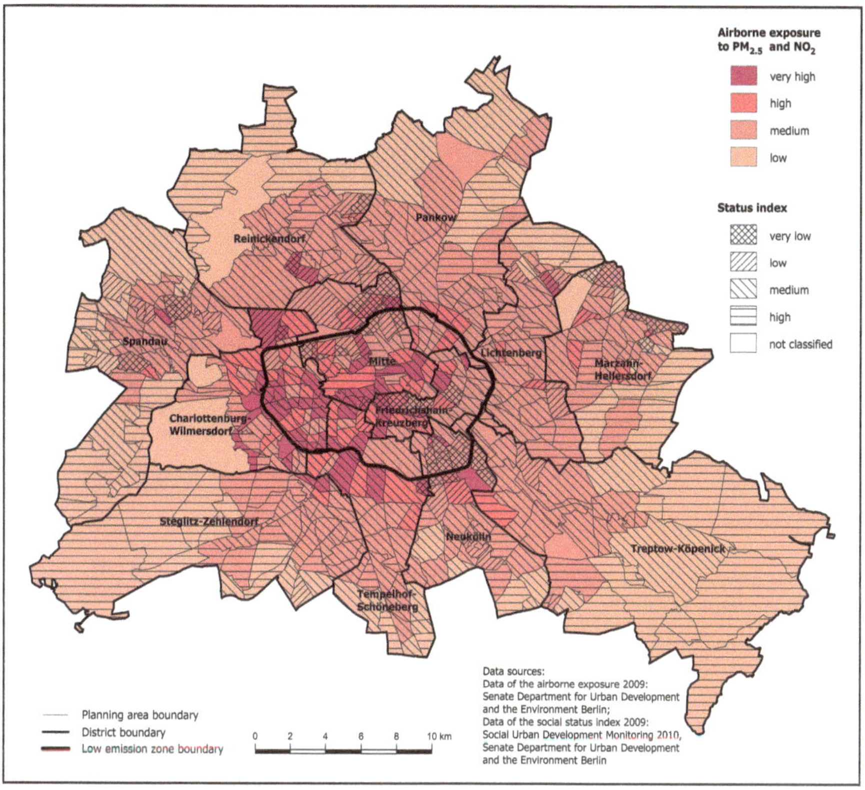

Fig. 4 Socio-spatial distribution of outdoor airborne exposure to $PM_{2.5}$ and NO_2 in the PLAs of Berlin in 2009

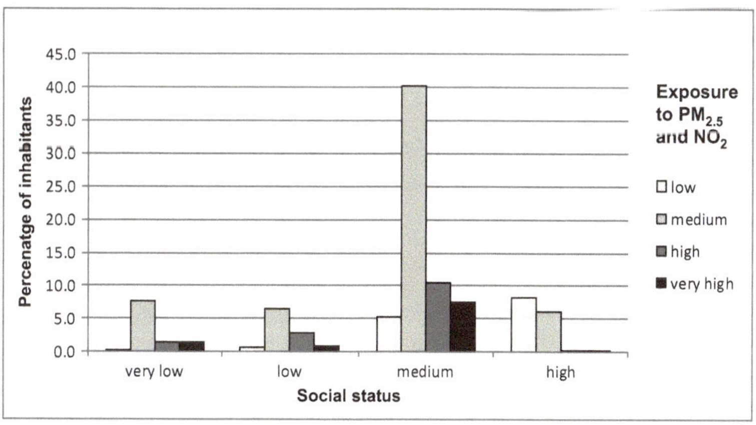

Fig. 5 Classification of Berlin's inhabitants in terms of their combined exposure to $PM_{2.5}$ and NO_2 and their social status index in 2009

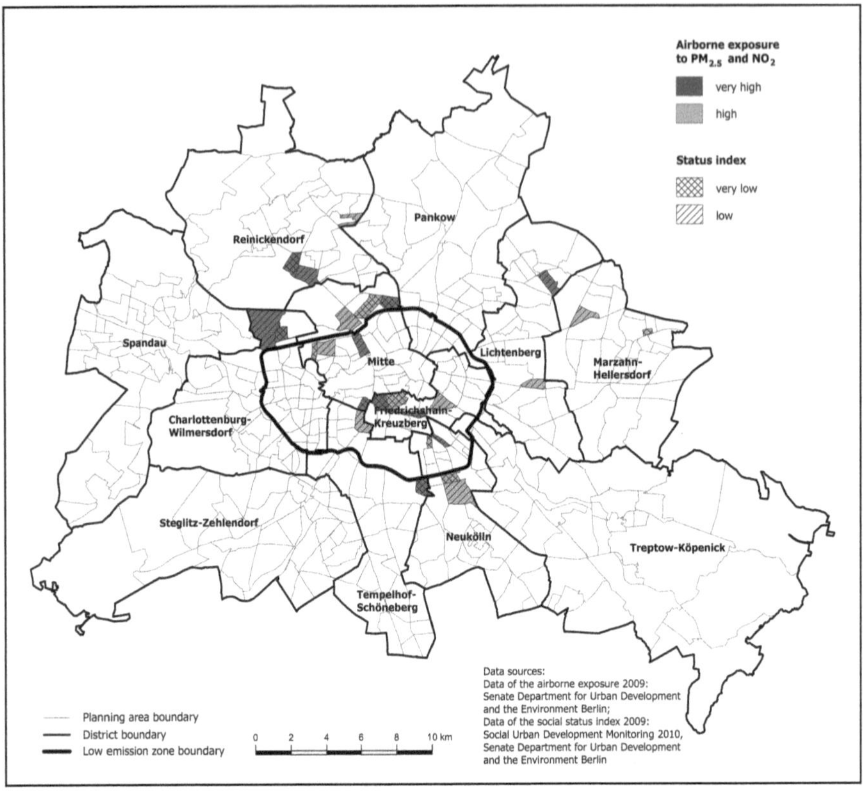

Fig. 6 Doubly disadvantaged PLAs with respect to high/very high exposure to $PM_{2.5}$ and NO_2 and very low/low social status in Berlin in 2009

Lichtenberg (2 PLAs), Marzahn-Hellerdorf (2 PLAs), and Reinickendorf (3 PLAs). From the perspective of environmental justice, these 27 PLAs are disadvantaged, compared to other PLAs of Berlin. The small-scale analysis of the exposure to $PM_{2.5}$ and NO_2 demonstrates in which parts of the city an urgent need for action exists.

7 Conclusion

In the case study of Berlin, the socio-spatial distribution of airborne outdoor exposure to $PM_{2.5}$ and NO_2 was analyzed according to the administrative division of the city into 447 PLAs. The PLAs are suitable spatial units to investigate the unequal socio-spatial distribution of airborne pollutants that adversely impact on environmental quality and, hence, well-being and quality of life. Referring to environmental justice, the immediate aim was to identify those PLAs with twofold discrimination in terms of the airborne exposure to $PM_{2.5}$ and NO_2 and the social status of their residents. It could be demonstrated that inner-urban areas of lower social status are more exposed to health-relevant environmental burdens than areas of higher social status. The results are quantitatively represented in statistics and visualized in maps.

Our research is practice-oriented. The results were applied to an up-dating of the clean air plan of Berlin (SenStadt 2013) and for decision-making support in urban planning. The method can be used as an additional instrument for environmental, social, and public health decision-making. Our results can help urban planners to develop management strategies and to implement suitable measures for improving air quality by mitigating outdoor airborne exposure, mainly in the doubly disadvantaged PLAs, but also in the other 82 PLAs with high/very high exposure to $PM_{2.5}$ and NO_2. The improvement of air quality in urban areas is a challenge and requires holistic solutions "that involve technological development, structural changes including the optimization of infrastructure and urban planning, and behavioral changes" (EEA 2015b, p. 7).The socio-spatial distribution of the outdoor airborne exposure to $PM_{2.5}$ and NO_2 could be an additional important indicator for the environmental justice index developed on the basis of noise pollution and green spaces for the PLAs in Berlin (Lakes et al. 2014). Furthermore, this approach is suitable for detecting important aspects of environmental injustice because it shows that the environmental quality and the quality of life are associated with socio-economic disparity. Both the scientific concept and the developed methodology can be transferred and adapted to other urban areas, depending on the availability of data.

Our results are helpful for decision makers such as politicians, mayors of single urban districts, urban planners, stakeholders, and the exposed population (if the results are made public), because they receive a detailed overview of the unequal socio-spatial distribution of air pollution by $PM_{2.5}$ and NO_2 in the PLAs in Berlin. Based on our results, the following research questions can be derived: In which PLAs is action needed to reduce the air pollution by $PM_{2.5}$ and NO_2 and to improve the air quality? Compared to noise, do the inhabitants perceive air pollution by $PM_{2.5}$ and NO_2 and do they know the health effects? How will urban air quality develop during the coming years? How can an increase in air pollution be prevented in PLAs of higher air quality? Which measures are suitable for reducing air pollution, in view of a population growth of about 50,000 inhabitants per year (Amt für Statistik Berlin-Brandenburg 2016b)? In addition to the low emission zone, which measures should be taken to reduce traffic-related air pollution? Does air pollution lead to an increasing socio-spatial differentiation of the population? How can environmental injustice be reduced within the city? How can the exposed population contribute to the improvement of air quality? Will it be possible to develop Berlin into a sustainable city with healthy living conditions, including good air quality, in the coming years? How does air pollution by $PM_{2.5}$ and NO_2 develop in cities of other European countries? What is the status of environmental justice in other European cities? Further investigations will also address the latest development in airborne pollution in the context of environmental justice in the city of Berlin.

Strengths and limitations of the study: The study aimed at investigating environmental justice and its violation. Particulate matter ($PM_{2.5}$) and nitrogen dioxide are the dominant air pollutants in many countries. Nevertheless, other air pollutants exist whose role with respect to human health has been recognized during recent years. For example, ultrafine and soot particles in the air can initiate and exacerbate cardio-respiratory diseases. Data on spatial variations of concentrations of these particles were not available.

Short-term periods of high pollutant concentrations may play a role in short-term health effects on humans. Therefore, they may be associated with environmental justice in a partially different way than yearly means of pollutant concentrations. Due to their availability, this study used only annual mean values of particle concentrations. On the other hand, tendencies of short-term changes of exposure concentrations may be relatively similar across the city, because they strongly depend on meteorological conditions. This means that differences between daily averages of pollutant concentrations in PLAs will typically show similar tendencies as yearly averages. Hence, it can be assumed that this study took the most important aspects of air quality associated with environmental quality and environmental justice into account.

This study is based on unique, space-resolved data on social parameters and mean ambient exposure concentrations of airborne pollutants, calculated for well-defined spatial units, the PLAs. This approach resulted in new insights into the interdependencies of social status and environmental exposure within a city with relatively clean air in a developed country. Air pollution and the availability of clean air are prominent indicators of environmental quality, quality of life, and environmental justice. All over the world, clean air, together with the access to clean drinking water, is particularly important for the protection of human health.

References

Amt für Statistik Berlin-Brandenburg (2016a) Einwohnerinnen und Einwohner im Land Berlin am 31. Dezember 2015. Statistischer Bericht A I5 – hj 2 / 15. https://www.statistik-berlin brandenburg.de/publikationen/stat_berichte/2016/SB_A01-05-00_2015h02_BE.pdf. Accessed 24 May 2016

Amt für Statistik Berlin-Brandenburg (2016b) Bevölkerungsstand Berlin 2011 bis 2015 jeweils zum 31.12. https://www.statistik-berlin-brandenburg.de/BasisZeitreiheGrafik/Zeit-Bevoelkerungsstand.asp?Ptyp=400&Sageb=12015&creg=BBB&anzwer=6. Accessed 3 Aug 2016

Anderson JO, Thundiyil JG (2012) Clearing the air: a review of the effects of particulate matter air pollution on human health. J Med Toxicol 8(2):166–175

Banzhaf E, de la Barrera F, Kindler A, Reyes-Paecke S, Schlink U, Welz J, Kabisch S (2014) A conceptual framework for integrated analysis of environmental quality and quality of life. Ecol Indic 45:664–668. doi:10.1016/j.ecolind.2014.06.002

Barcelo MA, Saez M, Saurina C (2009) Spatial variability in mortality inequalities, socioeconomic deprivation, and air pollution in small areas of the Barcelona Urban Region, Spain. Sci Total Environ 407(21):5501–5523

Bell ML, Ebisu K, Peng RD, Samet JM, Dominici F (2009) Hospital admissions and chemical composition of fine particle air pollution. Am J Respir Crit Care Med 179(12):1115–1120

Bolte G, Bunge C, Hornberg C, Köckler H, Mielck A (eds) (2012) Umweltgerechtigkeit. Chancengleichheit bei Umwelt und Gesundheit: Konzepte, Datenlage und Handlungsperspektiven. Huber, Bern

Brochu PJ, Yanosky JD, Paciorek CJ, Schwartz J, Chen JT, Herrick RF, Suh HH (2011) Particulate air pollution and socioeconomic position in rural and urban areas of the Northeastern United States. Am J Public Health 101(Suppl 1):S224–S230. doi:10.2105/AJPH.2011.300232

Brown AL (2003) Increasing the utility of urban environmental quality information. Landsc Urban Plan 65(1–2):85–93

Brunekreef B, Beelen R, Hoek G, Schouten L, Bausch-Goldbohm S, Fischer P, van den Brandt P (2009) Effects of long-term exposure to traffic-related air pollution on respiratory and cardiovascular mortality in the Netherlands: the NLCS-AIR study. Research Report Health Effects Institute 139. http://www.n65.nl/NCLS-AIR-Study-2009.pdf. Accessed 3 Aug 2016

Chen L, Mengersen K, Tong S (2007) Spatiotemporal relationship between particle air pollution and respiratory emergency hospital admissions in Brisbane, Australia. Sci Total Environ 373(1):57–67

Constanza R, Fisher B, Ali S, Beer C, Bond L, Roelof B et al (2007) Quality of life: An approach integrating opportuniteis, human need, and subjective well-being. Ecol Econ 61(2–3):267–276

Cooter EJ, Rea A, Bruins R, Schwede D, Dennis R (2013) The role of the atmosphere in the provision of ecosystem services. Sci Total Environ 448:197–208. doi:10.1016/j.scitotenv.2012.07.077

Crouse DL, Peters PA, Villeneuve PJ, Proux MO, Shin HH, Goldberg MS, Burnett RT (2015) Within- and between-city contrasts in nitrogen dioxide and mortality in 10 Canadian cities; a subset of the Canadian Census Health and Environment Cohort (CanCHEC). J Expo Sci Environ Epidemiol 25(5):482–489

Directive 2008/50/EC of the European Parliament and of the Council of 21 May 2008 On ambient air quality and cleaner air for Europe. Official Journal of the European Union L152/1 11.6.2008. http://eur-lex.europa.eu/eli/dir/2008/50/oj. Accessed 24 May 2016

EEA – European Environmental Agency (2015a) Air quality in Europe – 2015 report. EEA Report No. 5/2015. Luxembourg. http://www.eea.europa.eu/publications/air-quality-in-europe-2015. Accessed 3 Aug 2016

EEA – European Environmental Agency (2015b) SOER European briefings 2015, Air pollution. http://www.eea.europa.eu/soer-2015/europe/air. Accessed 5 May 2016

EEA – European Environment Agency (2016a) Environmental quality (definition). Glossary. Environmental Terminology and Discovery Service. http://glossary.eea.europa.eu/terminology/concept_html?term=environmental%20quality. Accessed 5 May 2016

EEA – European Environmental Agency (2016b) Air pollution http://www.eea.europa.eu/themes/air/intro. Accessed 10 Aug 2016

Elvers H-D (2011) Umweltgerechtigkeit. In: Groß M (ed) Handbuch der Soziologie. VS Verlag, Wiesbaden, pp 464–484

Elvers H-D, Gross M, Heinrichs H (2008) The diversity of environmental justice. Towards a European Approach. Eur Soc 10(5):835–856

EPA – US Environmental Protection Agency. (2016). Environmental justice. https://www.epa.gov/environmentaljustice. Accessed 17 May 2016

Eurostat (2015a) Quality of life Facts and views. Statistical books. Luxembourg. http://ec.europa.eu/eurostat/documents/3217494/6856423/KS-05-14-073-EN-N/742aee45-4085-4dac-9e2e-9ed7e9501f23. Accessed 24 May 2016

Eurostat (2015b) Eurostat Statistics Explained. Quality of life indicators – measuring quality of life. http://ec.europa.eu/eurostat/statistics-explained/index.php/Quality_of_life_indicators_-_measuring_quality_of_life. Accessed 24 May 2016

Faustini A, Rapp R, Forastiere F (2014) Nitrogen dioxide and mortality: review and meta-analysis of long-term studies. Eur Respir J 44(3):744–753

Forastiere F, Stafoggia M, Tasco C, Picciotto S, Agabiti N, Cesaroni G, Perucci CA (2007) Socioeconomic status, particulate air pollution, and daily mortality: differential exposure or differential susceptibility. Am J Ind Med 50(3):208–216

Foxon TJ, Reed MS, Springer LC (2009) Governing long-term socio-ecological change: what can the adaptive management and transition management approaches learn from each other? Environ Policy Gov 19(1):3–20

Franck U, Klimeczek H-J, Kindler A (2014a) Social indicators are predictors of airborne outdoor exposures in Berlin. Ecol Indic 36:582–593. doi:10.1016/j.ecolind.2013.08.023

Franck, U., Leitte, A. M., & Suppan, P. (2014b). Multiple exposures to airborne pollutants and hospital admissions due to diseases of the circulatory system in Santiago de Chile. Sci Total Environ, 468–469, 746–756 . doi:dx.doi.org/10.1016/j.scitotenv.2013.08.088.

Franck, U., Leitte, A. M., & Suppan, P. (2015). Multifactorial airborne exposures and respiratory hospital admissions — The example of Santiago de Chile. Sci Total Environ, 502, 114–121 . doi:dx.doi.org/10.1016/j.scitotenv.2014.08.093.

Franklin M, Zeka A, Schwartz J (2007) Association between PM2.5 and all-cause and specific-cause mortality in 27 US communities. J Expo Sci Environ Epidemiol 17(3):279–287

Garcia Diez S (2015) Indikatoren zur Lebensqualität. Vorschläge der europäischen Expertengruppe und ausgewählte nationale Initiativen. WISTA, 6. https://www.destatis.de/DE/Publikationen/WirtschaftStatistik/2015/06/IndikatorenLebensqualitaet_062015.pdf?__blob=publicationFile. Accessed 17 May 2016

Hornberg C, Bunge C, Pauli A (2011). Strategien für mehr Umweltgerechtigkeit. Handlungsfelder für Forschung, Politik und Praxis. Arbeitsgruppe 7 – Umwelt und Gesundheit, Fakultät für Gesundheitswissenschaften. Bielefeld: Universität Bielefeld

Hruba F, Fabianova E, Koppova K, Vandenberg JJ (2001) Childhood respiratory symptoms, hospital admissions, and long-term exposure to airborne particulate matter. J Expo Analysis and Environ Epidemiol 11(1):33–40

Janssen NA, Fischer P, Marra M, Ameling C, Cassee FR (2013) Short-term effects of $PM_{2.5}$, PM_{10} and $PM_{2.5\text{–}10}$ on daily mortality in the Netherlands. Sci Total Environ 463–464:20–26. doi:10.1016/j.scitotenv.2013.05.062

Johnson DL, Ambrose SH, Bassett TJ, Bowen ML, Crummey DE, Isaacson JS et al (1997) Meanings of environmental terms. J Environ Qual 26(3):581–589

Kabisch S, Kuhlicke C (2014) Urban Transformations and the Idea of Resource Efficiency, Quality of Life and Resilience. Built Environ 40(4):497–507

Klimeczek H-J (2011) Environmental justice in the Federal State of Berlin – development and implementation of a new a cross-cutting strategy. In C. Bunge, & K. Gebuhr (Ed.), Special Issue II. Environmental Justice, UMID: Umwelt und Mensch – Informationsdienst, No. 2/2011 (pp. 18–19). Federal Office for Radiation Protection (BfS), Federal Institute for Risk Assessment (BfR), Robert Koch Institute (RKI), Federal Environmental Agency (UBA)

Klimeczek H-J (2012) Umweltgerechtigkeit durch Chancengleichheit bei Umwelt und Gesundheit – Strategien auf Landesebene. In: Bolte G, Bunge C, Hornberg C, Köckler H, Mielck A (eds) Umweltgerechtigkeit. Chancengleichheit bei Umwelt und Gesundheit: Konzepte, Datenlage und Handlungsperspektiven. Huber, Bern, pp 205–218

Köckler H (2011) Ein Modell zur Analyse umweltbezogener Verfahrensgerechtigkeit. Umweltpsychologie 15(2):93–113

Lakes T, Brückner M, Krämer A (2014) Development of an environmental justice index to determine socio-economic disparities of noise pollution and green space in residential areas in Berlin. J Environ Plan Manag 57(4):538–556

Lall R, Ito K, Thurston GD (2011) Distributed lag analyses of daily hospital admissions and source-apportioned fine particle air pollution. Environ Health Perspect 119(4):455–460

Laurent E (2010) Environmental justice and environmental inequalities: a European perspective. Document de travail, N° 2010–05. Observatoire Francais des Conjunctures Économiques, Paris. http://www.ofce.sciences-po.fr/pdf/dtravail/WP2010-05.pdf. Accessed 24 May 2016

Marans RW, Stimson RJ (eds) (2011) Investigating quality of urban life. Theory, methods, and empirical research, Social indicators research series, vol 45. Springer, Dordrecht

Maschewsky W (2001) Umweltgerechtigkeit, Public Health und soziale Stadt. Frankfurt a. M. VAS

Maschewsky W (2004) Umweltgrechtigkeit – Gesundheitsrelevanz und empirische Erfassung. Veröffentlichungsreihe der Arbeitsgruppe Public Health. Forschungsschwerpunkt Arbeit. Sozialstruktur und Sozialstaat. Wissenschaftszentrum Berlin für Sozialforschung (WZB). https://bibliothek.wzb.eu/pdf/2004/i04-301.pdf. Accessed 9 May 2016

Maschewsky W (2006) "Healthy public policy" – am Beispiel der Politik zu Umweltgerechtigkeit in Schottland. In: Veröffentlichungen der Forschungsgruppe Public Health. Sozialforschung (WZB), Wissenschaftszentrum Berlin für. https://bibliothek.wzb.eu/pdf/2006/i06-304.pdf. Accessed 24 May 2016

Mills IC, Atkinson RW, Kang S, Walton H, Anderson HR (2015) Quantitative systematic review of the associations between short-term exposure to nitrogen dioxide and mortality and hospital admissions. BMJ Open 5(5):e006946. doi:10.1136/bmjopen-2014-006946

Minkos A, Dauert U, Schütze G, Feigenspan S, Himpel T, Kessinger S (2016). Luftqualität 2015. Vorläufige Auswertung. https://www.umweltbundesamt.de/sites/default/files/medien/376/publikationen/luftqualitaet_2015_vorlaeufige_auswertung.pdf. Accessed 24 May 2016

Noll H-H (2000) Konzepte der Wohlfahrtsentwicklung: Lebensqualität und "neue" Wohlfahrtskonzepte. WZB Discussion Paper, No. P 00–505. Wissenschaftszentrum Berlin. https://www.econstor.eu/dspace/bitstream/10419/50283/1/311841732.pdf. Accessed 9 May 2016

OECD – Organisation for Economic Co-operation and Development (2008) OECD Key Environmental Indicators. OECD Environment Directorate Paris, France. https://www.oecd.org/env/indicators-modelling-outlooks/37551205.pdf. Accessed 24 May 2016

Pacione M (2003) Introduction on urban environmental quality and human well-being. Landsc Urban Plan 65(1–2):1–3

Pascal M, Corso M, Chanel O, Declercq C, Badaloni C, Cesaroni G et al (2013) Assessing the public health impacts of urban air pollution in 25 European cities: Results of the Aphekom project. Sci Total Environ 449:390–400. doi:10.1016/j.scitotenv.2013.01.077

Perez L, Grize L, Infanger D, Kunzli N, Sommer H, Alt GM, Schindler C (2015) Associations of daily levels of PM10 and NO2 with emergency hospital admissions and mortality in Switzerland: Trends and missed prevention potential over the last decade. Environ Res 140:554–561. doi:10.1016/j.envres.2015.05.005

Pope CA 3rd, Burnett RT, Turner MC, Cohen A, Krewski D, Jerrett M, Thun MJ (2011) Lung cancer and cardiovascular disease mortality associated with ambient air pollution and cigarette smoke: shape of the exposure-response relationships. Environ Health Perspect 119(11):1616–1621

Prochaska JD, Nolen AB, Kelley H, Sexton K, Linder SH, Sullivan J (2014) Social determinants of health in environmental justice communities: examining cumulative risk in terms of environmental exposures and social determinants of health. Hum Ecol Risk Assess 20(4):980–994

Rice LJ, Jiang C, Wilson SM, Burwell-Naney K, Samantapudi A, Zhang H (2014) Use of segregation indices, townsend index, and air toxics data to assess lifetime cancer risk disparities in urban Charleston, South Carolina, USA. Int J Environ Res Public Health 11(5):5510–5526

Schwela D (2000) Air pollution and health in urban areas. Rev Environ Health 15(1–2):13–42

SenStadt (2005) Luftreinhalteplan und Aktionsplan für Berlin 2005–2010. Senatsverwaltung für Stadtentwicklung Berlin.

SenStadt (2007) Monitoring Soziale Stadtentwicklung Berlin 2007. Fortschreibung für den Zeitraum 2005–2006. Senatsverwaltung für Stadtentwicklung Berlin. http://www.stadtentwicklung.berlin.de/planen/basisdaten_stadtentwicklung/monitoring/download/2007/Endbericht-Monitoring2007.pdf. Accessed 24 May 2016

SenStadt (2010) Monitoring Soziale Stadtentwicklung 2010. Fortschreibung für den Zeitraum 2008–2009. Senatsverwaltung für Stadtentwicklung Berlin. http://www.stadtentwicklung.berlin.de/planen/basisdaten_stadtentwicklung/monitoring/download/2010/monitoring_soziale_stadtentwicklung_endbericht_2010.pdf. Accessed 24 May 2016

SenStadt (2013) Luftreinhalteplan 2011 bis 2017 für Berlin. Senatsverwaltung für Stadtentwicklung und Umwelt Berlin. http://www.stadtentwicklung.berlin.de/umwelt/luftqualitaet/de/luftreinhalteplan/. Accessed 3 Aug 2016

Shi L, Zanobetti A, Kloog I, Coull BA, Koutrakis P, Melly SJ, Schwartz JD (2016) Low-Concentration PM2.5 and Mortality: Estimating Acute and Chronic Effects in a Population-Based Study. Environ Health Perspect 124(1):46–52

Slaughter JC, Kim E, Sheppard L, Sullivan JH, Larson TV, Claiborn C (2005) Association between particulate matter and emergency room visits, hospital admissions and mortality in Spokane, Washington. J Expo Anal Environ Epidemiol 15(2):153–159

UBA – Umweltbundesamt (2015) Umweltgerechtigkeit im städtischen Raum – Entwicklung von praxistauglichen Strategien und Maßnahmen zur Minderung sozial ungleich verteilter Umweltbelastungen. Umwelt & Gesundheit 01/2015. Dessau-Roßlau: Umweltbundesamt

UBA – Umweltbundesamt (2016) Entwicklung der Luftqualität. https://www.umweltbundesamt.de/themen/luft/daten-karten/entwicklung-der-luftqualitaet. Accessed 3 Aug 2016

IVU Umwelt GmbH (2011) Fortschreibung des Berliner Luftreinhalteplans (LRP) 2009–2020. Endbericht. http://www.stadtentwicklung.berlin.de/umwelt/luftqualitaet/de/luftreinhalteplan/download/lrp_fortschreibung.pdf. Accessed 17 May 2016

UN – United Nations (2015). Transforming our world. The 2030 agenda for sustainable development. https://sustainabledevelopment.un.org/content/documents/21252030%20Agenda%20for%20Sustainable%20Development%20web.pdf. Accessed 3 Aug 2016

UNDP – United Nations Development Programme (2011) Human Development Report 2011. Sustainablity and Equity: A better Future for All. United Nations Development Programme (UNDP), 2011. http://hdr.undp.org/sites/default/files/reports/271/hdr_2011_en_complete.pdf. Accessed 3 Aug 2016

UNEP/UNECE – United Nations Environment Programme/United Nations Economic Commission for Europe. (2016). GEO-6 Assessment for the pan-European region. United Nations Environment Programme, Nairobi, Kenya. http://www.unep.org/roe/Portals/139/documents/GEO-6%20Assessment%20for%20the%20pan-European%20region%20EN.pdf. Accessed 3 Aug 2016

Vera J, Cifuentes L (2008) Association of Hospital Admissions for Cardiovascular Causes and Air Pollution (PM10, PM25 and O3) in Santiago, Chile. Epidemiology 19(6):S368. doi:10.1097/01.ede.0000340464.17551.d4

Walker G (2012) Environmental Justice: Concepts, Evidence and Politics. Routledge, London

WBGU – German Advisory Council on Global Change (2016) Humanity on the move: Unlocking the transformative power of cities. Berlin, WBGU. http://www.wbgu.de/fileadmin/user_upload/wbgu.de/templates/dateien/veroeffentlichungen/hauptgutachten/hg2016/hg2016_en.pdf. Accessed 4 Aug 2016

Welsch J, Bömermann H, Nagel H (2011) Data sources of Berlin pilot project: the Berlin Environmental Atlas and Social Urban Development Monitoring. In C. Bunge, & K. Gebuhr (Ed.), *Special Issue II. Environmental Justice*, UMID: Umwelt und Mensch – Informationsdienst, No. 2/2011 (pp. 20–24). Federal Office for Radiation Protection (BfS), Federal Institute for Risk Assessment (BfR), Robert Koch Institute (RKI), Federal Environmental Agency (UBA). https://www.umweltbundesamt.de/sites/default/files/medien/pdfs/umid0211-e.pdf. Accessed 24 May 2016

WHO – World Health Organisation (1997) WHOQOL Measuring Quality of Life. Programme on Mental Health and Prevention of Substance Abuse. http://www.who.int/mental_health/media/68.pdf. Accessed 9 May 2016

WHO – World Health Organisation (2006a) WHO Air quality guidelines for particulate matter, ozone, nitrogen dioxide and sulfur dioxide. Global update 2005. Summary of risk assessment. http://apps.who.int/iris/bitstream/10665/69477/1/WHO_SDE_PHE_OEH_06.02_eng.pdf. Accessed 24 May 2016

WHO – World Health Organisation (2006b) Air quality guidelines. Global update 2005. Particulate matter, ozone, nitrogen dioxide and sulfur dioxide. http://www.euro.who.int/__data/assets/pdf_file/0005/78638/E90038.pdf. Accessed 24 May 2016

WHO – World Health Organisation (2014) Burden of disease from Ambient Air pollution for 2012. Summary of results. http://www.who.int/phe/health_topics/outdoorair/databases/AAP_BoD_results_March2014.pdf. Accessed 3 Aug 2016

WHO – World Health Organisation (2016) Ambient Air Pollution Database, May 2016. http://www.who.int/entity/phe/health_topics/outdoorair/databases/WHO_AAP_database_May2016_v3web.xlsx. Accessed 3 Aug 2016

Yap PS, Gilbreath S, Garcia C, Jareen N, Goodrich B (2013) The influence of socioeconomic markers on the association between fine particulate matter and hospital admissions for respiratory conditions among children. Am J Public Health 103(4):695–702

Young GS, Fox MA, Trush M, Kanarek N, Glass TA, Curriero FC (2012) Differential exposure to hazardous air pollution in the United States: a multilevel analysis of urbanization and neighborhood socioeconomic deprivation. Int J Environ Res Public Health 9(6):2204–2225

What Really Matters in Green Infrastructure for the Urban Quality of Life? Santiago de Chile as a Showcase City

Ellen Banzhaf, Sonia M. Reyes-Paecke, and Francisco de la Barrera

1 Introduction

The built, green and social environment express the situation of a city and, to a large extent, indicate the development of the urban area. These components of the urban environment have a strong impact on the quality of life of citizens. Along with the concepts of resource efficiency and resilience in cities, the quality of life forms one of the three pillars on which our research on urban transformations is grounded. We approach the concept of quality of life from the environmental perspective and understand the human well-being as an integral part of the broader concept of quality of life. In this study we focus on green infrastructure (GI) as an indication for quality of life research. Here, we measure the extent to which people can access GI as a service and profit from this infrastructure for health-related and social dimensions (Scottish Executive 2005; Bognar 2005). Rapid urbanisation processes accelerate land-use changes that mostly go along with extensive urban land consumption and involve population developments. Such multi-dimensional changes in urban land use and land-consumption patterns are very dynamic and widely ramified.

E. Banzhaf (✉)
Department of Urban and Environmental Sociology, Helmholtz Centre for Environmental Research - UFZ, Leipzig, Germany
e-mail: ellen.banzhaf@ufz.de

S.M. Reyes-Paecke
Department of Ecosystems and Environment, Faculty of Agronomy and Forestry Engineering, Pontificia Universidad Católica de Chile. Center for Sustainable Urban Development (CEDEUS), Santiago, Chile
e-mail: sreyespa@uc.cl

F. de la Barrera
Institute of Geography, Centre for Urban Sustainable Development, Pontificia Universidad Católica de Chile, Santiago, Chile
e-mail: fdelabarrera@uc.cl

S. Kabisch et al. (eds.), *Urban Transformations*, Future City 10,
DOI 10.1007/978-3-319-59324-1_15

They can evoke unsustainable structures that entangle social-spatial differentiations which are discussed in the context of urban growth and shrinkage processes (see Haase et al., "From Shrinkage to Regrowth: The Nexus Between Urban Dynamics, Land Use Change and Ecosystem Service Provision", in this volume, dealing with urban dynamics, Seto et al. 2011; Kabisch and Kuhlicke 2014). As land is a limited and contested resource, it demands for infrastructural provision and, particularly with respect to urban quality of life and a sustainable urban development, for the provision of green infrastructure.

Urbanisation affects the development of land use in different ways and with varying consequences for different social strata even within the same urban area. Hence, land-use changes must be geared in a resource-efficient and resilient manner (UN 2016) to secure the environmental quality for all residents regardless their social strata. In this sense, urban land-use structure pictures environmental aspects of the quality of life, and relates to social-spatial disparities (see Kindler et al., "Socio-spatial Distribution of Airborne Outdoor Exposures – An Indicator for Environmental Quality, Quality of Life, and Environmental Justice: The Case Study of Berlin", in this volume). Regarding social-environmental aspects, land-use dynamics are the trigger to accelerate or decelerate urban transformations towards a sustainable urban development at various scales (Banzhaf et al. 2017).

A variety of commitments for a sustainable land use have been elaborated at both international and national level, e.g., the Rio 20 outcomes call for a neutral world without land degradation to solve the problem of land consumption (UN 2012). In this respect the EU has an objective of 'no net land take' by 2050 (EEA 2015), but Latin America is still far from adopting this type of political agreements because in many cities, major housing and infrastructure deficits persist and also steer further urban expansion (UNEP 2010). Beyond the high share of European and US citizens (82%, UN 2011; UN 2012), more than 90% of Latin Americans live in cities, thus making this continent to be the most urbanised on earth (UN 2012; World Bank 2014). Especially in Latin American countries such as Chile which have a strong socio-spatial differentiation environmental dimensions of quality of life, also depend on the social strata of the population. To illustrate these dimensions we exemplify our study on Santiago de Chile.

2 Tackling Questions of Green Infrastructure as a Major Component of Urban Land Use

Green infrastructure (GI) is defined as the interconnected network of natural and semi-natural areas with other environmental features to secure urban ecosystem services. Therefore it supports plant and animal species, maintains natural ecological processes in rural and urban areas, and contributes to the health and quality of life for human beings (CMP 2013; The Conservation Fund 2016; McDonald et al. 2005).

As mentioned above, Latin America lacks from taking on appropriate political conventions, and cities like Santiago de Chile are exposed to further acceleration of urban processes such as urban land use change accompanied by shortcomings in environmental quality. In contrast, in Europe, the policy of GI has been elaborated under the EEA guise of 'Science for Environment Policy' (EEA 2011). The European Commission (EC 2013) emphasizes that GI policy has the capability to offer win-win, or 'no-regret' solutions (see Knapp et al., "Do Urban Biodiversity and Urban Ecosystem Services Go Hand in Hand, or Do We Just Hope It Is That Easy?", in this volume) and could promote integrated spatial planning by identifying multi-functional zones and by incorporating habitat restoration measures into land use plans and management strategies (EEA 2011). Hence, we want to shed light to the relation between urban quality of life and GI in the Latin American context.

In our view, it is important to analyse the various characteristics of GI for the cornerstones of urban transformations, foremost its role for a resource-efficient handling of urban land and a resilient and well-adapted vegetation cover to secure a sustainable quality of life in cities. Therefore we want to tackle the following question on different spatial scales to understand prospective land-use options related to GI: How can green infrastructure contribute to residents' well-being and thus maintain or even enhance urban quality of life in changing urban environments? In this sense it is important to evaluate the multiple contributions of GI for ecosystem services, against different socioeconomic conditions, and in different urban and suburban settings by using a refined set of indicators. Comprehending the facets of GI's regional and local specifications we will shed light to their effects at each particular scale for Santiago de Chile as case study which is characterised by a very fast urban growth and an extreme social-spatial differentiation.

GI covers all kinds of vegetated urban spaces, with diverse forms and structures with contributing multiple functions (Jim et al. 2015) and is one of the major suppliers of urban ecosystem services. Typically, GI refers to an interconnected network of such multifunctional green spaces, gardens, street aligned trees which is strategically planned and managed to supply a range of such services (Benedict and McMahon 2006; Wright 2011). Its development has been driven by changes in local demand and urban form over time, but in the past decade it has attracted burgeoning interest, including notably its potential as a climate change intervention (Hansen et al. 2016; Naumann et al. 2011a). It corresponds to the spatial structure of areas covered by vegetation may it be natural or semi-natural, adjacent to water bodies or to buildings and public infrastructure. Therefore GI is the basis for the provision and regulation of urban ecosystems to serve as multiple benefits for citizens (EC 2016). The high level of return over time gained from healthy GI supplies a higher environmental quality of life and is complementary to the so-called grey infrastructure. Depending on its size, form, location, internal facilities and configuration, the emanating GI can be of advantage for human well-being on the local and regional level of cities.

Citizens have strong needs for GI and related ecosystem services and, at the same time, they produce far-reaching environmental implications. Urban GI is a specific type of goods that supplies services by natural structures designed or conserved (Naumann et al. 2011b), and needs to be constantly maintained in cities. As land consumption is a critical issue regarding urban sustainability, the maintenance, connectivity and distribution of GI are essential parameters to secure resource-efficient urban land use. Beyond its capacity to amplify resilience of cities (e.g., micro-climate regulation, capture of pollutants, flood regulation) GI can also improve the urban quality of life by ensuring cultural services (e.g., recreation, aesthetics), especially in densely populated cities. Urban ecosystem services are generally understood as the direct and indirect contributions of ecosystems to human well-being (TEEB 2010, p. 33) for which GI in cities accounts for environmental health (De Groot et al. 2010; Haines-Young and Potschin 2010). Being recognized and promoted differently by people and local governments according their priority needs and values, urban GI offers a great diversity of tangible and intangible benefits for citizens.

3 Methodological Considerations

3.1 GI from Different Spatial Perspectives

As GI delivers essential ecosystem services in a multi-faceted way our goal is to understand its regional and local characteristics and their impact better at the respective scale. In a multi-scale approach we assess the quantity and quality of different kinds of GI to get findings on its sensitivity to environmental pressures related to dynamic urban processes (see Fig. 1). Hence, we give spatially explicit evidence on urban GI as a supplier for ecosystem services to preserve or enhance the environmental quality of life. Keeping multifunctional aspects of GI in mind, we refer to the urban quality of life from environmentally driven considerations on human well-being in urban areas.

To comprehend the role of GI as a proxy for environmentally induced quality of life we elaborate a differentiated picture on multiple scales. Key aspects are the urban structure which may result in (un)evenly distributed GI, and potential social inequalities in relation to this structure. Thus, the various facets of GI may be a superior indicator for different urban contexts with different environmental goods expressing different levels of quality of life in the entire urban area of Santiago de Chile. For a differentiated comprehension of its significance, we perform a down-scaling from the urban central municipalities to suburban development, from one municipality to another, and even from one neighbourhood to the adjacent one.

Figure 1 depicts our scale-dependent concept starting from the entire urban area on the regional scale, continuing with comparisons of different municipalities on the intermediate local scale and finally focussing on one suburban municipal neighbourhood. Dependent on the respective scale we evaluate the GI to first illustrate different urban and suburban configurations (urban scale), then discuss green spaces in socio-spatially differentiated urban environments and their corresponding role for ecosystem services, and finally shed light on neatly differentiated vegetation in a newly developed suburban neighbourhood to explain the plant-soil complex.

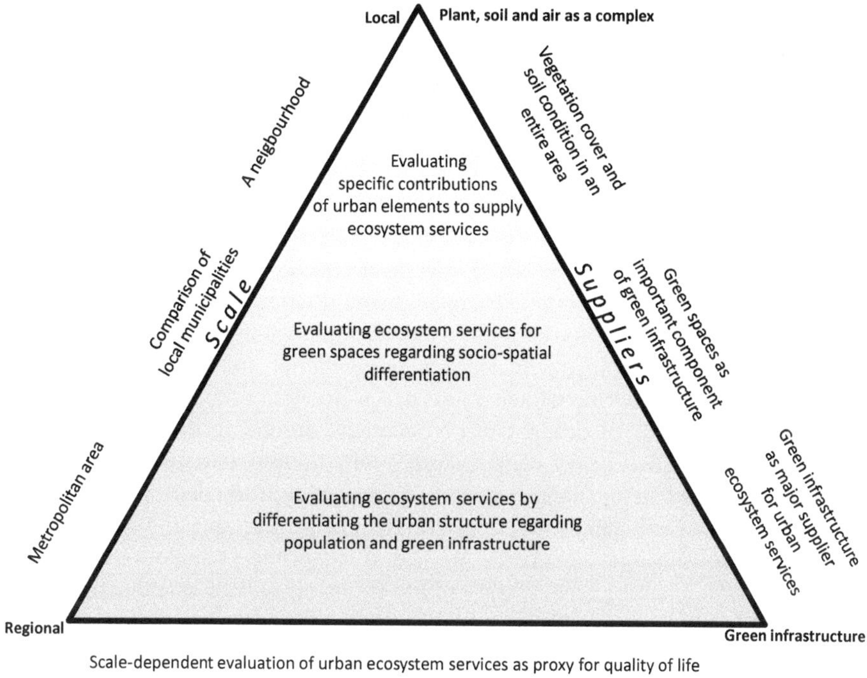

Fig. 1 Scale-dependent evaluation of green infrastructure (GI) as main constituent for urban quality of life

Distinctions between GI, green spaces and vegetation cover need to be pointed out. As defined above, GI refers to the performance of green components as functional elements, implies connections and thus establishes a network. Public green spaces are one of the most important components of GI, they are considered as public goods which allow free access to all citizens and represent pockets of nature for all residents, but their coverage of vegetation can strongly differ. Vegetation cover, however, is a broad concept with a diverse structural pattern. It comprises root penetration, ramification, foliation, and adheres to different types of grass, flowers, shrublands, and trees (De la Barrera et al. 2016a, b; Lehmann et al. 2014).

3.2 *Santiago de Chile as a Showcase City*

Many growing cities are characterised by their high heterogeneity and assemble a vast number of large residential and commercial developments reaching from core areas far into the urban fringe, sometimes with much environmentally and economically degraded space in-between (Loibl and Toetzer 2003; Wu 2010). As a complex phenomenon, suburbanisation is thus emerging as intertwined trajectories with locally pronounced developments. The sprawl into suburban areas is not a single urban feature and does not steadily proceed outwards from a single centre. Urban and suburban dynamics entail multifarious problems, which include various combinations of environmental implications such as superimposed soil sealing, inadequate configuration with GI, and socio-spatial inequalities amongst others.

Like so many other (emerging) megacities, Santiago de Chile has high urban growth, a tremendous increase in population, suburbs and leap-frog development which transforms land use. Such unrestricted urban expansion evolves a high pressure on the urban region, converts open spaces into urbanised land-use types and produces unbalanced distribution and quality of GI due to increasing demand on housing and public infrastructure. There are many reasons to investigate GI in this city. Amongst others our showcase city is located in a valley framed by the Andes and the Pacific Coastal Range with a lack of air ventilation and subsequent episodes of severe air pollution during winter. Santiago is a city with a Mediterranean climate implying long periods of drought in hot summers producing intense urban heat island effects, and short heavy winter rainfalls with flood hazards. The urban landscape comprises a heterogeneous pattern of urbanisation, from densely built to disperse urban areas with a high structural diversity, being surrounded by agricultural land and native vegetation remnants. As such, it is a prime example for a city that can enhance the environmental quality of life for its citizens by establishing a sustainable GI and, consequently, providing and regulating ecosystem services (e.g., cooling effects; Inostroza et al. 2016). Like the strong socio-spatial differentiation in so many soaring cities, Santiago de Chile can be characterised by high urban diversity in terms of (1) population and housing densities, with its highly dynamic processes of urbanisation by encroaching upon the surrounding landscapes; and (2) GI and its extremely different expression in the residential areas. From urban to municipal to neighbourhood scale the different components of GI need to be investigated twofold: in terms of their composition and structure, and their distribution and accessibility (Reyes-Paecke and Figueroa 2010).

3.3 *Elaborated Methods*

In our approach we carry out an enhanced multiple scale analysis by means of remote sensing techniques, spatial analysis, population statistics, field mapping and social survey. This mixed method approach allows us to comprehend the significance of GI for urban quality of life at various scales and for selected municipalities and neighbourhoods. For the entire urban area of Santiago de Chile and their respective 34 municipalities we quantify land-use and population dynamics in space and time as well as their share in green spaces. To do so, we need to know the amount, distribution, share, and configuration of GI which we gain on the basis of the following performance and sustainability indicators: fraction of *population density* for the entire administrative units compared to the fraction for the particular *built-up areas*, densities *of built-up areas* regarding their *degree of imperviousness*, as well as *public green spaces*, their *dynamics*, and their ratio *per capita*. To gain all these pieces of information we combine remote sensing techniques with GIS exploiting satellite data, census data and additional population statistics. Based on that knowledge we then cluster the urban area into *urban* and *suburban* municipalities for which we get findings about their dichotomy and similarities regarding structural characteristics of their environmental quality. Based on this detailed spatial analysis we extract three socioeconomically contrasting municipalities for a further and much more refined picture of the urban and suburban situation regarding their green spaces' *quality*, *quantity*, *spatial distribution* and *accessibility*. For the latter we evaluate these municipalities through social survey techniques. By in-situ observation of people's use of green spaces on a random sample size and further interviews with various users of public green spaces, we gain a deeper understanding about such qualitative aspects as their attitude, perception of safety, social interactions and appreciation. Finally, a special attention is exemplarily paid to a newly constructed suburban area and its respective neighbourhoods, to derive a sophisticated knowledge on its implication for urban ecosystem services within and beyond such a neighbourhood. Here, its vegetation composition is recorded by field mapping, with added value from remotely sensed data to map the dynamics of vegetation cover from the time before the development took place and in the presence of the newly developed suburbs.

4 Results: GI Analysis at Multiple Scales

4.1 Mirroring Urban with Suburban Municipalities Auditing Dichotomy and Similarities

To understand the impact of land-use changes and population dynamics on the urban quality of life we investigate the urban area of Santiago de Chile in the juxtaposition of urban and suburban municipalities. Location and built-up densities form the criteria for this subdivision at municipal level on which we analyse changes in population densities, development of the built-up areas, and of GI. As major drivers of land-use changes we identify *population, population density* of urban and suburban municipalities and for the particular *built-up area* to understand local and regional environmental pressures on land. Areas of lower and higher supply by GI are gained from *public green spaces*, especially their *amount*, *distribution* and *dynamics*, to gather a sophisticated insight into their multifunctionality and their configuration regarding the socio-spatial differentiation of Santiago de Chile. These indicators help to measure and compare the urban and suburban patterns and their impact on the urban quality (Briggs 2003; OECD 1997; UNEP 2010). The size and shape of the green areas, as well as the density and diversity of vegetation present therein are key factors to the ability to provide ecosystem services. The larger green areas allow for greater plant diversity, they contribute more effectively to the microclimatic regulation and to reduce stormwater runoff. Therefore the structural attributes are good proxies for estimating ecosystem services provision (Cilliers et al. 2013; Inostroza et al. 2016; Lehmann et al. 2014).

According to its urban structure, we divided the city into 21 urban and 13 suburban municipalities covering an area of 2274 km^2. In 2009, Santiago was home to approximately 6.2 million inhabitants, which is equivalent to a population density of 2734 inh./km^2 (Encuesta CASEN 2009). Romero et al. (2012, p. 81) point to the high urbanisation process. They show that the built-up area almost doubled between 1975 and 2009 from approximately 341 km^2 to about 616 km^2. In this respect urban sprawl has developed in an uneven suburban pattern over space and time being home to low and middle class population in dense settlements and offering space for disperse high-class areas.

Despite a generally very positive population growth, the 21 urban municipalities are characterised by a residential decline of −6.5% (1992–2002) and −10% (2002–2009). Here, approximately 90% of the space is built-up, leaving little area for GI to supply the citizens with ecosystem services. The dynamics of public green spaces is not well pictured by statistics only. Although they increased by 1.5% in the urban municipalities between 2001 and 2006, their spatial distribution varies greatly. In the urban area, parks were created for low-income families and scarce green spaces due to a program of the Ministry of Housing (called Urban Parks Program). In contrast, the 13 suburban municipalities have gained tremendously in the number of residents by 40% and 37% in the respective intervals. Although, in average, suburbs have less than 50% of their surface sealed by building activities, the tendency for further densification is observed if the natural setting (on foothills or along valleys) allow for. Here, public green spaces gain 0.4% in the same period, documenting that in suburban municipalities the values are lower while the built-up areas grow faster (Banzhaf et al. 2013). Differences in municipal policies are reflected in individual positive or negative changes, since some municipalities allocated resources to increase green spaces, other municipalities did not set the same priority. In some suburban municipalities least public green spaces occur where private green spaces are biggest, mirroring the high-income sectors. In these few suburban municipalities private green spaces and the related ecosystems are well supplied and the quality of life secured without much public space needed.

Predominantly though, the spatial analysis shows the necessity to improve environmental quality in GI twofold: in the urban core area most space is being built-up, and most green spaces refer to historical parks. Hence, the environmental quality of life could be improved by more vegetation cover (e.g., street trees, green façades and roofs), so that different types of GI provide benefits for residents. Suburban municipalities of lower to middle classes are densely built-up whereas other municipalities in the outskirts with higher socio-economic classes signify low-density residential areas. Therefore GI varies more in the suburban municipalities and generally provides very little green spaces. Less affluent municipalities offer smaller patches of land to residents with less open areas and have less financial resources to irrigate green spaces. The municipal effort to increase the number of green spaces is a positive planning aspect where urban areas still have a scarce amount of green spaces. Likewise do more affluent municipalities establish few public green spaces, but their local dwellers own large private gardens and have less need for GI on public grounds. They can provide irrigation for public and private spaces for which the supply with ecosystem services is ensured at local level. Such diverse socio-spatial structures have different physical settings, some of them do not meet the greater needs for GI. For a regional comparison of large agglomerations as exemplified here, the amount of public green spaces per capita serves well to give evidence on socio-ecological priority and deficit areas. Applying this indicator (minimum of 9–11 m^2 of green spaces per inhabitant to contribute to an adequate quality of life) we can estimate the capacity of public green spaces as ecosystem service supplier that serves as a standard quality of life measure (UNEP 2010, p. 157). During the last decade (2001–2006), most of Santiago de Chile showed an increase, but the amplitude differs: in the urban municipalities GI rises from 5.7 to 7.9 m^2/inh. (+38%) and from 4.0 to 5.7 m^2/inh. in the suburbs (+44%). The differences could continue to lessen if this trend goes on. Individual distinctions are not related to urban or suburban conditions, since municipalities with very low values (less than 5 m^2/inh.) exist in both areas. Only seven municipalities exceed 10 m^2/inh. in 2006, five of which are urban and two are suburban. In urban municipalities the increase is due to two simultaneous processes: the creation of new public green spaces, and the decrease in residents (for further details of this study see Banzhaf et al. 2013).

Other suburban areas of Santiago fulfil the commonly significant and negative relationship between built-up areas and vegetation cover (Banzhaf et al. 2013). From the perspective of GI as supplier for ecosystem services, most of these suburbs are densely built-up, offer little space for vegetation and therefore experience a shortage in regulating, providing and cultural ecosystem services. Although having increased their public green spaces after 2000, the overall vegetation cover has remained little in the built-up area of these municipalities (Banzhaf et al. 2013, p. 188). Only those newly developed suburbs along the Andean piedmont differ substantially for which we discuss their value in a subsequent section (see Sect. 4.4 in this chapter).

4.2 *Inter-Municipal Inequalities in the Supply of Green Spaces*

For the socio-ecological situation complementary indicators must still be considered to get a more comprehensive view when testing the relationship between environmental and socio-spatial inequalities. While GI are suppliers for ecosystem services and especially trees are relevant at the regional and even global level for carbon sequestration, other services are also important at less aggregated levels, where GI provides microclimate regulation, flood and landslide mitigation and cultural services such as recreation and beautification of the urban landscape. The municipal level is defined here as spatial units smaller than the urban area, according the administrative organization of each city.

Inequalities in the spatial distribution of GI generate uneven ecosystem services (in quantity and quality) throughout the urban area. The magnitude of these inequalities in Santiago de Chile have been analysed by applying a set of spatially explicit indicators for measuring green spaces' quality, quantity, spatial distribution and accessibility (De la Barrera et al. 2016a). The study results by Banzhaf et al. (2013) provides the prerequisite to select three socioeconomically differentiated municipalities (see Fig. 2) for which the elaborated indicators are applied. The municipality with the lowest income level (Cerro Navia) has the least area of green spaces per inhabitant (2.6 m^2/inh.) although these spaces occupy a larger proportion to the built-up area (4.3%) compared to the municipalities with middle-income population (2.8 m^2/inh. and 3.1% of built-up area) and high income population (7.7 m^2/inh. and 3% of the built-up area). These seemingly contradictory values are due to the marked difference in population density whose highest value is 166 inh./ha in the poorest municipality and reaches only 37 inh./ha in the municipality (Vitacura) with highest incomes. The differences for GI are magnified on the municipal surface in municipalities of low (15%), medium (25%) and high income (41%) respectively. As family income increases, the share of private spaces also increases and with it the amount of GI and thus ecosystem services supply, while in low-income municipalities public green spaces are the major suppliers (for further details see De la Barrera et al. 2016a). The comprehensive results are depicted in Table 1. They show strong differences associated with the different income levels of the population.

In Table 1 the quantity, quality and spatial distribution of public green spaces refers to the respective municipality. In Santiago de Chile, public green spaces are publicly assigned areas belonging to the respective municipality. It is of greatest relevance that their furnishing with plants and their maintenance underlies the municipal budget which must settle the score of high water prices. Significantly, vegetation cover of the poorer municipality Cerro Navia is mainly found on public green spaces, which allows for the conclusion that there is hardly any vegetation outside public green spaces, but these spaces are not well covered by vegetation either. Coincidently, the urban structure limits the available private space for vegetation to small gardens. In relation to the high number of inhabitants there are obviously disproportionately few public green spaces in this representative municipality. Therefore the share of population supplied by green spaces reaches 70% in general, but this share decreases to 41% if we define the size of a green space to be larger than 0.5 ha. Vitacura, in contrast, has rather few inhabitants but large public green spaces, which are well covered with vegetation. Hence, the share of population supplied by green spaces reaches almost 82%, and even if the size of a green space is larger than 0.5 ha the supply for inhabitants still reaches 67%. For the latter it can be observed that vegetation cover of this municipality is mainly on private grounds, explaining for the share of *public* green spaces per entire vegetation cover being comparatively low (De la Barrera et al. 2016a, Table 3).

Regulating and supporting ecosystem services can be supplied by public and private spaces, e.g., tree cover helps to mitigate the urban heat island and provides a suitable habitat for wildlife (Soule 1991). Further references support the findings of this study in which neighbourhoods with highest incomes show higher vegetation cover in general, thus lower temperatures in summer situations and even increased presence of native birds than lower-income ones (Díaz and Armesto 2003; Romero

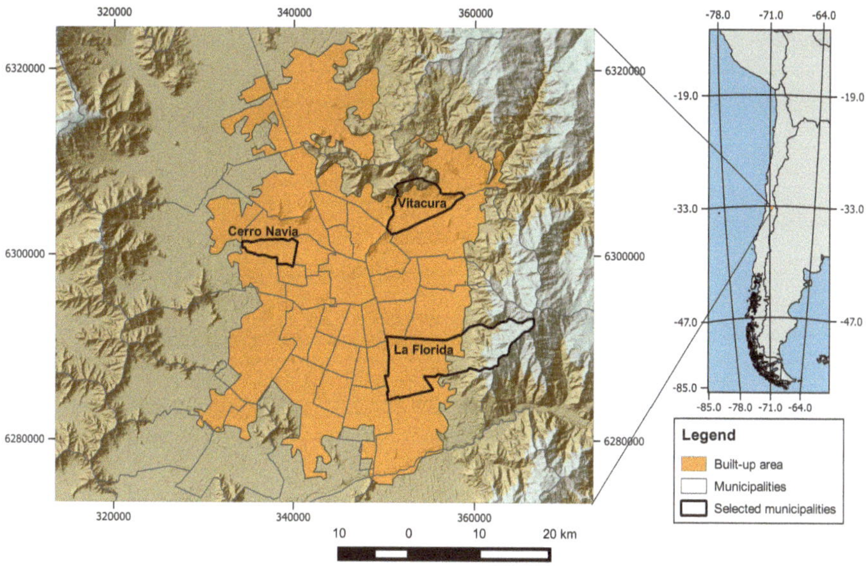

Fig. 2 Location of the urban area of Santiago de Chile (*grey* feature is the built-up area). For the study of inter-municipal inequalities three contrasting municipalities are selected: Cerro Navia (low incomes), La Florida (middle incomes), and Vitacura (high incomes) – highlighted in *bold lines* (Source: Own sources, De la Barrera et al. (2016a))

et al. 2012). Most cultural services require direct contact of citizens with nature through recreation, nature trails, socialising amongst others. As a consequence, private spaces (residential gardens, private parks) supply these services solely or mainly to their owners or to those with access to such places. For this reason public green spaces are relevant for supplying cultural services even if there are private green spaces in the vicinity. Hence, public green spaces are even more important in densely urbanised areas where many houses have no gardens. Shortages of a municipality in (public and private) GI cannot be compensated by some vegetation cover on green spaces only, because the urban structure limits any available space to be converted into new green spaces. Central urban municipalities, being poorer or richer, are mainly built-up, although some of them possess large historical parks or some vacant land. Generally speaking, they are much less flexible when it comes to developing new GI. Therefore we made our selection of contrasting municipalities along the eastern suburban fringe and in the urbanised central west.

Beyond, specific GI such as tree canopies on public land has a significant role for social, aesthetic and environmental benefits for residents (Troy et al. 2012; Salmond et al. 2016). Due to the need to build housing and infrastructure, growing cities such as the study area face a constant pressure on land. The combination of population and economic growth (which facilitates real-estate investment) and high density of urbanisation restricts the presence of environmental amenities (Livert Aquino and Gainza 2014) which include GI, and even threatens existing green spaces.

Table 1 Set of indicators for public green spaces (GS) evaluated in three contrasting urban settings

Indicators	Low income municipality Cerro Navia	Medium income municipality La Florida	High income municipality Vitacura
Quantity of GS			
GS per inhabitant			
GS per vegetation cover			
Quality of GS			
Mean size of GS			
Mean vegetation cover per GS			
Vegetation cover on GS per inhabitant			
Spatial distribution and accessibility to GS			
Share of population supplied by GS			
Share of population supplied by GS > 0,5 ha			

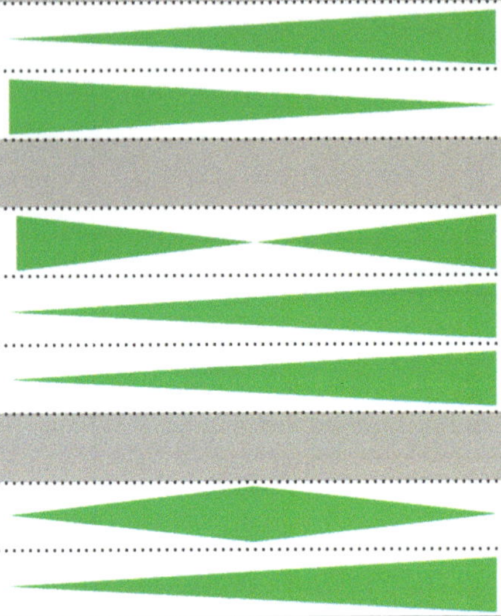

The positive correlation between the household incomes and the abundance of vegetation has been demonstrated in other cities, e.g., city of Tampa, Florida, USA (Landry and Chakraborty 2009) and Montreal, Canada (Pham et al. 2012). In Santiago de Chile, poorer and richer municipalities invest a similar share (4.1–4.7%) of their budget to tree management on green spaces (Escobedo et al. 2006) achieving different results, not only because their municipal budgets are quite different. This study shows how a high income municipality keeps an abundant vegetation cover in private gardens and on green spaces of public use that would not survive unattendedly in such a semi-arid environment where urban vegetation is dominated by exotic species requiring irrigation and maintenance. In cities located in arid and semiarid regions the correlation between income level and vegetation cover is rather evident (Jenerette et al. 2007; Halper et al. 2012). In these municipalities the low-income population faces a double constraint in regards to the maintenance of vegetation. Those contrasts are land and water scarcity. The lowest income groups live in smaller properties which again reduce the area that could be planted, and cannot devote a lot of water to irrigate the plants, due to the cost of drinking water. It also emphasizes the demand of lavishly planted public green spaces for poorer municipalities to balance out the deficiency of vegetation cover on private spaces. The mentioned evidences also suggested that the higher the household incomes, the richer the structural diversity (e.g., tree, shrub and herbs) (Reyes-Paecke and Meza 2011).

Generally speaking, the supply of multiple ecosystem services depends on the amount and diversity of GI and especially tree cover (Dobbs et al. 2011). As mentioned before, many cultural services such as recreation, contact with nature, rest and social interactions depend on the actual use of green spaces by people. The use of green spaces ensures that these services are actually delivered, but it is greatly influenced by structural attributes such as size, quantity and quality of equipment and vegetation, lighting and cleanliness, accessibility and proximity to the population (Giles-Corti et al. 2005; Lapham et al. 2016). In addition, it also depends on residential characteristics such as age, gender, physical limitations (disability), and their perception regarding the safety of these spaces (Parra et al. 2010; Lapham et al. 2016).

4.3 *The Benefits of Green Spaces for Different Neighbourhoods*

It is important to evaluate the use of green spaces in neighbourhoods of different socioeconomic status in order to know which factors are more decisive or better explain their use in different urban contexts. If these factors are known, they could contribute to better planning and maintenance of green spaces, which is a very important issue for local governments. For several reasons, Santiago is a representative for complex cultural perceptions occurring in many big and diverse cities. First, despite the differences between the three analysed neighbourhoods described above, in all cases the people that effectively use green spaces positively value their existence and their environmental and social role. Most of them consider them to be especially important for children, and also as spaces to be in contact with nature, for sports and leisure. Regarding the physical characteristics of green spaces, users consider the spaces with more trees and grass more attractive, with playgrounds, benches and lighting and those where they feel safe. To conclude, residents benefit from green spaces as cultural ecosystem services in cities.

4.3.1 The Perception of Safety

The sense of security appears to be linked to the knowledge residents have of other users, i.e., they are not perceived as strangers, but as part of the community. Insecurity seems to correspond to certain types of activities of other external users, factors such as the use of alcohol and drugs, and the absence of people during extended periods of time (De la Barrera et al. 2016c). In high income neighbourhoods the presence of security guards is frequent. According with studies conducted in other Latin-American cities (Bogotá, Colombia and Santa Cruz, Bolivia) the perception of safety is also influenced by cleanliness, lighting, the presence of guards and other visitors (Parra et al. 2010; Wright-Wendel et al. 2012). Meanwhile Lapham et al. (2016) analyse the interactions between parks equipment and perceived safety in four urban areas of the U.S. They find that parks with several different facilities – as playgrounds, tennis courts, basketball courts - were perceived as rather safe and report less incivilities which give evidence about perceived quality and safety as key factors for the quality of green spaces.

4.3.2 Socially Explicit Usage Categories

Santiago shows interesting differences in the intensity of the use of public green spaces between different neighbourhoods. In the lowest income neighbourhoods different types of users and uses are observed simultaneously. In these spaces a mixed age structure with children, youth, adults and seniors performing different planned and spontaneous activities coexist even if green spaces are smaller than 1000 m^2. In contrast, in higher income neighbourhoods the users are almost exclusively children accompanied by caring adults (parents or nannies), with little presence of youth, adults and seniors, not accompanying children (De la Barrera et al. 2016c). Consequently, the use of green spaces contrasts between planned meetings or chance encounters of neighbours in low income neighbourhoods and very low interaction in green spaces among residents of high income neighbourhoods.

4.3.3 Perception of Ecosystem Services by Residents

Despite that citizens are not explicitly aware of the concept of ecosystem services, they clearly perceive social and environmental benefits provided by green spaces, most prominently the contact with nature, physical activity, leisure, socialising with friends or neighbours, and the aesthetic value it brings to the neighbourhood. Furthermore, regulating services such as pollution abatement and temperature mitigation are appreciated by those interviewees regularly using the analysed green spaces and even by residents infrequently visiting the respective sites. Both, air pollution and high temperatures are serious environmental problems affecting Santiago at different times of the year with a rising awareness of its residents.

4.4 The Importance of GI in Neighbourhoods of Newly Constructed Suburban Areas

The suburban landscape changes fast in expanding cities like Santiago where the new urban settings have different forms, densities and landscaping depending on the concept of their development. In suburbs, population and built-up densities vary more than in central parts thus affecting the urban ecosystem services supplied by the newly planted and designed GI. In such newly built suburban neighbourhoods design options could take advantage of the GI concept to bring ecological and social systems closer together leaving soils unsealed to offer more available spaces for conservation of remnants of vegetation (Margules and Pressey 2000). Options exist to select adaptive species and to design the spatial pattern of built and green infrastructures in a more sustainable way.

The suburban area of Santiago has experienced a replacement of vegetated areas with natural and semi-natural vegetation or crops by residential uses (Romero and Vásquez 2005; Romero 2007; Pavez et al. 2010). Fuentes et al. (1984) and Holmgren (2002) also describe the replacement of native dense vegetation by a low dense open savannah of thorny vegetation. The latter represents natural but degraded vegetation, typical of agricultural abandoned lands or landscapes after fires. It used to have annual exotic grasses which increase the risks because such grasses are easy to ignite. In addition, the open savannahs facilitate the presence of exotics herbivores. All these factors obstruct the establishment and conservation of native vegetation less resistant to fires and herbivory. In addition, it can provide much less ecosystem services because of its lower biomass (Holmgren 2002; De la Barrera et al. 2016b). Thus, all these land-use changes suggest a critical loss in the composition and quality of GI, if they assume urbanisation being just sealing of soils and newly developed real estates.

Exemplified in this study is a suburban area located in the Andean piedmont where the current built-up area was a mosaic of active and abandoned agricultural lands about 60 years ago with only 68.4 ha of urbanised area, having been transformed into 1916.6 ha of suburban area by 2010 (De la Barrera et al. 2016b). Meanwhile local population increased from 13,092 to 106,187 inhabitants between 1970 and 2016 according to official data. In 2009 its population density was only 60 inh./ha in the urbanised part of the municipality thus evincing a much lower figure than the average of suburban and urban municipalities of Santiago (97 and 100 inh./ha respectively; Banzhaf et al. 2013).

These fast land-use changes brought an increase in an index referred to healthy vegetation and quantity not only at the extension of the new suburban area but also in the entire mostly rural watershed. In general, the trend shows that more urbanisation renders more vegetation cover. This trend is explained by the sustained increase of vegetation cover by replacing bare soils by a structural diversity of GI, such as street trees, green spaces and private gardens. Those changes result in an increase of three ecosystem services. First, the service of maintaining or improving air quality is reinforced because there is more biomass to capture pollutants. Secondly, rainwater infiltrates better because areas with low vegetation cover are replaced by densely vegetated areas, offsetting the increase in sealed soils. Thus, the hazard of landslides decreases in this hilly area making the area more resilient against such a natural hazard. Finally, microclimate regulation is favoured by newly planted trees providing shade. These types of GI supply the currently new suburban developed area along the foothills in a higher quantity compared to the previous historical suburban situation. Beyond, accumulated anthropogenic disturbances in rural conditions produce poor vegetation cover with less abundant biomass such as firewood, charcoal production, agricultural activities, pasture (De la Barrera et al. 2016b).

As a consequence, only those low density new suburban areas developed along the Andean piedmont mostly coincide with a high GI coverage which is configurated by trees and other native plants to secure high benefits of ecosystem services. These affluent neighbourhoods have more available private spaces to maintain vegetation and their socio-economic situation allows keeping their gardens lush. On the one side this observation confirms the simple suggestion that some specific types of urbanisations constituting of affluent residents are well supplied with provisioning of ecosystem services. On the other hand the privately established GI serves as a regulating ecosystem service reducing flooding, preventing landslides or diminishes transmitting of pollution, waste or exotic species from urban to rural settings beyond their own privileged neighbourhoods with a positive effect on the piedmont region. In particular, suburban areas can serve as a buffer between the rural surroundings and the urban core helping to mitigate these adverse effects.

5 Conclusions

Our multi-scale investigation gives empirical evidence on the differentiated evaluation of GI for the urban quality of life, exemplified by one of the fastest growing cities in Latin America. In Santiago de Chile, different urban environments can sustain the high value that GI provides for the supply of ecosystem services like increasing the cooling effects, maintaining or enhancing cultural and aesthetical values, decreasing inundations and landslides to name the most important. But these services can only be secured at neighbourhood and municipal level. Besides, GI can enhance resilience, ensuring the human well-being, and thus securing the urban quality of life at neighbourhood scale.

As a showcase city, Santiago de Chile reveals that GI can greatly contribute to the urban quality of life valued in neighbourhoods. Restrictions refer to decisions on GI that depend on the authorities at different levels of governance and the municipal budget. When we sum up our answers to the question "*What really matters in green infrastructure for the urban quality of life?*" we can state the following points of view: The most obvious implication is that the creation and conservation of public green spaces really matters, because these places are one of the most important and equitable contributors of GI and, additionally, they are highly appreciated by users. Secondly, the urban structure (i.e. patterns of spatial heterogeneity and land use) and the dynamics in urbanisation are significant factors when creating or maintaining GI as it requires available space. Finally, the structure of GI really matters, and as land is limited, this vegetation composition must ensure the quality and supply of ecosystem services in a resource-efficient way, for instance by including native plants with low water requirements. In this regard, there is a strong recommendation to other cities to consider vegetation species that are naturally adapted to the local climate which has strong advantages. Only urban planning that takes a well-adapted GI into account provides a more resilient environment to citizens and cares for a more sustainable urban development.

Cities like Santiago de Chile are facing multiple challenges: they must sort out the deficit in social infrastructures, improve their urban governance and also incorporate the scientific knowledge advances into decision making and, in parallel, they must adapt to a changing climate scenario. Regarding GI, these challenges can be solved only with a greater integration of scientific research and citizen participation in decision-making. Changes are required in current maintenance practices in order to ensure the ecosystem services provision without increasing unsustainable levels such as by high irrigation water consumption. Governance constraints become obvious as changes must be supported by the municipalities for their successful implementation. For example, in urban landscaping sustainable GI involves an important change that must be accepted by the community, understanding that this is an imperative measure for sustainability of both green spaces and the entire urban environment.

In the developing world, urban transformations have some particular complexities. What is striking in Santiago de Chile like in so many other large agglomerations (e.g., Mexico City, Buenos Aires, Pearl River Delta, Calcutta, etc.), it either does not have a central authority or the administrative boundaries do not include the urban area. The lack of this level of management makes a normative approach difficult, as a central authority could design an integrated land-use planning. Impeding is the individual land-use planning of each municipality, and the lack of strategic land-use planning for the entire urban area. This situation causes larger differences between municipalities, since each of them defines its environmental and land-use policies independently. Although there are some regional institutions (Regional Secretariat of Housing and Urbanism, Regional Government) that try to regulate regional policies (in an indicative, not in a normative way, and only by suggestions), they are not always mandatory for municipalities; one reason for the heterogeneity found in several of the studied dimensions. Therefore the role of the urban area as a political decision actor would be important for territorial choices and a sustainable development to safeguard the current and future urban quality of life.

The numerous civil society initiatives seeking to improve GI and public policies going in the same direction allow setting up the basis to design a more liveable environment and thus a consistent or even better quality of life. It must be based on the recognition that today we are building the GI that people will need throughout the next century.

References

Banzhaf E, Reyes-Paecke S, Müller A, Kindler A (2013) Do demographic and land-use changes contrast urban and suburban dynamics? A sophisticated reflection on Santiago de Chile. Habitat Int 39:179–191

Banzhaf E, Kabisch S, Knapp S, Rink D, Wolff M (2017) Integrated research on land-use changes in the face of urban transformations – an analytic framework for further studies. Land Use Policy 60:403–407

Benedict MA, McMahon ET (2006) Green infrastructure: linking landscapes and communities. Island Press, Washington, DC

Bognar G (2005) The concept of quality of life. Soc Theory Pract 321(4):561–580

Briggs D (2003) Making a difference: indicators to improve children's environmental health. World Health Organization, Geneva

Cilliers S, Cilliers J, Lubbe R, Siebert S (2013) Ecosystem services of urban green spaces in African countries – perspectives and challenges. Urban Ecosyst 16(4):681–702

De Groot RS, Alkemade R, Braat L, Hein L, Willemen L (2010) Challenges in integrating the concept of ecosystem services and values in landscape planning, management and decision making. Ecol Complex 7:260–272

De la Barrera F, Reyes-Paecke S, Banzhaf E (2016a) Indicators for green spaces in contrasting urban settings. Ecol Indic 62:212–219

De la Barrera F, Rubio P, Banzhaf E (2016b) The value of vegetation cover for ecosystem services in the suburban context. Urban Forestry & Urban Greening 16:110–122

De la Barrera F, Reyes-Paecke S, Harris J, Bascuñán D, Farías JM (2016c) People's perception influences on the use of green spaces in socio-economically differentiated neighborhoods. Urban Forestry & Urban Greening 20:254–264

Díaz IA, Armesto JJ (2003) La conservación de aves silvestres en ambientes urbanos de Santiago (wild birds conservation in urban environments of Santiago). Ambiente y Desarrollo (Chile) XIX:31–38

Dobbs C, Escobedo FJ, Zipperer WC (2011) A framework for developing urban forest ecosystem services and goods indicators. Landsc Urban Plan 99:196–206

EC – European Commission (2013) Green infrastructure (GI)—enhancing Europe's natural capital. European Commission, Brussels

EC – European Commission (2016) Environment, nature & biodiversity. Green infrastructure. European Commission, Brussels. http://ec.europa.eu/environment/nature/ecosystems/index_en.htm. Accessed Sept 2016

EEA – European Environmental Agency (2011) Green infrastructure and territorial cohesion. The concept of green infrastructure and its integration into policies using monitoring systems. Technical report 18. European Environment Agency, Copenhagen, Denmark.[online] URL: http://www.eea.europa.eu/publications/green-infrastructure-and-territorial-cohesion. Accessed Sept 2016

EEA – European Environmental Agency (2015) SOER 2015 – The European environment – state and outlook 2015. A comprehensive assessment of the European environment's state, trends and prospects, in a global context. European Environmental Agency, Copenhagen. http://www.eea.europa.eu/soer. Accessed Sept 2016

Encuesta CASEN (2009) National socio-economic characterization survey. Ministry of Planning, Government of Chile (MIDEPLAN), Santiago de Chile

Escobedo FJ, Nowak DJ, Wagner JE, De la Maza CL, Rodríguez M, Crane DE et al (2006) The socioeconomics and management of Santiago de Chile's public urban forests. Urban Forestry & Urban Greening 4(3):105–114

Fuentes E, Espinoza G, Fuenzalida I (1984) Cambios vegetacionales recientes y percepción ambiental: El caso de Santiago de Chile. Revista de Geografía Norte Grande 11:45–53

Giles-Corti B, Broomhall MH, Knuiman M, Collins C, Douglas K et al (2005) Increasing walking: how important is distance to, attractiveness, and size of public open space? Am J Prev Med 28(2S2):169–176

Haines-Young RH, Potschin M (2010) The links between biodiversity, ecosystem services and human well-being. In: Raffaelli D, Frid C (eds) Ecosystem ecology: a new synthesis, BES ecological reviews series. Cambridge University Press, Cambridge

Halper EB, Scott CA, Yool SR (2012) Correlating vegetation, water use, and surface temperature in a semiarid city: a multiscale analysis of the impacts of irrigation by single-family residences. Geogr Anal 44(3):235–257

Hansen R, Werner R, Santos A, Luz AC, Száraz L, Tosics I et al. (2016) Advanced urban green infrastructure planning and implementation – innovative approaches and strategies from European cities. Technical report, April 2016, p 205. doi: 10.13140/RG.2.1.3948.9680
Holmgren M (2002) Exotic herbivores as drivers of plant invasion and switch to ecosystem alternative states. Biol Invasions 4:25–33
Inostroza L, Palme M, De la Barrera F (2016) A heat vulnerability index: spatial patterns of exposure, sensitivity and adaptive capacity for Santiago de Chile. PLoS One 11(9):e0162464. doi:10.1371/journal.pone.0162464
Jenerette GD, Harlan SL, Brazel A, Jones N, Larsen L, Stefanov WL (2007) Regional relationships between surface temperature, vegetation, and human settlement in a rapidly urbanizing ecosystem. Landsc Ecol 22(3):353–365
Jim CY, Lo AY, Byrne JA (2015) Charting the green and climate-adaptive city. Landsc Urban Plan 138:51–53
Kabisch S, Kuhlicke C (2014) Urban transformations and the idea of resource-efficiency, quality of life and resilience: first conceptual considerations for an interdisciplinary research program. Built Environ 40(4):497–507
Landry S, Chakraborty J (2009) Street trees and equity: evaluating the spatial distribution of an urban amenity. Environ Plann A 41:2651–2670
Lapham SC, Cohen DA, Han B, Williamson S, Evenson KR, McKenzie TL et al (2016) How important is the perception of safety to park use? A four city survey. Urban Stud 53(12):2624–2636. doi:10.1177/0042098015592822
Lehmann I, Matheya J, Rößler S, Bräuer A, Goldberg V (2014) Urban vegetation structure types as a methodological approach for identifying ecosystem services – application to the analysis of micro-climatic effects. Ecol Indic 42:58–72
Livert Aquino F, Gainza X (2014) Understanding density in an Uneven City, Santiago de Chile: implications for social and environmental sustainability. Sustainability 6(9):5876–5897
Loibl W, Toetzer T (2003) Modeling growth and densification processes in suburban regions – simulation of landscape transition with spatial agents. Environ Model Softw 18(6):553–563
Margules CR, Pressey RL (2000) Systematic conservation planning. Nature 405:243–253
McDonald L, Allen W, Benedict M, O'Connor K (2005) Green infrastructure plan evaluation frameworks. J Conserv Plann 1:12–43
Naumann S, Anzaldua G, Gerdes H, Frelih-Larsen A, Davis M, Berry P et al (2011a) Assessment of the potential of ecosystem-based approaches to climate change adaptation and mitigation in Europe. Final report to the European Commission, DG Environment, Brussels
Naumann S, Davis M, Kaphengst T, Pieterse M, Rayment M (2011b) Design, implementation and cost elements of green infrastructure projects, Final report. European Commission, Brussels, p 138
OECD – Organisation for Economic Co-Operation and Development (1997) Better understanding our cities. The role of urban indicators. Organisation for Economic Co-Operation and Development, Paris, Report, 94 pp
Parra DC, Gomez LF, Fleischer NL, Pinzon JD (2010) Built environment characteristics and perceived active park use among older adults: results from a multilevel study in Bogotá. Health Place 16:1174–1181
Pavez E, Lobos G, Jaksic F (2010) Cambios de largo plazo en el paisaje y los ensambles de micromamíferos y rapaces en Chile central. Rev Chil Hist Nat 83:99–111
Pham T, Apparicio P, Séguin A, Landry S, Gagnon M (2012) Spatial distribution of vegetation in montreal: an uneven distribution or environmental inequity? Landsc Urban Plan 107(3):214–224
Reyes-Paecke S, Figueroa I (2010) Distribución, superficie y accesibilidad de las áreas verdes urbanas en Santiago de Chile. EURE 36(109):89–110
Reyes-Paecke S, Meza L (2011) Jardines residenciales en Santiago de Chile: extensión, distribución y cobertura vegetal (*Residential gardens in Santiago de Chile: extent, distribution and vegetation cover*). Rev Chil Hist Nat 84:581–592

Romero H (2007) Cambio climático y crecimiento urbano de las metrópolis chilenas. Mesa redonda sobre aspectos urbanos. Sao Paulo: III conferencia regional sobre cambios globales en América del Sur. 06 de Noviembre de 2007
Romero H, Vásquez A (2005) Evaluación ambiental del proceso de urbanización de las cuencas del piedemonte andino de Santiago de Chile. Revista EURE 94:97–18
Romero H, Vásquez A, Fuentes C, Salgado M, Schmidt A, Banzhaf E (2012) Assessing urban environmental segregation (UES): the case of Santiago de Chile. Ecol Indic 23:76–87
Salmond JA, Tadaki M, Vardoulakis S et al (2016) Health and climate related ecosystem services provided by street trees in the urban environment. Environ Health 15(Suppl 1):36. doi:10.1186/s12940-016-0103-6
Scottish Executive (2005) Quality of life and well-being: measuring the benefits of culture and sport: a literature review and thinkpiece. Scottish Executive Social Research, Edinburgh. http://www.scotland.gov.uk/socialresearch. Accessed Nov 2016
Seto KC, Fragkias M, Günerralp B, Reilly MK (2011) A meta-analysis of global urban land expansion. PLoS One 6(8):1–9
Soule ME (1991) Land use planning and wildlife maintenance: guidelines for conserving wildlife in an urban landscape. J Am Plan Assoc 57:313–323
TEEB (2010) The economics of ecosystems and biodiversity: mainstreaming the economics of nature: a synthesis of the approach, conclusions and recommendations of TEEB. http://www.teebweb.org
The Conservation Fund (2016) Strategic conservation planning. Green Infrastructure. http://www.conservationfund.org/what-we-do/strategic-conservation-planning. Accessed Sept 2016
The Conservation Measures Partnership – CMP (2013) Open standards for the practice of conservation. Version30 / April 2013. http://www.iai.int/wp-content/uploads/2015/08/CMP_Open_Standards_Version_3.0_April_2013.pdf. Accessed Sept 2016
Troy A, Grove JM, O'Neil-Dunne J (2012) The relationship between tree canopy and crime rates across an urban-rural gradient in the greater Baltimore region. Landsc Urban Plan 106:262–270
UN – United Nations (2011) World population prospects: the revision 2010. United Nations publications ST/ESA/SER.A/313 and ST/ESA/SER.A/317. http://www.un.org/en/development/desa/population/publications/pdf/urbanization/WUP2011_Report.pdf. Accessed Sept 2016
UN – United Nations (2012) World urbanization prospects the 2011 revision. United Nations, Department of Economic and Social Affairs (DESA), Population Division, Population Estimates and Projections Section, New York
UN – United Nations (2016) Global sustainable development report 2016. United Nations, Department of Economic and Social Affairs, New York
UNEP – United Nations Environment Programme (2010) Global environmental outlook: Latin America and the Caribbean (GEO LAC 3) http://www.unep.org/pdf/GEOLAC_3_English.pdf. Accessed 20 Apr 2016
World Bank (2014) The World Bank annual report 2014. © World Bank, Washington, DC. https://openknowledge.worldbank.org/handle/10986/20093 License: CC BY-NC-ND 3.0 IGO
Wright H (2011) Understanding green infrastructure: the development of a contested concept in England. Local Environ 16(10):1003–1019
Wright-Wendel H, Zarger RK, Mihelcic JR (2012) Accessibility and usability: green space preferences, perceptions, and barriers in a rapidly urbanizing city in Latin America. Landsc Urban Plan 107:272–282
Wu F (2010) Gated and packaged suburbia: packaging and branding Chinese suburban residential development. Cities 27(5):385–396

Do Urban Biodiversity and Urban Ecosystem Services Go Hand in Hand, or Do We Just Hope It Is That Easy?

Sonja Knapp, Dagmar Haase, Stefan Klotz, and Nina Schwarz

1 Introduction

Interest in urban ecosystem services has grown steadily over recent years (Haase et al. 2014), with policies promoting the enhancement of ecosystem service provision in urban areas (e.g. DG Environment 2012). Indeed, the increasing share of humans living in urban areas worldwide and environmental threats such as climate change urge us to consider ways to maintain and improve the well-being of urban dwellers. In this context, ecosystem services, which are the benefits humans receive from ecosystems (Millennium Ecosystem Assessment 2005b), are a promising supplement to technical solutions.

Urbanization often takes place in areas that are characterized by high levels of biodiversity, thus threatening global biodiversity (Kühn et al. 2004; Seto et al. 2012). At the same time, urban biodiversity is highly relevant to urban dwellers because, first, the majority of people worldwide lives in cities and urban biodiversity is thus that part of global biodiversity they can experience every day. Second,

S. Knapp (✉) • S. Klotz
Department Community Ecology, Helmholtz Centre for Environmental Research-UFZ, Halle (Saale), Germany
e-mail: sonja.knapp@ufz.de; stefan.klotz@ufz.de

D. Haase
Geography Department, Humboldt-Universität zu Berlin, Berlin, Germany
e-mail: dagmar.haase@geo.hu-berlin.de

N. Schwarz
Department of Computational Landscape Ecology, Helmholtz Centre for Environmental Research-UFZ, Leipzig, Germany
e-mail: nina.schwarz@ufz.de

S. Kabisch et al. (eds.), *Urban Transformations*, Future City 10,
DOI 10.1007/978-3-319-59324-1_16

urban biodiversity has an important value for education because it connects people who live in highly artificial systems to nature and consequently creates sufficient reasons for actions in biodiversity conservation (Haase et al. 2014). Third, diverse ecosystems are assumed to be more stable and resilient, compared to monotonous sites, to environmental and man-made hazards such as storms, pests, or climate warming, and thus could contribute positively to well-being in urban areas (Elmqvist et al. 2013). Fourth, biodiversity has an intrinsic value that makes it worth protecting. As a consequence, several initiatives and strategies currently target biodiversity conservation within urban areas (e.g., the Berlin Strategy for Biodiversity; Senatsverwaltung für Stadtentwicklung und Umwelt 2012).

Urban biodiversity is often considered fundamental to the generation of urban ecosystem services (UES) and enhancing UES is considered beneficial for protecting biodiversity (Mace et al. 2012). The Millennium Ecosystem Assessment (2005b, p. vi), for example, depicts biodiversity as the basis of all ecosystem services – a representation referred to by many publications in science and policy-making. The European Commission, in its report on "The Multifunctionality of Green Infrastructure", emphasizes that the role of green infrastructure "in protecting biodiversity is highly dependent on its role in promoting ecosystem services and vice versa." (DG Environment 2012, p. 2). If these assumptions were true, there would be win-win situations facilitating urban transformations towards sustainable development. Current research, however, reveals that there are also trade-offs among biodiversity and UES. As Haase et al. (2014, p. 414) stated, "ecosystems do not necessarily provide more or better UESs when the level of biodiversity is changed". Similarly, the Millennium Ecosystem Assessment acknowledges that ecosystem service delivery can improve, even when biodiversity declines (Millennium Ecosystem Assessment 2005b). The recent attempts to enhance the provision of ecosystem services in urban areas might thus conflict with urban biodiversity conservation.

Here, we present and discuss the state of knowledge on the relationships between biodiversity and UES. Because 'biodiversity' is often erroneously used as synonym for 'nature' (Jax and Heink 2015), we first define the term 'biodiversity'. For the term 'ecosystem services', we refer to the definition provided by Kabisch et al., "Introduction: Urban Transformations – Sustainable Urban Development Through Resource Efficiency, Quality of Life, and Resilience", in this volume. Second, we briefly outline how urbanization affects biodiversity. Third, we discuss the assumption that biodiversity is fundamental to the generation of ecosystem services. This is followed by a survey of current knowledge about relationships between biodiversity and ecosystem services in urban areas. Finally, we conclude with recommendations related to urban transformations.

2 Biodiversity

Biodiversity is the diversity of genes, populations, species, communities, and ecosystems (Millennium Ecosystem Assessment 2005a; United Nations 1992). It can be quantified using measures of (1) richness (e.g., the number of ecosystems in a landscape or the number of species in an ecosystem); (2) abundance (e.g., the frequency of species within an ecosystem) or (3) evenness (e.g., uniformity in the distribution of ecosystems in a landscape or of species in an ecosystem). One of the simplest and most popular measures of biodiversity is species richness. Just counting species, however, does not distinguish between different types of species such as trees, shrubs, and herbs or herbivorous, omnivorous and carnivorous mammals. Such differences in the anatomical, behavioral, biochemical, morphological, phenological, physiological, or structural characteristics of species are called functional differences and the characteristics themselves are called traits. Traits represent the adaptation of species to their environment; they thus affect an organism's growth, reproduction, and survival (Violle et al. 2007). The flying seeds of dandelion, for example, are an adaptation to dispersal by wind, and therefore, support spread and reproduction. Moreover, traits affect ecosystem functioning (Lavorel and Garnier 2002). An example is the lignin content of leaves, which affects the degradability of leaves and, therefore, soil formation.

Consequently, 'biodiversity' is neither synonymous with 'nature' nor with the presence of a green space per se, such as the presence of a park within an urban area. Rather, always more than one element (ecosystem, species, trait, etc.) is required to create diversity. The wealth and differing meanings of biodiversity measures show that, when investigating the relevance of biodiversity for the provision of ecosystem services, it is insufficient to restrict biodiversity to species richness. We need to consider genetic, species, and ecosystem diversity, as well as traits and functional diversity to fully understand how biodiversity is related to ecosystem services.

3 Effects of Urbanization on Biodiversity

Urbanization is currently increasing faster than ever, supporting global biodiversity loss (CBD 2012; Seto et al. 2012). Thus, a key question with respect to urban transformations is whether we can design and manage urban areas in a biodiversity-friendly way. To answer this question, we briefly outline how biodiversity responds to urbanization.

Urbanization, on the one hand, can drive native species to extinction and, on the other hand, promote an increasing number of non-native (potentially invasive) as well as native generalist species (Hahs et al. 2009; Lososová et al. 2012; Sukopp 2004). In total, urban areas often harbor more species (both native and non-native) than rural areas (Kühn et al. 2004). This high species richness results from (1) the non-random distribution of urban areas across the world, within or close to areas of high biodiversity (Seto et al. 2012) and (2) geologically heterogeneous landscapes

(Kühn et al. 2004), as well as (3) the internal heterogeneity of urban areas (which usually comprise a range of different ecosystems, especially when compared to intensively used homogeneous agricultural landscapes; Niemelä 1999), and (4) the high number of non-native species.

Urban areas can even harbor rare species (Ives et al. 2016) but such rare species, especially those specialized to a narrow range of environmental conditions or on few interaction partners (e.g., pollinators), are often the ones that are eradicated first (Duncan et al. 2011). One example is the green-veined orchid, *Orchis morio* L., a plant species dependent on insect pollinators, which has become rare in central Europe but which may occur in remnants of nutrient-poor grasslands, e.g., at the urban fringe (Stolle and Klotz 2004). Moreover, species populations within urban areas are not necessarily self-sustaining: Urban bird populations, for example rely on the migration of individuals from the rural surroundings into the urban landscape, as shown for the city of Dunedin, New Zealand by van Heezik and Ludwig (2012) – if birds did not migrate into the city, urban bird populations could not persist, because predation by cats alone exceeds their regeneration rate.

Whether single species respond to urbanization with increasing or decreasing abundance depends on their traits. Species that thrive in urban areas usually share traits that represent adaptations to typical urban environmental conditions, such as dry to mesic, nutrient rich and alkaline soils, high degrees of isolation, high disturbance intensities, high temperatures and a wealth of anthropogenic food resources (Bateman and Fleming 2012; Kuttler 2008; Williams et al. 2015). Urban plant communities are often characterized by species with, e.g., small seeds and by short-lived species, dispersed by humans, that are highly competitive when nutrient loads are high (Knapp et al. 2008; Williams et al. 2015). Animals that can exploit urban environments are often omnivores (Bateman and Fleming 2012), indoor breeders (Møller 2010), or able to adapt their communication strategy to urban noise (Nemeth et al. 2013; Parris et al. 2009). Although not all trait responses to urbanization have been identified, yet (Williams et al. 2015), it is clear that "urbanization causes shifts in species' trait state frequencies" (Knapp et al. 2008, p. 375) – with likely consequences for ecosystem functions and services.

From the facts presented so far, we can conclude that urban landscape heterogeneity may promote both species richness and functional diversity (because varying environmental conditions can support species characterized by different functional traits) but also that urbanization usually goes along with favoring a particular set of trait values. A second crucial point arises from the species area relationship. This basic ecological rule states that large areas will host more species than small ones. The species-area relationship applies to ecosystems worldwide, including urban ecosystems. Designing and managing urban areas in a way that is beneficial to biodiversity therefore requires large and diverse green spaces with intact vegetation within and close to urban areas (Aronson et al. 2014; Beninde et al. 2015).

4 Biodiversity and Ecosystem Services

Biodiversity is regarded as a regulator of ecosystem processes (such as nutrient cycling or primary production) and it has been classified as an ecosystem service itself (Mace et al. 2012). As a consequence, biodiversity is usually considered fundamental to the generation of ecosystem services and human well-being, regardless of the rarity of empirical evidence for such statements (cf. Haase et al. 2014; Ziter 2016). The positivistic point of view – that ecosystem services and biodiversity in cities go hand in hand (e.g. DG Environment 2012) – tends to neglect four arguments: (1) that nature has an intrinsic value beside its utility for humans; (2) that nature can also produce disservices (Lyytimäki and Sipilä 2009), such as allergies; (3) that ecosystem service provision might be independent of the actual level of biodiversity; and (4) that there is ample evidence for the fact that the promotion of a particular ecosystem service within an ecosystem may actually harm biodiversity (as does the promotion of high agricultural yields; Meyer et al. 2013). All these considerations lead to a general conclusion that demonstrates the limited and anthropocentric character of the ecosystem service concept: Nature does not exist for our benefit and therefore, we cannot expect that everything in nature is arranged for our comfort or meets our interests.

In addition, several methodological gaps in ecosystem service research remain: Studies comparing services and disservices are still rare (Dunn 2010; but see Sladonja et al. 2015), but we have to consider both the costs and benefits provided by ecosystems. Monitoring schemes for ecosystem services are rare and not well developed: The International Long Term Ecological Research network (ILTER), for example, showed that only scattered monitoring data are available (Vihervaara et al. 2013). In addition, Rasmussen et al. (2016), focusing on provisioning services, emphasized that a combination of methods (e.g., plot monitoring, collection diaries, interviews, and participant observation) is necessary to compare the availability and use of ecosystem services (see Hansjürgens et al., "The TEEB Approach Towards Sustainable Urban Transformations: Demonstrating and Capturing Ecosystem Service Values", in this volume for methods to monitor and evaluate ecosystem services).

Irrespective of these gaps, biodiversity and ecosystem services are often discussed together, also with respect to urban areas: Urban green spaces and their plant, bird, or mammal species are supposed to contribute significantly to the liveability of cities by providing cultural, provisioning, and regulating ecosystem services such as physical health, food, or local climate regulation (e.g. Elmqvist et al. 2013; Keniger et al. 2013), although they often differ in terms of structural diversity and species richness (Shanahan et al. 2014; Strohbach et al. 2009).

In total, surprisingly little empirical evidence is available about biodiversity-ecosystem service (BES) relationships in urban systems (Gómez-Baggethun et al. 2013; Kowarik 2011; Ziter 2016). Instead, empirical evidence mostly stems from rural contexts (e.g. Cardinale et al. 2012; Verheyen et al. 2016); however, it cannot be easily extrapolated to urban areas for a number of reasons. Whilst the focus in

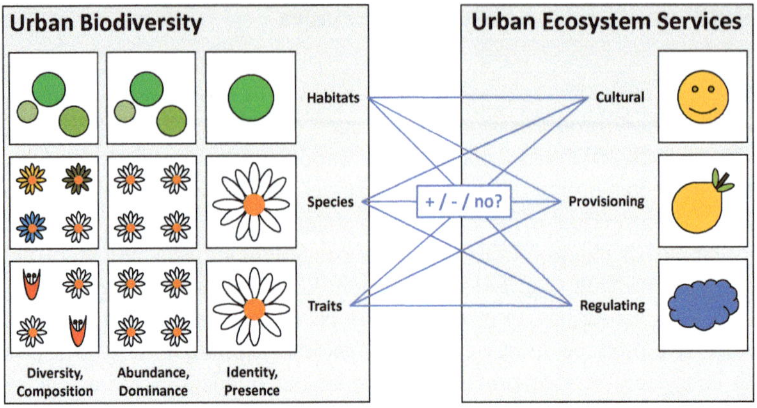

Fig. 1 Conceptual framework illustrating the question of how various biodiversity indicators (e.g., the number of habitats or species or traits) are related to cultural, provisioning, and regulating ecosystem services in urban areas. Relationships might be positive (+), indicating synergies among the protection of biodiversity and the enhancement of ecosystem service delivery. They might also be negative (−), indicating trade-offs among biodiversity and UES, or even non-significant ("no"). Figure adpated from Schwarz et al. 2017

rural areas is mainly on provisioning services (especially food production), cultural and regulating services play a major role in urban areas (Haase et al. 2014). Moreover, urban areas differ from rural contexts in terms of habitat structure and abiotic conditions (Kowarik 2011); for example, higher temperatures and artificial lighting in the former (Russ et al. 2015). Urban green spaces are also largely planned and consist of artificial assemblages of species, for instance in parks or gardens. These differences result, amongst other aspects, in changes of biodiversity along gradients of urbanization, as summarized above. It follows that, if the assumption that the provision of ecosystem services is correlated to biodiversity is true, this provision should change with changes in species richness and functional composition along urbanization gradients. Thus, empirical results are needed to clarify whether biodiversity indicators (e.g., the number of ecosystems or species or traits, etc.) are positively (equalling synergies among biodiversity and UES), negatively (equalling trade-offs among biodiversity and UES) or not at all related to cultural, provisioning, and regulating ecosystem services in urban areas (Fig. 1).

5 The Relationship Between Biodiversity and Ecosystem Services in Urban Areas

A review by Ziter (2016) for BES relationships in urban areas showed that the majority of studies analysed biodiversity at the species level. A review of urban BES studies, which the authors contributed to (Schwarz et al. 2017), assessed various categories of biodiversity indicators (based on a systematic review of publications

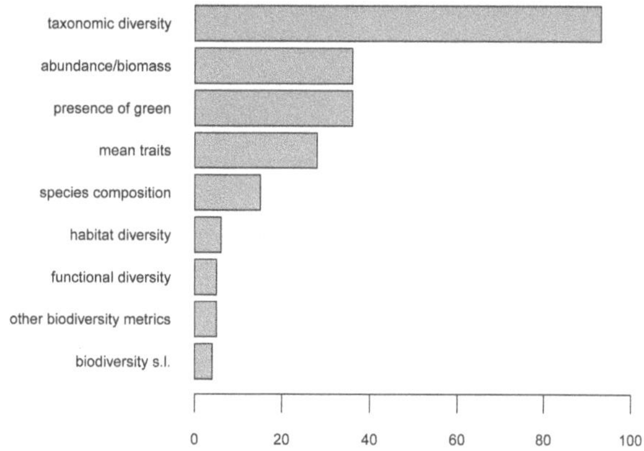

Fig. 2 Biodiversity categories and the number of cases in which a link among these categories and ecosystem services was tested (data from Schwarz et al. 2017). "Biodiversity s.l." stands for biodiversity *sensu latu*, and refers to studies that did not categorize biodiversity any further

listed in the ISI Web of Science that covered studies from cities across the globe). They found that most BES links (within 35 publications that did not only suggest but really tested such links) were assessed using taxonomic diversity, followed by metrics of abundance or biomass and metrics quantifying the presence of green, mean traits (i.e., the mean value of a trait within a species assemblage, such as the mean number of seeds produced by all species in a grassland system), and metrics of species composition (Fig. 2). The majority of BES links was tested for regulating services (115 cases), whilst there were 87 cases for cultural services, 18 cases for provisioning services, and eight cases in which ecosystem services were not further distinguished.

Within these 228 tested cases, far more BES relationships were positive (52%) rather than negative (11%). However, 28% of all tested cases did not reveal evidence for a BES link. Thus, in total, roughly half of all tested cases demonstrated a synergy between biodiversity and UES but the other half did not. To date, claims for specific BES links are often not reliable, because most were either not tested, tested only once, or studies showed inconsistent results. For example, taxonomic diversity was shown to positively affect human health, recreation and wellbeing in eight cases but negatively in two cases. Another eight cases showed no relationship at all.

To sum up, the scientific literature shows that the idea of biodiversity enhancing UES holds in a number of cases – but far from always. Some links among biodiversity and UES are even negative. Pollinator efficiency, for example, was lower in urban than in natural areas in California, probably as a result of the high number of introduced plant species in urban areas (Leong et al. 2014). Similarly, models for an English city showed that short-rotation coppice plantings composed of the tree species *Eucalyptus gunnii* Hook F. and *Populus tremula* L. sequester 12 times more carbon than the current, more diverse urban tree stock does (McHugh et al. 2015) – which, at the same time, means that enhancing carbon sequestration might come at the cost of restricting tree diversity.

What is more, the relationship between biodiversity and urban ecosystem disservices (Lyytimäki and Sipilä 2009) is largely unexplored. Such disservices sometimes are linked to the presence of urban green per se (e.g., perceived safety in dense urban forests). In other cases, a possible relationship to biodiversity needs to be further explored. For example, allergy potential is, on the one hand, clearly linked to the presence or absence of certain species that emit allergenic pollen (and therefore to species' traits). However, higher biodiversity has a positive effect on the human immune system, leading to lower overall allergic dispositions (Hanski et al. 2012). Consequently, there is no general synergy among UES and biodiversity that implies that more biodiversity always means more services and greater human well-being.

In addition, access to biodiversity and ecosystem services is often not distributed uniformly across a city, thus reducing environmental justice (Fuller et al. 2007; Stigsdotter and Grahn 2011; also see Kindler et al., "Socio-Spatial Distribution of Airborne Outdoor Exposures – An Indicator for Environmental Quality, Quality of Life, and Environmental Justice: The Case Study of Berlin", in this volume). Even if urban green spaces are available and accessible to everyone, only part of the urban population regularly uses these spaces. Visitation frequency is strongly shaped by gender, age, green space properties, the cultural background of urban dwellers, preferences, and socio-economics, to name the most important relevant variables (Lin et al. 2014; Shan 2014; Shanahan et al. 2015). Up until now, no studies consider the question of which level (or scale) of biodiversity, from whole ecosystems to genes, including the taxonomic, structural and trait aspects of biodiversity, benefits which socio-cultural groups, because we assume that human-species relationships are more specific and not universal across the variety (age, gender, size, niche, appearance, rareness) of both groups: humans and non-human (flora/fauna) organisms. For example, evidence suggests that the perception of biodiversity and its elements differs between human cultures. Some species, for example, are seen as pests in one country but are used as sources of pharmaceuticals and other raw materials in other countries – such as the tree of heaven, *Ailanthus altissima* (Mill.) Swingle, which originated in China and has now invaded temperate urban areas in Europe (Sladonja et al. 2015).

6 Conclusions: Developing Biodiversity and Ecosystem Services in the Face of Urban Transformations

Cities can be seen as a cultural and biological "rendezvous" that provide many situations for associations between urban dwellers and biodiversity (Aronson et al. 2014; Elmqvist et al. 2013). A range of studies has been undertaken to assess the importance of biodiversity conservation in urban areas (e.g. Godefroid and Koedam 2003; Kowarik 2011) and of all the benefits people receive from ecosystems, such as climate regulation, fresh air, food, spiritual enrichment, cognitive development, or recreation (Millennium Ecosystem Assessment 2005b). Urban biodiversity and urban ecosystem services, however, do not always go hand in hand. Consequently, there is no "one-size-fits-all" solution to enhance both biodiversity and UES that can be applied in cities worldwide and across different ecosystem services.

Moreover, it should not be forgotten that nature can also produce disservices (Dunn 2010). Consequently, biodiversity conservation, services, and disservices have to be balanced when designing and managing urban green infrastructure. Karl et al. (2009), for example, provide a list of European tree species and their species-specific potential to emit biogenic volatile organic compounds that increase the concentration of ozone in the atmosphere – a disservice with implications for human health. Similarly, databases for the selection of trees with respect to their suitability for urban areas and for withstanding climate warming have been created recently (e.g. Gillner et al. 2016). Such overviews can support decisions about species selection and related UES and disservices.

Generally, cities should provide space for the provision of ecosystem services, space for biodiversity and – as space is a limited resource – especially multifunctional space. Lundholm (2015), for example, showed that green roofs provide multiple ecosystem services – and that multifunctional service provision improves with increasing species diversity of roof planting mixes. Furthermore, urban planning should not only consider the prevailing forms of biodiversity conservation, but also man-made habitats and their biodiversity (e.g., urban parks, urban gardens, and brownfields; see Banzhaf et al., "Potentials of Urban Brownfields for Improving the Quality of Urban Space", in this volume, dealing with potentials of urban brownfields). This should include green architecture, e.g., green roofs and green walls in which plants not only serve ornamental purposes but also assist in microclimate management, maintain suitable habitats for bird and insect species, and capture pollutants, thus supporting human health (Lee and Maheswaran 2011). The spatial layout of cities has the potential to positively influence both biodiversity and ecosystem services – even in cases where the two are independent of each other: Urban biodiversity increases with the size of available green spaces (Beninde et al. 2015), as does the provision of several ecosystem services (e.g. microclimate regulation; Jaganmohan et al. 2016; or groundwater retention).

Consequently, urban planning has to be case-specific, depending on (1) the main aim of biodiversity conservation, (2) the ecosystem services in focus, and (3) the socio-cultural groups that will supposedly benefit from UES.

References

Aronson MFJ, La Sorte FA, Nilon CH, Katti M, Goddard MA, Lepczyk CA et al (2014) A global analysis of the impacts of urbanization on bird and plant diversity reveals key anthropogenic drivers. Proc R Soc B Biol Sci 281(1780). doi:10.1098/rspb.2013.3330

Bateman PW, Fleming PA (2012) Big city life: carnivores in urban environments. J Zool 287(1):1–23

Beninde J, Veith M, Hochkirch A (2015) Biodiversity in cities needs space: a meta-analysis of factors determining intra-urban biodiversity variation. Ecol Lett 18(6):581–592

Cardinale BJ, Duffy JE, Gonzalez A, Hooper DU, Perrings C, Venail P et al (2012) Biodiversity loss and its impact on humanity. Nature 486(7401):59–67

CBD – Secretariat of the Convention on Biological Diversity (2012) Cities and Biodiversity Outlook. CBD, Montreal

DG Environment – European Commission's Directorate-General Environment (2012) The Multifunctionality of Green Infrastructure. European Commission's Directorate-General Environment, Bristol, pp 1–37

Duncan RP, Clemants SE, Corlett RT, Hahs AK, McCarthy MA, McDonnell MJ et al (2011) Plant traits and extinction in urban areas: a meta-analysis of 11 cities. Glob Ecol Biogeogr 20(4):509–519

Dunn RR (2010) Global Mapping of Ecosystem Disservices: The Unspoken Reality that Nature Sometimes Kills us. Biotropica 42(5):555–557

Elmqvist T, Fragkias M, Goodness J, Güneralp B, Marcotullio PJ, McDonald RI et al (2013) Urbanization, Biodiversity and Ecosystem Services: Challenges and Opportunities: A Global Assessment. Springer, Dordrecht

Fuller RA, Irvine KN, Devine-Wright P, Warren PH, Gaston KJ (2007) Psychological benefits of greenspace increase with biodiversity. Biol Lett 3(4):390–394

Gillner S, Hofmann M, Tharang A, Vogt J (2016) Development of a database for urban trees. In: Roloff A (ed) Urban Tree Management – for a Sustainable Development of Green Cities. Wiley-VCH, Oxford, pp 196–210

Godefroid S, Koedam N (2003) How important are large vs. small forest remnants for the conservation of the woodland flora in an urban context? Glob Ecol Biogeogr 12(4):287–298

Gómez-Baggethun E, Gren Å, Barton DN, Langemeyer J, McPhearson T, O'Farrell P et al (2013) Urban Ecosystem Services. In: Elmqvist T, Fragkias M, Goodness J, Güneralp B, Marcotullio PJ, McDonald RI et al (eds) Urbanization, Biodiversity and Ecosystem Services: Challenges and Opportunities, A Global Assessment. Springer, Dordrecht, Heidelberg, New York, London

Haase D, Larondelle N, Andersson E, Artmann M, Borgstrom S, Breuste J et al (2014) A Quantitative Review of Urban Ecosystem Service Assessments: Concepts, Models, and Implementation. Ambio 43(4):413–433

Hahs AK, McDonnell MJ, McCarthy MA, Vesk PA, Corlett RT, Norton BA et al (2009) A global synthesis of plant extinction rates in urban areas. Ecol Lett 12(11):1165–1173

Hanski I, von Hertzen L, Fyhrquist N, Koskinen K, Torppa K, Laatikainen T et al (2012) Environmental biodiversity, human microbiota, and allergy are interrelated. PNAS 109(21):8334–8339

Ives CD, Lentini PE, Threlfall CG, Ikin K, Shanahan DF, Garrard GE et al (2016) Cities are hotspots for threatened species. Glob Ecol Biogeogr 25(1):117–126

Jaganmohan M, Knapp S, Buchmann CM, Schwarz N (2016) The bigger, the better? The influence of urban green space design on cooling effects for residential areas. J Environ Qual 45(1):134–145

Jax K, Heink U (2015) Searching for the place of biodiversity in the ecosystem services discourse. Biol Conserv 191:198–205. doi:10.1016/j.biocon.2015.06.032

Karl M, Guenther A, Koble R, Leip A, Seufert G (2009) A new European plant-specific emission inventory of biogenic volatile organic compounds for use in atmospheric transport models. Biogeosciences 6(6):1059–1087

Keniger LE, Gaston KJ, Irvine KN, Fuller RA (2013) What are the Benefits of Interacting with Nature? Int J Environ Res Public Health 10(3):913–935

Knapp S, Kühn I, Wittig R, Ozinga WA, Poschlod P, Klotz S (2008) Urbanization causes shifts in species' trait state frequencies. Preslia 80:375–388

Kowarik I (2011) Novel urban ecosystems, biodiversity, and conservation. Environ Pollut 159(8–9):1974–1983

Kühn I, Brandl R, Klotz S (2004) The flora of German cities is naturally species rich. Evol Ecol Res 6(5):749–764

Kuttler W (2008) The urban climate – basic and applied aspects. In: Marzluff JM, Shulenberger E, Endlicher W, Alberti M, Bradley G, Ryan C et al (eds) Urban ecology: an international perspective on the interaction between humans and nature. Springer, New York, pp 233–248

Lavorel S, Garnier E (2002) Predicting changes in community composition and ecosystem functioning from plant traits: revisiting the Holy Grail. Funct Ecol 16(5):545–556
Lee ACK, Maheswaran R (2011) The health benefits of urban green spaces: a review of the evidence. J Public Health 33(2):212–222
Leong M, Kremen C, Roderick GK (2014) Pollinator Interactions with Yellow Starthistle (Centaurea solstitialis) across Urban, Agricultural, and Natural Landscapes. Plos One 9(1):e86357
Lin BB, Fuller RA, Bush R, Gaston KJ, Shanahan DF (2014) Opportunity or Orientation? Who Uses Urban Parks and Why. Plos One 9(1):e87422
Lososová Z, Chytrý M, Tichý L, Danihelka J, Fajmon K, Hájek O et al (2012) Native and alien floras in urban habitats: a comparison across 32 cities of central Europe. Glob Ecol Biogeogr 21(5):545–555
Lundholm JT (2015) Green roof plant species diversity improves ecosystem multifunctionality. J Appl Ecol 52(3):726–734
Lyytimäki J, Sipilä M (2009) Hopping on one leg – the challenge of ecosystem disservices for urban green management. Urban Forestry & Urban Greening 8(4):309–315
Mace GM, Norris K, Fitter AH (2012) Biodiversity and ecosystem services: a multilayered relationship. Trends Ecol Evol 27(1):19–26
McHugh N, Edmondson JL, Gaston KJ, Leake JR, O'Sullivan OS (2015) Modelling short-rotation coppice and tree planting for urban carbon management – a citywide analysis. J Appl Ecol 52(5):1237–1245
Meyer S, Wesche K, Krause B, Leuschner C (2013) Dramatic losses of specialist arable plants in Central Germany since the 1950s/60s – a cross-regional analysis. Divers Distrib 19(9):1175–1187
Millennium Ecosystem Assessment (2005a) Ecosystems and human well-being: current state and trends: findings of the Condition and Trends Working Group. Island Press, Washington
Millennium Ecosystem Assessment (2005b) Ecosystems and Human Well-being: Synthesis. Island Press, Washington, DC
Møller AP (2010) The fitness benefit of association with humans: elevated success of birds breeding indoors. Behav Ecol 21(5):913–918
Nemeth E, Pieretti N, Zollinger SA, Geberzahn N, Partecke J, Miranda AC et al (2013) Bird song and anthropogenic noise: vocal constraints may explain why birds sing higher-frequency songs in cities. Proceedings of the Royal Society B-Biological Sciences 280(1754):20122798
Niemelä J (1999) Is there a need for a theory of urban ecology? Urban Ecosyst 3(1):57–65
Parris KM, Velik-Lord M, North JMA (2009) Frogs Call at a Higher Pitch in Traffic Noise. Ecol Soc 14(1):25
Rasmussen LV, Mertz O, Christensen AE, Danielsen F, Dawson N, Xaydongvanh P (2016) A combination of methods needed to assess the actual use of provisioning ecosystem services. Ecosyst Serv 17:75–86. doi:10.1016/j.ecoser.2015.11.005
Russ A, Ruger A, Klenke R (2015) Seize the night: European Blackbirds (Turdus merula) extend their foraging activity under artificial illumination. J Ornithol 156(1):123–131
Schwarz, N, Moretti, M, Bugalho, M, Davies, Z, Haase, D, Hack, J, Hof, A, Melero, Y, Pett. T, Knapp, S (2017) Understanding biodiversity-ecosystem service relationships in urban areas: a comprehensive literature review. Ecosyst Serv 27: 161–171
Senatsverwaltung für Stadtentwicklung und Umwelt (2012) Berlins biologische Vielfalt: Berliner Strategie zur biologischen Vielfalt; Begründung, Themenfelder und strategische Ziele. Senatsverwaltung für Stadtentwicklung und Umwelt, Berlin
Seto KC, Güneralp B, Hutyra LR (2012) Global forecasts of urban expansion to 2030 and direct impacts on biodiversity and carbon pools. PNAS 109(40):16083–16088
Shan XZ (2014) Socio-demographic variation in motives for visiting urban green spaces in a large Chinese city. Habitat Int 41:114–120
Shanahan DF, Lin BB, Bush R, Gaston KJ, Dean JH, Barber E et al (2015) Toward Improved Public Health Outcomes From Urban Nature. Am J Public Health 105(3):470–477

Shanahan DF, Lin BB, Gaston KJ, Bush R (2014) & Fuller, R.A. (2014). Socio-economic inequalities in access to nature on public and private lands: A case study from Brisbane, Australia. Landsc Urban Plan 130:14–23. doi:10.1016/j.landurbplan.2014.06.005
Sladonja B, Sušek M, Guillermic J (2015) Review on Invasive Tree of Heaven (Ailanthus altissima (Mill.) Swingle) Conflicting Values: Assessment of Its Ecosystem Services and Potential Biological Threat. Environ Manag 56(4):1009–1034
Stigsdotter UK, Grahn P (2011) Stressed individuals' preferences for activities and environmental characteristics in green spaces. Urban Forestry & Urban Greening 10(4):295–304
Stolle J, Klotz S (2004) Flora der Stadt Halle (Saale). Calendula, hallesche Umweltblätter, Halle (Saale)
Strohbach MW, Haase D, Kabisch N (2009) Birds and the City: Urban Biodiversity, Land Use, and Socioeconomics. Ecol Soc 14(2):31
Sukopp H (2004) Human-caused impact on preserved vegetation. Landsc Urban Plan 68(4):347–355
United Nations (1992) Convention on Biological Diversity. UNEP, Rio de Janeiro
van Heezik Y, Ludwig K (2012) Proximity to source populations and untidy gardens predict occurrence of a small lizard in an urban area. Landsc Urban Plan 104(2):253–259
Verheyen K, Vanhellemont M, Auge H, Baeten L, Baraloto C, Barsoum N et al (2016) Contributions of a global network of tree diversity experiments to sustainable forest plantations. Ambio 45(1):29–41
Vihervaara P, D'Amato D, Forsius M, Angelstam P, Baessler C, Balvanera P et al (2013) Using long-term ecosystem service and biodiversity data to study the impacts and adaptation options in response to climate change: insights from the global ILTER sites network. Curr Opin Environ Sustain 5(1):53–66
Violle C, Navas ML, Vile D, Kazakou E, Fortunel C, Hummel I et al (2007) Let the concept of trait be functional! Oikos 116(5):882–892
Williams NSG, Hahs AK, Vesk PA (2015) Urbanisation, plant traits and the composition of urban floras. Perspect in Plant Ecol Evol Syst 17(1):78–86
Ziter C (2016) The biodiversity – ecosystem service relationship in urban areas: a quantitative review. Oikos 125(6):761–768

Part IV
Urban Risks and Resilience

Outline

Kerstin Krellenberg and Christian Kuhlicke

The discourse on urban risks stemming from natural hazards and disasters has been adopted not only by researchers from various disciplines, but has also been employed in municipal programs, plans, and strategies. Given the increase in extreme events related to climate change and unbroken urbanization processes on the global scale, both a better understanding of the underlying drivers of urban risks as well as how to respond to the anticipated negative consequences have become vital questions for many cities and regions around the world. This section of the book therefore presents a variety of perspectives on how urban areas respond to urban risks, in terms of adaptation, resilience, and transformation.

In the light of urban transformations, the contributions of this section are both conceptual and empirical, and linked by the overall objective of enhancing, building, or developing capacities in organizations, communities, and entire urban systems in order to better anticipate, cope with, and recover from the impacts of climate change. The paper of **Kuhlicke et al.** introduces the concepts of resilience, adaptation, and transformation to discuss, using two case studies from Germany and Chile, how they help to understand different management trajectories for flooding in urban areas. Staying in the Latin-American region, **Welz et al.** present empirical research from Santiago de Chile and discuss the potential of adapting built-up areas to heat hazard as an urban response to the resilience and quality of life dimensions of urban transformations. The results, which stem from modelling and household surveys, support the combination of green and white interventions. The contribution of **Reese** explores how climate proofing can be fostered within the legal framework of urban development in Germany, focusing on environmental impact and risk assessment procedures and adaptation planning instruments. He presents a set of general key requirements for steering urban development towards climatic resilience. **Meyer** discusses the suitability of decision heuristics in flood management to support decision makers when reaching decisions under conditions of high uncertainty and complexity.

The contributions within this section evolve along both conceptual as well as empirical considerations. What becomes apparent is that urban transformations towards more resilience in cities must consider a combination of strategies, in order to achieve fundamental changes. Small-scale interventions such as, for example, planting trees along streets and open spaces, whitening roofs, or applying decision heuristics approaches in flood risk management should become part of an overall picture or strategy. Accordingly, Reese's contribution complements the chapter of Kuhlicke et al., by giving additional insights into legal instruments of risk assessment and how they should be further enhanced in order to better manage anticipated risks in urban areas.

Resilience, Adaptation and Transformation: Conceptual and Empirical Insights from Two Case Studies in Germany and Chile

Christian Kuhlicke, Kerstin Krellenberg, and Juliane Welz

1 Introduction

This chapter engages with three different concepts shaping current attempts to deal with the anticipated negative consequences of climate change, namely resilience, adaptation, and transformation. These concepts are not only the basis for many policy-initiatives; they are also constitutive for more specific plans of actions and strategies, defining the future development of cities across the globe.

A closer look, however, reveals that the individual concepts represent a great plurality of meanings; meanings that overlap in some cases and, in others, contradict each other. The Intergovernmental Panel on Climate Change (IPCC), for instance, defines resilience in its 5th assessment report as "*[t]he capacity of a social-ecological system to cope with a hazardous event or disturbance, responding or reorganizing in ways that maintain its essential function, identity, and structure, while also maintaining the capacity for adaptation, learning, and transformation*" (IPCC 2014, p. 3). Obviously, this definition not only includes a reference to the concept of adaptation and transformation, it also implies an antagonism between a rather conservative interpretation of change, as suggested by terms such as "reorganizing" and "maintaining". This echoes classical, engineering-based definitions of resilience as the capacity of a system to bounce back (e.g., Pimm 1984). It also includes more progressive and fundamental notions of change, as suggested by the reference to "learning" and "transformation" (e.g., Pelling 2010). As a practical implication, the IPCC definition suggests, paradoxically, that cities, in order to become resilient, should be able to both bounce back effectively and maintain or restore the status quo ante after a crisis while, at the same time, they should be able to change, learn, and transform, and hence radically alter their structures and functions.

C. Kuhlicke (✉) • K. Krellenberg • J. Welz
Department of Urban and Environmental Sociology, Helmholtz Centre for Environmental Research-UFZ, Leipzig, Germany
e-mail: christian.kuhlicke@ufz.de; kerstin.krellenberg@ufz.de; juliane.welz@ufz.de

S. Kabisch et al. (eds.), *Urban Transformations*, Future City 10,
DOI 10.1007/978-3-319-59324-1_17

This contribution therefore, first, argues for greater conceptual clarity and thus distinguishes between the concepts of resilience, adaptation, and transformation more precisely. Second, it introduces two case studies, in order to unravel how different urban actors respond to the anticipated challenges of future climate-related risks and how these can be understood through the conceptual lenses of resilience, adaptation, and transformation. By placing a specific focus on cities and flood hazards, this chapter pleads for a more thorough, empirically grounded examination of underlying similarities and differences buried in the three concepts this chapter addressed and calls for a more careful interpretation.

In Sect. 2, the concepts are introduced and defined in more detail and thereafter, two empirical case studies, one from Germany and one from Chile, are presented to explore different, empirically grounded pathways for managing past and potential future flood events. The case study from Germany sketches out that resilience and transformation are the dominant framing concepts for responding to a recent flood event and for anticipating possible future flood events; the case study from Chile is more focused on incremental changes and, hence, is framed as adaptation. The final section then links both case studies back to the context of urban transformations.

2 Resilience, Adaptation, and Transformation – A Clarification

In a first approximation, we understand resilience, adaptation, and transformation as both future-oriented concepts that engage with the question of how to respond to more or less uncertain anticipated future developments and, at the same time, as concepts that shape current practices and decision-making processes (Anderson 2010). However, although they share a forward-looking orientation, they differ profoundly with regard to their openness toward change, as we describe in more detail below (for a more comprehensive overview see O'Brien 2012; Pelling 2010; Pelling et al. 2015).

Although a broad range of definitions and understandings of resilience exist (Alexander 2013, Weichselgartner and Kelman 2014), we agree with Brand and Jax (2007, p. 1) that "both conceptual clarity and practical relevance of the concept of resilience" is at risk, because it is overloaded with too many meanings. The term resilience has undergone a considerable reframing, from a rather descriptive concept in ecology to a rather messy and vague interpretation of resilience in various other scientific disciplines. Although "increased conceptual vagueness" (ibid.) can have quite positive implications, because it allows communication across disciplinary boundaries, it also has its downsides: the concept is quite often overloaded with normative meanings that tend to describe how things should be instead of looking at empirical phenomena in a more descriptive and analytical sense.

We therefore argue that resilience is a concept that refers to action, decisions, and processes that aim at withstanding or effectively recovering from the negative consequences of environmental risks.

The term *adaptation* refers, according to our understanding and in line with Pelling et al. (2015, p. 116), to "incremental adjustments that preserve systems integrity when conditions change". This includes processes, actions, or measures that aim at coping with, or adjusting to, changing conditions, external stresses, dangers, risks, and even opportunities (Smit and Wandel 2006). Usually, adaptation is framed as a response either to changes in the physical environment or to internal stimuli, such as economic, demographic, or socio-political changes. However, a closer look reveals that it is less an actual change in the environment that triggers adaptation but rather, in many cases, an anticipated and expected potential change that may lead to some kind of adaptive process. Usually, adaptation also requires an understanding of underlying vulnerabilities of societies and ecosystems to the impacts of change and whether established capacities are sufficient to respond to these changes and also whether the socio-economic cost of adaptation are bearable (Klein et al. 2007).

The term *transformation* is understood here as actions, or measures for strategies that are associated with more fundamental changes, in order to be better able to cope with and respond to anticipated future risks (Pelling et al. 2015, p. 116). This represents a narrowed down definition of the general transformation definition given in the introduction of this book. If a system, for instance, has already moved into an undesirable regime in which adaptation to the negative consequences of an anticipate threat is no longer feasible, a system transformation might become a vital option. In this sense, transformation differs from adaptation because the latter predominantly refers to incremental changes that aim at increasing the effectiveness or efficiency of an existing management or planning system, whilst transformative activities engage with more fundamental changes. Transformation describes a way of overcoming symptoms of crisis and entails relatively profound alterations of a given system that can cross system borders and deal with multiple as well as uncertain development options and aim at changing the very structures and functions of a system (McCormick et al. 2013). We therefore understand transformation as processes that aim at "altering fundamental attributes of a system (including value systems; regulatory, legislative, or bureaucratic regimes; financial institutions; and technological or biological systems)" (IPCC 2012, p. 564).

In the following chapter, we lay out in more detail how the individual concepts build on each other and how, through associated policy choices, they also block alternative choices and pathways.

3 Two Empirical Examples

Empirical evidences from different case studies in Europe and South America are given to underpin the previous conceptual specifications on adaptation, resilience, and transformation. While the case-studies originate from very different starting points regarding frequency and intensity of experienced flood events, they help to understand the different trajectories urban settings can take in order to prepare for and cope with future hydro-climatic extreme events and how they are shaped by the different concepts of resilience, adaptation and transformation.

3.1 Resilience, Transformation and the Experience of Repeated Flood Events in Germany

Since, in 1997, the Odra inundated parts of Poland and Germany, Central Europe has undergone a series of devastating flood events. In Germany, the flood of 2002 went down in German history as the most costly flood event to date, with damages summing up to 11.5 billion € (DKKV 2015, pp. 32 + 44). The 2002 flood was unprecedented in most people's memory and caused enormous damage, but it was then followed by a series of smaller flood events such as the 2006 flood in the upper parts of the Elbe River, the 2010 flood of the Neiße River, and finally, by the 2013 flood, resulting again in approx. 6 billion € of financial damage in Germany (ibid.).

What are the consequence of experiencing repeated flood events in a relatively short time-span? Do people and organisations simply bounce back to where they were before the flood (resilience), do they initiate smaller adaptive steps to prevent future flood losses (adaptation), or do they start questioning the very place they are located in and where they face repeated flood events by starting thinking about moving out of harm's way by relocating to less flood-prone areas (transformation)? This chapter sheds some light on the consequences of repetitive flood events for affected households and communities. It is based on an empirical study (Kuhlicke et al. 2015) that includes a series of interviews conducted between 2004 and 2013, as well as the results of a large-scale household survey conducted in the aftermath of the 2013 flood.

3.1.1 The Phase from 2002 to 2013: Bouncing Back and Adapting to Future Flood Risks

Generally, the 2002 flood represents a turning point in flood risk management, both in Saxony as well as in Germany, because, as a consequence, new laws were established, the organisational capacity to respond to and manage flood events was enhanced, and a large-scale information campaign was initiated (DKKV 2015).

Furthermore, for the majority of people, the 2002 flood is still remembered as a complete surprise that caught municipalities and citizens unprepared and demonstrated their vulnerability. In order to restore people's relationship with their immediate environment and particularly with the rivers inundating people's property to an unknown extent, most residents placed great hope on the restoration and improvement of the technical flood protection system. Technical flood protection systems were a dominant topic that was addressed in most of the interviews we conducted. Their presence is furthermore difficult to ignore, visibly, when one walks through many communities along larger rivers such as the Elbe or the Mulde. People placed great hope, in the aftermath of the 2002 flood, on the levees and walls built and rebuilt in their community and hoped they would restore their feeling of security. Carsten Felgentreff's subtitle of a case study on the Odra Flood of 1997 captures this perspective pointedly: "The Dyke Broke, but People's Trust in Technical Solutions Remained Unbroken" (2000).

In this sense, there was a great desire to restore the status quo before the flood by improving the technical protection systems along the rivers, which resulted in many adaptive actions in the aftermath of 2002. Most prominently, huge investments and improvements were undertaken with regard to flood protection measures and warning systems. Between December 2002 and April 2005, a total of 47 flood protection concepts, including 1600 individual measures, were developed in Saxony and ranked through a risk-based priorisation scheme. This required financial investments of about 2 billion € (SMUL 2005, p. 10) and resulted in the new construction of hundreds of kilometres of new dikes and dams as well as detention reservoirs. For example, in the City of Eilenburg, situated on the Mulde River, a total of 13 km of flood protection walls and dikes have been constructed, at a total cost of about 35 million € (LTV 2008).

Many disaster management units have also been established on the local level and so-called dike guards have been introduced – trained personnel who watch the dikes and dams 24 h a day during a flood event. Some cities now use SMS-warning systems. Others have technically improved their siren systems. And most cities now use the internet for information purposes, for example, to inform residents about water levels and actions to be taken. In addition, the planning system on the urban level was reinforced with regard to flood prevention. Future flood risks are, meanwhile, considered as an important criterion within local planning processes. This criterion has been acknowledged to be of greater relevance, after 2002, than it was in the past. Development and land-use plans have been changed or adapted to local flood risk; some have been even annulled. The promotion and implementation of flood-adapted construction designs also became part of urban planners' work.

Adaptation took place not only on the organisational or the municipality level; it also played a more prominent role on the level of single households than before the 2002 flood. According to the empirical study by Kuhlicke et al. (2015) about half of the respondents have implemented some kind of adaptive measure after the 2002 flood (n = 941). Figure 1 demonstrates that most of the private actions that have been taken occurred after a flood event.

The types of measures used most often are behavioural measures (e.g., moving furniture and valuables to higher floors in the building, using sandbags and other methods to seal doors and windows, organizing a generator and pump, and moving vehicles to higher ground). These actions are usually taken shortly before a flood event. Less often, households tend to take more costly technical measures to make building more flood-resilient. With regard to insurance, 65.7% of respondents (n = 883) reported having an insurance policy that covers natural hazards. Those who reported not being insured believe that such insurance is too expensive (11%). Some respondents also reported not being able to receive insurance for natural hazards or that their policy was cancelled by the insurance provider (9.6%). Other respondents reported not being satisfied or wanting to be insured (6.7%). Some also reported having taken other private measures, instead (3.1%).

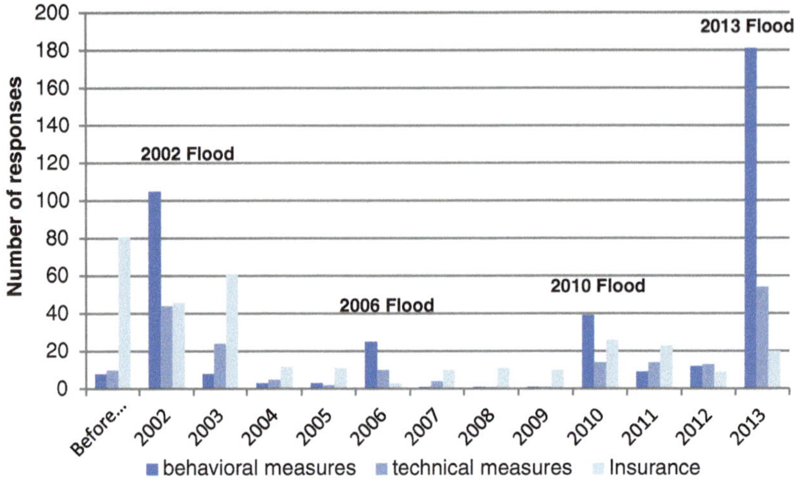

Fig. 1 Types of private actions taken by households between 2002 and 2013

3.1.2 After the 2013 Flood: Initiation of Transformative Processes?

The 2013 flood, which had, in some localities, an extent similar to the 2002 flood, was, to a certain extent, a test for the adaptive measures that had been implemented after the 2002 flood. Many interviewees of the empirical study (Kuhlicke et al. 2015) were generally quite positive with regard to the effectiveness of the measures. Basically, most case study sites have had consistently positive experiences during the 2013 event, with regard to the adaptive steps taken, at least on the operational level. Local actors felt much more independent and could intervene and respond faster and better to the needs of the residents, for example, during the evacuation process, as the following quotes from interviews with local practicioners demonstrate: "We still have to practice a bit, but as I see it, you learn from every disaster. And, in any case, the performance was much better organized than in 2002, definitely". Additionally, the cooperation between the different responsible administrative and operational bodies "has worked well, has worked better than in 2002". However, the improved effectiveness of adaptive steps was clearly related to the operational and technical procedures within existing institutional structures and hence, based on incremental changes.

After the 2013 flood, more fundamental changes in existing structures, values, and norms were also observable than after 2002. In general, the perception of the threat potential of flood risk has changed considerably as a consequence of the 2013 event. Whilst, after 2002, the focus was on improving existing flood management systems (e.g., new and better dikes, improved warning systems, improved emergency management), the 2013 flood substantially shattered the idea that increased effectiveness will reduce the risk of flooding, at least on the local level. Many communities, quite openly, admit that the risk of flooding is not reducible to zero through improving the established approach, as this quote underlines: “It is not preventable that there will be future damage caused by flooding. [...] An absolute protection is not possible”. On the contrary, flood events such as those in 2002 or 2013 can happen on a regular basis: “There will be more floods in the future, for sure”.

In this sense, the reflection and learning processes in the aftermath of the 2013 flood are more fundamental than after the 2002 flood event. This is clearly indicated not only by the questioning of the ‘safety promise’ of technical measures, but also by discussions about more self-responsibility of exposed residents and businesses to take appropriate actions in individual flood protection, such as obtaining insurance against natural hazards or taking private mitigation measures in buildings and on sites, as one local decision-maker argues: “In any case, it makes sense to protect themselves. In any case! Even if you get a central flood protection.”

Moreover, there are on-going debates about the relocation of residents at risk.

Many communities would support citizens willing to move, but the financial assistance for recovery promotes and regulates the rehabilitation and reconstruction of buildings only at the original location, as underlines by this quote: “There have been areas in the city that were flooded three, four times since 2002. And of course there was a request from affected citizens to resettle. [...] But what is offered, as support, is far from sufficient to make a resettlement economically feasible. [...] And so we hoped that the State of Saxony shows more commitment. But the Flood Aid reconstruction fund considers only one-to-one compensation”.

The German case studies emphasize how resilience and adaptation-based strategies are increasingly questioned as a result of the 2013 flood and, therefore, more wide-ranging transformative processes were observable in the case study.

3.2 Adaptation to Flood Hazard in the Urban Area of Santiago de Chile

In the Urban Area of Santiago de Chile (MAS), urban and river flooding is an annual winter phenomenon. Here, flood occurrence is heavily influenced by the combination of recent urbanization patterns, the growing demand for land, and climate change (Krellenberg et al. 2013). Urban expansion has led to the conversion of important amounts of green space and agricultural land to built-up areas, generating loss of important infiltration and retention areas and, therefore, aggravating the preconditions for flood hazard generation. In some parts of the urban area, flooding tends to inundate streets and ground floors of dwellings (Müller and Höfer 2014). Thus the construction of water collectors in the city over the last decades has significantly contributed to diminishing the overall damage of flood events. Nevertheless, certain areas of the city are still exposed to and affected by annual flooding.

In a city like MAS, with its profound patterns of social inequality and on-going processes of socio-spatial differentiation, it is particularly interesting to know who suffers from the annual flooding, why and how well individuals can respond, and whether institutions intervene or not. Empirical evidence, to be presented in the following, was gathered between 2010 and 2015, derived from workshops, expert interviews, household surveys, and field mapping, and contributed to in-depth vulnerability assessments and recommendations for adaptation measures in response. The data came from a variety of international research projects that aimed at developing, collaboratively and at the interface of science, policy and practice, decisive adaptation measures (Krellenberg and Barth 2014). Prior to these projects, the topic of climate change and, in particular, attempts to react through adaptation were not yet on the political agenda (Barton et al. 2015). Resilience, on the other hand, has not yet been discussed.

Although one could assume, for a segregated city, that residents of the lower socio-economic strata are those most exposed to flood hazard, the picture in MAS is more sophisticated: the better-off are more prone to floods, often as the result of deliberate choice, for panorama or prestige purposes (Welz et al. 2014). Notwithstanding, a large number of households from the middle and lower socio-economic strata were also found to be exposed to flood hazards (see Fig. 2).

In this context, results from household surveys show that socio-structural dimensions of the household conditions (e.g., employment status, age structure of households, etc.) are eased by the likelihood of being affected by flood hazard. Individuals' flood hazard coping capacities are, in general, very low in the case study areas under review, and highly influenced by homogeneous social networks that are activated, or not, in case of an emergency, as well as perception, preparedness, and awareness of the hazardous incident (Krellenberg and Welz 2016). However, empirical findings reveal that experience with a flood event in the past clearly influences the perception of preparedness, although affected residents have, so far, only rarely implemented adaptation measures in response. This might be related to the fact that information regarding the implementation of prevention measures is still only rarely available.

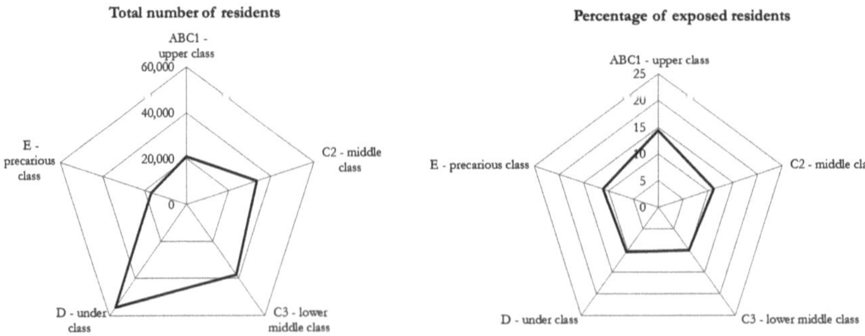

Fig. 2 Residents exposed to annual flood events in Santiago de Chile according to their socio-economic status (Source: Welz and Krellenberg 2016)

Overall, the findings from the vulnerability assessment show that institutional adaptation is needed, because the individual coping capacities of flood-exposed households are not sufficient to reduce their vulnerability. Here, an analysis of the institutional adaptive capacity (assessed through expert interviews and by the Adaptive Capacity Wheal of Grupta et al. 2010) provides insights into strengths and weaknesses of the institutional structure of selected municipalities of the MAS, which to date, rarely promotes adaptation activities. The general organisational structure of the municipalities, legal restrictions, and the lack of knowledge about adaptation options hamper the response to flood events. The legal framework determines, by and large, the way institutions function. It goes without saying that this reduces spontaneity and makes responses to newly emerging challenges, such as climate change and associated hazards, difficult. On the other hand, a certain propensity to respond to the needs of the people exists; this is one reason why priority is often given to visible problems and in such a way that short-term improvement is achieved.

In general, human resources play a key role, in terms of factors that limit the adaptive capacity of institutions, which relates to knowledge gaps and personal experience, as well as the capacity to create and build social capital. Without building capacity and leadership among municipality staff, it cannot be expected that advanced transformative action in response to flooding will occur.

3.2.1 From Adaptation to Transformation?

Current activities to adapt to climate change in MAS can be interpreted as starting points for more fundamental transformations. The activities can be described as incremental changes that aim at increasing the effectiveness or efficiency of the existing management or planning system. Nevertheless, they are important, because it was observed that there is an increasing awareness of the linkages between specific flood events and more general conditions of climate change over the past years and there is hope that this might become the starting point for more fundamental transformations. A national adaptation plan has recently been established and several sectoral adaptation plans have been initiated. At the urban level, an adaptation plan for the Urban Region of Santiago de Chile was developed (Krellenberg and Hansjürgens 2014) that influenced the decision to also elaborate sectoral adaptation plans for cities at the national level. Nevertheless, regional and local planning instruments currently at the disposal of the Urban Region of Santiago de Chile have not yet considered climate change specifically. Only during the last 5 years has more detailed information emerged on the potential impacts and vulnerabilities of climate change (e.g., Krellenberg and Hansjürgens 2014; Cortés et al. 2012; Romero et al. 2010). Hence the general debate on climate change response, notably on adaptation, is a comparatively recent phenomenon in Chile. Furthermore, as stated above, institutional capacities at the local, municipal level are still limited and it cannot be expected that in the short-run, transformations will occur in response to, e.g., flood events.

In view of the current trend to urbanize hazard-exposed locations, MAS is likely to experience an increase in the exposure of its population to hazardous events in the future. However, this will depend on climate change adaptation policies and on whether urban development does, in fact, strengthen capacities to protect the hazard-exposed population, diminish its vulnerability and, importantly, prevent a surge of new exposures. Adaptation, thus, must be seen as a social process and how transformative action is addressed and launched is key factor in this scenario. The Chilean case, therefore, confirms that adaptation can be seen as an important starting point for more fundamental changes in terms of urban transformations.

4 Discussion and Conclusion

This chapter argued for greater conceptual and terminological clarity in recent discussion on urban attempts to prepare for the anticipated consequences of future climate-induced changes. Whilst resilience refers to actions, decisions, and processes that aim at withstanding or effectively recovering from the negative consequences of environmental risks, adaptation is a concept that describes incremental changes and measures aiming at mitigating the anticipated consequences of climate change; transformation comprises more fundamental changes that go beyond adaptation or resilience.

With the aid of two empirical examples, it was shown that strategies and actions based on the concept of resilience and adaptation are prevailing, at least in the case studies this contribution focuses upon. In the German case study, a series of repeated flood events since 2002 had devastating consequences for many communities. However, it was only the most recent flood that initiated more fundamental changes aimed at addressing the underlying root causes of communities' exposure to flood risk. Previously, the dominant way of preparing for future flood risks was a mixture of restoring the status quo ante after flood events and increasing the performance of flood risk management, in order to mitigate future damages. Similarly, the case study of the urban area of Santiago de Chile indicates that "adaptation" is the dominant way of managing future flood risks, because measures have been proposed and, to some extent, already implemented, that are based on relatively incremental changes that can lead, but have not led yet, to fundamental transformations. The reasons therefore are grounded, above all, in the established institutional and organisational framework as well as in the lack of human capital; these undermine a more comprehensive approach for managing the anticipated consequences of future flood risks.

Whether strategies that are predominantly based on the idea of resilience and adaptation are sufficient to tackle the long-term consequences of climate change is a matter of speculation; however, better understanding of the societal tipping points at which the limits of established practices become apparent, and fundamental changes more acceptable, should be a matter of further research, in order to better understand the underlying causes for urban transformations with regard to mitigation of the consequences of climate change.

References

Alexander DE (2013) Resilience and disaster risk reduction: an etymological journey. Nat Hazards Earth Syst Sci 13:2707–2716

Anderson B (2010) Preemption, precaution, preparedness: anticipatory action and future geographies. Prog Hum Geogr 34:777–789

Barton JR, Krellenberg K, Harris JM (2015) Collaborative governance and the challenges of participatory climate change adaptation planning in Santiago de Chile. Climate Dev 2:175–184

Brand FS, Jax K (2007) Focusing the meaning(s) of resilience: resilience as a descriptive concept and a boundary object. Ecol Soc 12, online

Cortés G, Schaller S, Rojas M, Garcia L, Descalzi A, Vargas L, McPhee J (2012) Assessment of the current climate and expected climate changes in the Metropolitan Region of Santiago de Chile. UFZ-Discus Pap, Leipzig

DKKV – Deutsche Komitee für Katastrophenvorsorge (2015) Das Hochwasser im Juni 2013: Bewährungsprobe für das Hochwasserrisikomanagement in Deutschland. DKKV-Schriftenreihe Nr. 53, Bonn

Felgentreff C (2000) Impact of the 1997 Odra flood on flood protection in Brandenburg (FRG): the dyke broke, but the people's trust in technical solutions remained unbroken. In: Bronstert A, Bismuth C, Menzel L (eds) European conference on advances in flood research (PIK-report 65). PIK, Potsdam, pp 614–626

Gupta J, Termeer C, Klostermann J, Meijerink S, Van den Brink M, Jong P, Nooteboom S, Bergsma E (2010) The adaptive capacity wheel: a method to assess the inherent characteristics of institutions to enable the adaptive capacity of society. Environ Sci Pol 13(6):459–471

IPCC – Intergovernmental Panel on Climate Change (2012) Glossary of terms. In: Field CB et al (eds) Managing the risks of extreme events and disasters to advance climate change adaptation. Cambridge University Press, Cambridge

IPCC – Intergovernmental Panel on Climate Change (2014) Climate change: impacts, adaptation, and vulnerability. Part a: global and sectoral aspects. Contribution of working group II to the fifth assessment report of the Intergovernmental Panel on Climate Change. Cambridge University Press, Cambridge, UK/New York, USA

Klein RJ, Eriksen SE, Næss LO, Hammill A, Tanner TM, Robledo C, O'Brien KL (2007) Portfolio screening to support the mainstreaming of adaptation to climate change into development assistance. Clim Chang 84:23–44

Krellenberg K, Barth K (2014) Inter- and transdisciplinary research for planning climate change adaptation responses-the example of Santiago de Chile. Interdiscip Sci Rev 39(4):360–375

Krellenberg K, Hansjürgens B (eds) (2014) Climate Adaptation Santiago. Springer, Heidelberg

Krellenberg K, Welz J (2016) Assessing urban vulnerability in the context of flood and heat Hazard: pathways and challenges for indicator-based analysis. Soc Indic. doi:10.1007/s11205-016-1324-3

Krellenberg K, Müller A, Schwarz A, Höfer R, Welz J (2013) Flood and heat hazards in the urban region of Santiago de Chile and the socio-economics of exposure. Appl Geogr 38:86–95

Kuhlicke C, Begg C, Kunath A, Dressler G, Beyer M, Callsen I (2015) Resilience and river floods in Central Europe. Deliverable 5.2 of the embrace-project. https://drive.google.com/file/d/0BySVyEajWokmOXJDck9tQXExOUE/view. Last accessed 02 Dec 2016

LTV – Landestalsperrenverwaltung (2008) Hochwasserschutz für Eilenburg https://publikationen.sachsen.de/bdb/artikel/15599. Last accessed 28 Feb 2017

McCormick K, Anderberg S, Coenen L, Neij L (2013) Advancing sustainable urban transformation. J Clean Prod 50:1–11

Müller A, Höfer R (2014) The impacts of climate and land-use change on flood and heat hazards. In: Krellenberg K, Hansjürgens B (eds) Climate Adaptation Santiago. Springer, Berlin, pp 107–126

O'Brien KL (2012) Global environmental change (2): from adaptation to deliberate transformation. Prog Hum Geogr 36(5):667–676

Pelling M (2010) Adaptation to climate change: from resilience to transformation. Routledge, London/New York

Pelling M, O'Brien K, Matyas D (2015) Adaptation and transformation. Clim Chang 133:113–127

Pimm SL (1984) The complexity and stability of ecosystems. Nature 307:321–326

Romero H, Salgado M, Smith P (2010) Cambios climáticos y climas urbanos: Relaciones entre zonas termales y condiciones socioeconómicas de la población de Santiago de Chile. Revista INVI 70(25):151–179

Smit B, Wandel J (2006) Adaptation, adaptive capacity and vulnerability. Glob Environ Chang 16:282–292

SMUL - Sächsisches Staatsministerium für Umwelt und Landwirtschaft (2005) Ergebnisse der landesweiten Priorisierung von Hochwasserschutzmaßnahmen. Sächsisches Staatsministerium für Umwelt und Landwirtschaft. Dresden, Germany. [online]. http://www.umwelt.sachsen.de/umwelt/download/wasser/051206_HwskMaListe_GU_HswskRang_051206.pdf. Accessed 23 Feb 2017

Weichselgartner J, Kelman I (2014) Geographies of resilience: challenges and opportunities of a descriptive concept. Prog Hum Geogr 39(3):249–267

Welz J, Krellenberg K (2016) Vulnerabilidad frente al cambio climático en la Región Urbana de Santiago de Chile: posiciones teóricas versus evidencias empíricas. Revista EURE 42(125):251–272

Welz J, Schwarz A, Krellenberg K (2014) Understanding hazard exposure for adaptation in a climate change context. In: Krellenberg K, Hansjürgens B (eds) Climate Adaptation Santiago. Springer, Berlin, pp 127–147

Adapting Built-Up Areas to Climate Change: Assessment of Effects and Feasibility of Adaptation Measures on Heat Hazard

Juliane Welz, Daniel Hertel, Kerstin Krellenberg, and Uwe Schlink

1 Introduction

Looking at future trends in urbanization, with more than two thirds of the world's population living in cities by the 2050s (UN 2014), and at global trends of climate change, heat waves are expected to become more frequent and more intense in cities (Perkins et al. 2012). This calls particularly for local adaptation strategies to anticipate the consequences in urban areas.

In general, adaptation to climate change in cities has received considerable attention within the political and academic discourse in recent years (Measham et al. 2011; Eriksen et al. 2011). In this context, adaptation refers to processes, actions, or outcomes from individual households up to regions and countries that allow coping with, managing, or adjusting to changing conditions, external stresses, dangers, risks, and even opportunities (Smit and Wandel 2006). Many different ways of dealing with climate change can be conceived that cover a broad range of forms. Urban transformations towards climate adapted cities are coming to the fore when fundamental changes are addressed in an adequate way. Far-reaching changes in the urban context refer to energy, transportation, water use, land use, ecosystems, and associated growth patterns, consumption, and lifestyles.

Adapting built-up areas to heat hazard, as an urban response to the resilience and quality of life dimensions of urban transformations (see Kabisch et al., "Local Residential Quality from an Interdisciplinary Perspective: Combining Individual Perception and Micrometeorological Factors", in this volume) will be investigated

J. Welz (✉) • D. Hertel • K. Krellenberg • U. Schlink
Department of Urban and Environmental Sociology, Helmholtz Centre for Environmental Research-UFZ, Leipzig, Germany
e-mail: juliane.welz@ufz.de; daniel.hertel@ufz.de; kerstin.krellenberg@ufz.de; uwe.schlink@ufz.de

S. Kabisch et al. (eds.), *Urban Transformations*, Future City 10,
DOI 10.1007/978-3-319-59324-1_18

in this chapter by discussing empirical findings stemming from research in the Urban Area of Santiago de Chile (MAS).

In the future, the MAS will have to deal with increasing median temperatures (Mc Phee et al. 2014), and consequently, heat stress will become a serious problem (Müller and Höfer 2014). With a specific focus on heat hazard, we use micro-scale modelling of urban climate to identify feasible adaptation measures at the neighborhood level. We employ an interdisciplinary methodological approach that combines climate modelling with household survey data in order to assess the effects of different adaptation interventions, such as greening and reflective materials, on temperature distribution. Based on the empirical findings, we discuss how these interventions can contribute to fundamental urban transformations regarding the increase of resilience and quality of life.

2 Green and White Adaptation – Urban Responses to Climate Change in the Light of Urban Transformations?

Adaptation in the face of climate change is conceived as a dynamic phenomenon and there are many ways of characterizing adaptation. In general, adaptation is about confronting the societal context in which climate change occurs rather than simply accommodating its impacts (O'Brien et al. 2015). According to Pelling (2011), adaptation includes changes in the scope and range of values, institutions, behavior, and assets, with transformation being the deepest form of adaptation. This kind of adaptation implies systemic changes in over-arching political/economy regimes and the associated cultural discourses on development, security, and risk (O'Brien 2012). In the literature on adaptation to climate change, two different ways to approach transformation are discussed:

(i) A transformational adaptation that directly responds to the impacts of climate change and involves actions or interventions altering the nature, composition, and/or location of systems on a larger scale, with greater intensity and over longer time-periods (Kates et al. 2012).
(ii) A transformational adaptation that is concerned with the wider and less visible roots of vulnerability and that alters the risks for global development and human security posed by climate change (Pelling 2011; O'Brien 2012).

Both approaches highlight the idea that transformation of developmental pathways towards a more resilient and livable society stands out as one of the most important aims of adaptation to climate change.

Turning to urban transformations, urban planning and urban design have a critical role to play. Modifying the existing form and layout of buildings and urban neighborhoods can support cooling and ventilation, which, in turn, can contribute to reducing energy demand and allowing citizens to cope with higher temperatures. Therefore, we argue that applying adaptation measures in response to heat hazard in cities can push processes of urban transformations, because physical factors (e.g., radiation, elevation, wind) and land use interact with urban structures such as housing orientation, construction materials, ventilation, and other heat protection measures (Fernandez and Creutzig 2015; Coseo and Larsen 2014) and result in differences in the local climate (Klok et al. 2012).

In general, great importance is attached to the potential of green infrastructure and reflective materials, in particular for minimizing the impacts of heat hazard on people, directly and indirectly (Demuzere et al. 2014). Green urban infrastructure is related to urban forests, wetlands, parks, green roofs and walls, bioswales, innovative 'green street' techniques, permeable pavements, or rain gardens (Santamouris 2014; Li et al. 2013; Kraus and Spafford 2013, see also Banzhaf et al. in this volume dealing with green infrastructure). Depending on the shape, quality, size or type of the green infrastructure, its cooling effects may differ. It depends on evapotranspiration, shade, absorption of short wave solar radiation, and emitted thermal radiation (Bowler et al. 2010; Mentens et al. 2006; Stovin 2009). Effective management and maintenance can help to avoid excessive absorption of daytime solar energy and night time longwave radiation, as well as extensive watering, i.e., irrigation. In particular, in cities where water scarcity is a central issue, irrigation can be demanding and should be carefully considered because its side effects may frustrate the overall benefits that green infrastructures involve.

Another way of adapting built-up areas to heat hazard involves reflective materials (e.g., cool or white roofs) (Georgescu et al. 2014). By minimizing solar absorption through higher albedo, the warming of roofs and surrounded air can be reduced. However, the overall positive impacts of cool roofs on regional climate are still debated and are uncertain. In general, the adoption of cool roofs is seen as an effective tool for reducing building energy use in hot climates, for reducing urban heat islands, and for lowering regional air temperatures (Zhang et al. 2016). The need for cooling energy in buildings decreases (Akbari and Matthews 2012) as the flow of heat from the roof into the building decreases (Levinson and Akbari 2010). White roofing costs are comparable to those of conventional roofs, thus the energy that can be saved results in saving money, which makes such roofs appropriate for residential areas.

A precondition for the successful implementation of both adaptation strategies in built-up areas – greening and reflective materials – is the understanding of the local context, the urban structure of built-up areas, and the population affected by heat hazard.

3 The Urban Area of Santiago de Chile

The urban area of Santiago de Chile (MAS) currently consists of about 6.6 million inhabitants (INE 2002). According to projections, the population will exceed eight million people by 2030 (MINVU 2008). The MAS is located between the central Andean and coastal cordilleras; the climate is Mediterranean (Csb, according to the Köppen classification) with dry summers. Recent decades have shown an increase in median temperatures, a decline in average precipitation rates, a greater number of days with air temperatures above 30 °C, and a concentration of rainfall in extreme events. For the future, results from regional models predict higher median temperatures and less precipitation as a consequence of climate change (Mc Phee et al. 2014). Ongoing changes in land use and land cover are leading to a significant loss of environmental services such as storm water infiltration, heat mitigation, and biodiversity conservation in certain areas of the city (Ebert et al. 2010; Müller and Höfer 2014).

Already today, changing land-use and climate conditions in the MAS are reinforcing heat hazards, and are likely to amplify the intensity of potential hazardous events, leading to exposure patterns that vary by location, physical housing conditions, and standards. According to Welz et al. (2014), dwellings with deficient or precarious housing conditions, which indicate a very high likelihood of heat exposure, are of particular interest.

The case study area Villa Peru is located in the municipality of La Florida, in the south-east of the MAS (Fig. 1). The selected neighborhood is predominantly residential, with medium population density, and was developed at the beginning of the 1970s. The residents of Villa Peru are mainly elderly and belong to the lower socioeconomic strata. The residential structure is organized in a modular way and was built by the Public Housing Corporation (CORVI). Villa Peru shows a mixed structure of housing conditions, mostly characterized by detached houses of concrete and masonry with metal plates used for roof material. Thus, the indoor thermal behavior of these dwellings is not optimal because they readily accumulate heat.

The neighborhood contains only a very small amount of green infrastructure. Although the available green spaces (small squares and street greening) regularly

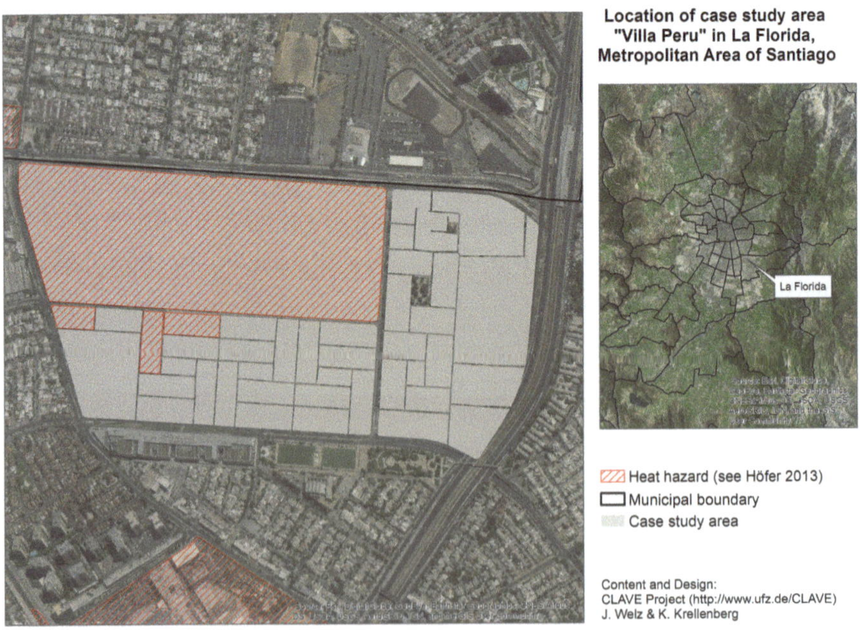

Fig. 1 Study region "Villa Peru" (La Florida municipality)

maintained, their cooling behavior is poor (Krellenberg et al. 2017). Consequently, the degree of imperviousness is high and, thus, favors the accumulation of heat in the urban environment. This calls for context-specific measures for diminishing heat hazard. In order to evaluate if either green or reflective surfaces have the potential to bring forth the necessary cooling effects, or if a combination of both produces them, modelling and survey approaches have been applied in the case study area.

4 Methodological Approach to Assess Green and White Adaptation

In order to analyze or simulate possible effects of adaptation measures for the cooling of urban areas, ENVI-met (version 4), a micrometeorological model, has been applied (http://www.envi-met.de). ENVI-met is designed for micro-scale analysis and allows the investigation of small-scale interactions between the building structure, surfaces, green infrastructure, and the urban atmospheric boundary layer. The model is therefore increasingly used to assess the effectiveness of urban planning measures (e.g., adaptation measures) for the local microclimate (Emmanuel and Loconsole 2015; Bruse and Fleer 1998). ENVI-met is a three-dimensional non-hydrostatic model consisting of a vegetation model, an atmosphere model including a radiative transfer model, and a one-dimensional soil model (Bruse and Fleer 1998).

For the analysis of the case study area, surface characteristics, buildings, and vegetation structure are mapped on a grid of 60× 65 pixels (each pixel representing 4 m×4 m). Meteorological boundary conditions (such as temperature and relative humidity) are retrieved from a weather station in La Florida (geo coordinates: UTM 352504 E 6290304 N) operated by the Environmental Ministry of Chile (Sistema de Información Nacional de Calidad del Aire – SINCA: www.sinca.mma.gob.cl). With the aim of analysing heat hazard, a hot weather situation in summer was simulated, based on temperature records from the SINCA weather station on 22 January 2014. In addition, model input parameters, such as specific humidity at 2.500 m height (derived from radio sounding), wind speed, and wind direction at 10 m height (both from the weather station in La Florida), cloud coverage (default value 0 according to the weather data on the simulated day), albedo of roofs and walls (values for typical materials, based on the literature, and different scenarios in the study area), are specified for the total simulation area.

The current situation in Villa Peru presents a roof albedo of 0.5; from the overall grid (3900 pixels), 79 pixels represent hedges, and 289 pixels represent trees (Fig. 2). In order to assess the efficiency of green infrastructures and reflective materials in the case study area, five different scenarios are simulated by:

1. Positioning of 60 additional trees,
2. Siting 150 additional trees,
3. Increasing the roof albedo to 0.9,
4. Increasing the roof albedo to 0.8, and
5. Combining scenario 1 and 3.

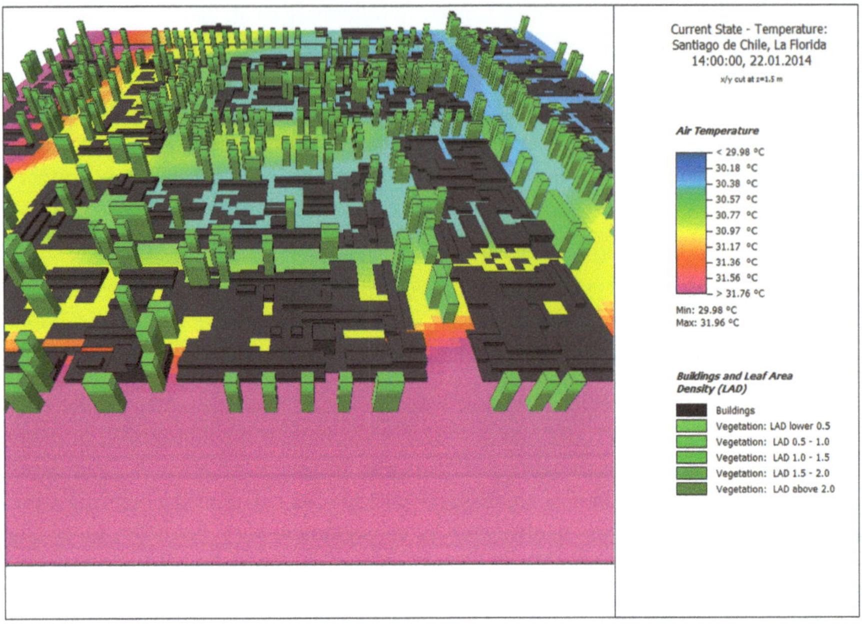

Fig. 2 Air temperature in 1.5 m height above ground and urban structure of Villa Peru; current state

Each scenario is compared with the modelled temperature field of the current situation of the study area. This allows the assessment of the effects of green and reflective interventions on the surrounding temperature and the evaluation of the most appropriate adaptation strategy for the neighborhood in La Florida.

Furthermore, the ENVI-met simulations are combined with residents' perception of heat hazard. This makes it possible to depict real-world conditions in terms of residents' perception and helps to underpin the most pressing action needed, in terms of heat hazard in the case study area. Data from a household survey, conducted in April 2014, were used to gain information on how the hot summer situation was perceived by the residents. Considering a confidence level of 95%, a sample of 347 from the 3548 target households in the study area was selected. In total, 129 households had completed the survey questionnaire that corresponds to a response rate of 37%.

5 Current Climate Situation and Residents' Perception of Heat Hazard in the Study Area

According to data from the SINCA weather station in La Florida, the average weather situation during the summer season 2013/2014 revealed temperatures between 13.5 °C (at night) and 31.5 °C (at midday) with relative humidity between

75% (at night) and 24% (at midday). The modelled microscale temperature field characterizes a representative, average weather situation of this summer season. High temperatures occur in the central square of the area that is only very sparsely covered by trees (see Fig. 2). The shadow provided by the loose vegetation is low, causing heating of the ground that can lead to temperatures above 40 °C. Similar to the central square, the surrounding bituminized streets are hotspots of heat accumulation, which is transferred to the ground level air. The verge of the buildings has well-shadowed areas, with temperatures varying between 27 and 29 °C at midday, whereas roof temperatures exceed all others.

The results of the household survey underpin the ENVI-met simulations. Heat perception is highly influenced by the building structure (dense structure of detached houses) and the roof material, because almost all houses have galvanized corrugated roofs. Households that feel most affected by extreme heat generally belong to lower socioeconomic status groups and display very poor living conditions. Moreover, these households are poorly prepared for heat hazard, and knowledge about protection measures is scant. When a heat hazard occurred, two-thirds of all respondents opened the windows to ventilate their dwellings (72%) and almost half of them turned on an electric fan (49%). The first measure, in particular, contributes to an increase of indoor temperature. Moreover, the majority of the respondents suffered transpiration and needed to take a shower several times a day (70%) and also drank more water than usual (68%). Only 15% of all respondents use green urban infrastructure for cooling effects. In general, although they have been affected by heat, residents have not changed their lifestyles or their residential environment after the last hot spell, to be better prepared in the future (Krellenberg et al. 2017). Against this backdrop, the following section will shed light on the question of whether green infrastructure and/or white roofs present feasible adaptation strategies to reduce air/surface temperature within the study area and thereby minimize heat hazard and its impacts for the inhabitants.

6 Effects of Green and White Adaptation in Built-Up Areas on Heat Hazard

Depending on the applied adaptation scenario, different levels of temperature reduction can be achieved (Table 1). Calculating, for air temperature at 1.5 m above ground, the absolute differences between the adapted scenario and the current

Table 1 Reduction of air temperature at 2 p.m., compared to the current state (see Fig. 2)

Scenario	Description	Max. temp. change at 1.5 m height [K]	Range of temp. change at all heights [K]
1	60 additional trees	−1.37	0.14 to −1.45
2	150 additional trees	−1.35	0.14 to −1.44
3	Increased roof albedo (0.9)	−1.45	−0.14 to −1.62
4	Increased roof albedo (0.8)	−1.45	−0.14 to −1.62
5	Combination of scenario 1 and 3	−2.35	−0.19 to −2.49

situation can be identified, leading to conclusions about locations that benefit most from the adaptation measures.

According to the ENVI-met simulations of scenarios 1 and 2, where the amount of vegetation cover in the central square as well as the principal street in the south of the residential area was enhanced, we observe that the resulting temperature reduction is not necessarily proportional to the number of additional trees in the study area (Table 1). This supports other studies in which it was found that not only the amount of urban green is important for the regulation of the local climate, but also the type, form, arrangement, and irrigation of the plants (Emmanuel and Loconsole 2015). Furthermore, green infrastructure does not result in temperature reduction in all simulated locations of the study area. What is more, whereas temperature reductions of up to 1.45 Kelvin are achieved at some locations, other sites show an increase of the air temperature by up to 0.14 Kelvin (Table 1, rightmost column).

In contrast, simulation of reflective materials (scenarios 3 and 4) shows a much more rigorous cooling effect than green adaptation; temperature reduction occurs at all locations of the study area with a greater magnitude (see Table 1). Temperature reduction by 1.45 Kelvin is much stronger than the reduction by 1.37 Kelvin resulting from green adaptation in scenario 1.

The concerted application of both green and white adaptation measures in scenario 5 (Fig. 3) results in an overall temperature reduction of 0.45–2.35 Kelvin, which is the highest of all scenarios (Table 1). In terms of spatial distribution, strongest cooling effects occur in the central square (Fig. 4), where several trees were

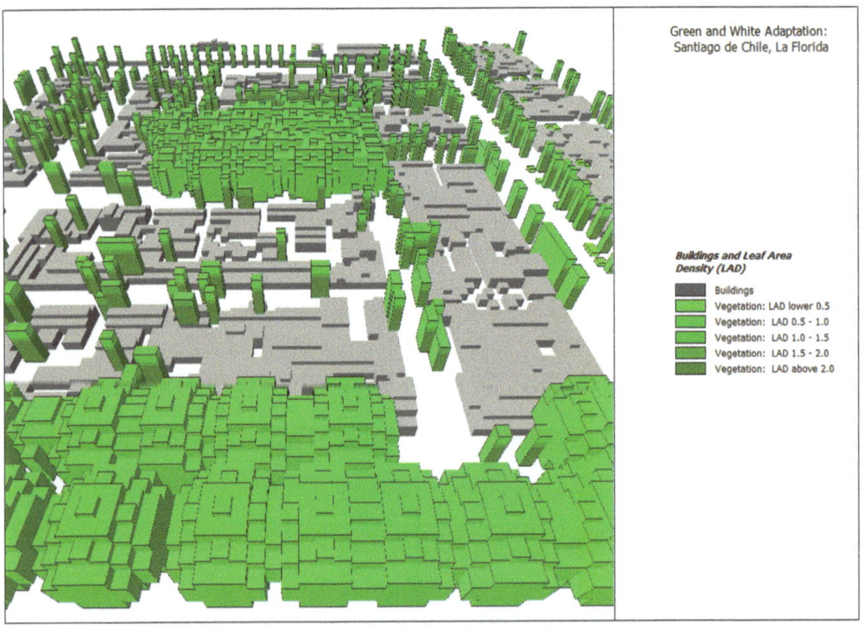

Fig. 3 Urban structure of Villa Peru after greening (60 additional trees in the central square and main roads) and white adaptation (increased roof albedo up to 0.9, illustrated by light gray buildings)

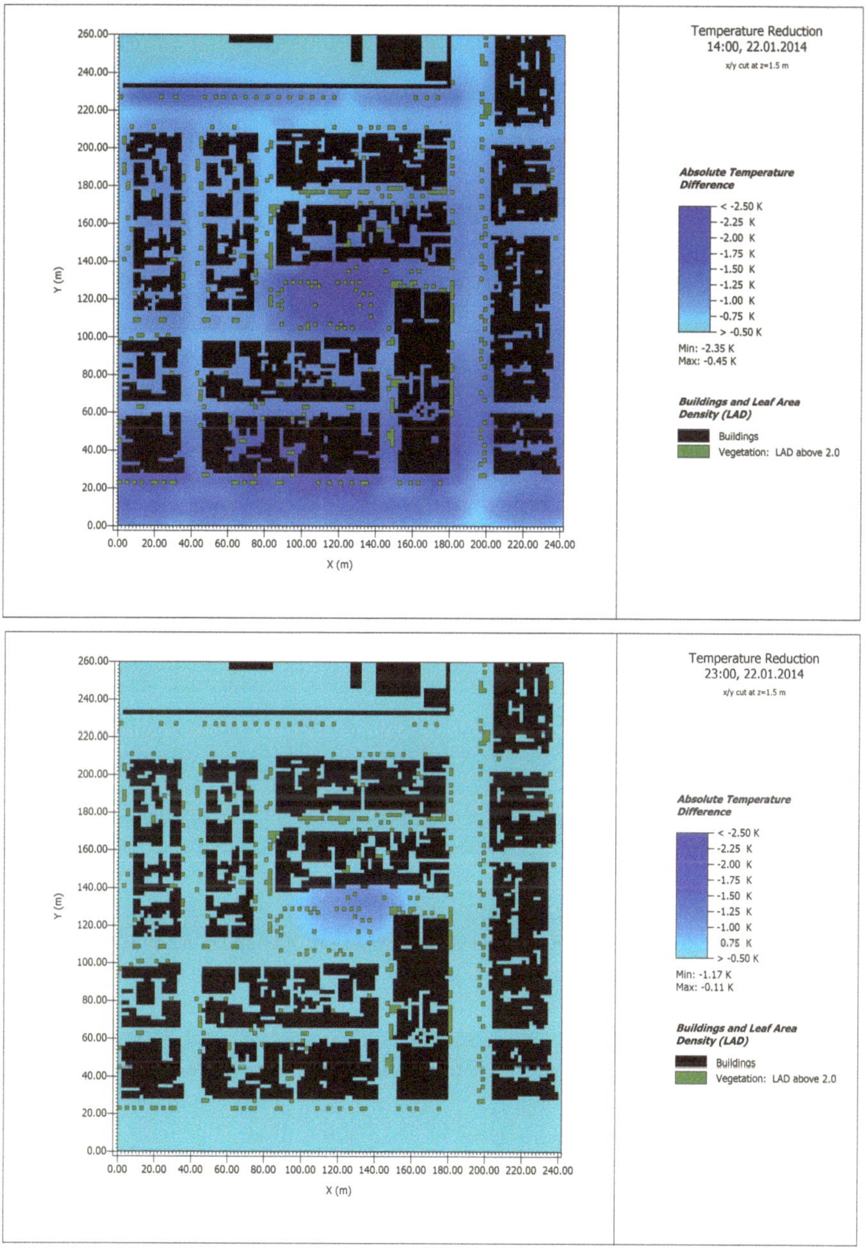

Fig. 4 Temperature reduction at noon and at midnight, due to green and white adaptation

simulated for shadowing effects. This demonstrates that a combination of green and white adaptation measures can promote an overall local cooling effect within the neighborhood of up to 1.1 K.

7 Conclusion

With the urban area of Santiago de Chile (MAS), an area was selected that suffers from the impacts of climate change and increased urbanization in terms of heat hazard. This paper aimed at presenting empirical findings that could support possible interventions, in terms of adaptation measures, as a step towards urban transformation. Our findings are twofold: methodological and content-related.

In terms of methodology, we strengthen the usefulness of micrometeorological simulations as a detailed database for the evaluation of various adaptation measures. Notwithstanding, the ENVI-met technique is limited because it only refers to one particular day and does not represent climatic averages, for example, of the prevailing wind, temperature, etc., situation. This means that simulations can only be undertaken for a single situation and not for long-term heat burdens over the year. The latter would require the integration of information on changing weather conditions according to the geographic location of the considered region.

Following this up, and in terms of content-related conclusions, the simulations for the heat-adapted neighborhood of Villa Peru in the MAS indicate that the most effective interventions comprise the combined application of green infrastructures and reflective (white) materials. Thereby, we highlight that the extent of green infrastructure and reflective materials plays a significant role in adapting urban neighborhoods to extreme heat. Green infrastructure was identified as a vital component in lowering the heat effect as well as the albedo of the urban structure (Gill et al. 2007). In contrast, white adaptation measures are most effective during the summer time and can also result in visual discomfort and glare, and as such, their application is limited to specific urban areas (e.g., not in the vicinity of airfields or air lanes) (Al-Obaidi et al. 2014).

What remains open is the question of who, in practice, and under real-world conditions, will take the responsibility for implementing the proposed interventions. Here, we see a clear role that municipalities have to play. However, research on the adaptive capacity of the municipalities in MAS shows that, in general, they do not yet have the necessary conditions for adapting to climate change, because the institutional structure itself does not promote transformational adaptation. In particular, the legal framework appears as an obstacle because it reduces spontaneity of the system and makes it very difficult to respond to new challenges imposed by climate change. Additional obstacles are related to human resources, knowledge, and experience in the face of adaptation to climate change. Thus, the simulation data on the micro-scale also point to interventions that could, to a certain extent, also be undertaken by the inhabitants themselves: painting their roofs white or planting trees in their front or backyards (Krellenberg et al. 2017).

According to our empirical findings, we argue that the transformation towards more resilience and quality of life in cities must consider a combination of strategies in order to achieve fundamental changes. Small-scale interventions, such as planting trees along streets and in open spaces, as well as roof whitening, should be part of the overall picture. This follows a transformational adaptation in cities as understood by Pelling (2011) that starts from local and small-scale interventions and relates to the wider and less visible roots of vulnerability.

Acknowledgements Our thanks go to D. Dobersalske who was involved in ENVI-met modelling. D. H. (AZ:20,015/373) gratefully acknowledges the financial support given by the Deutsche Bundesstiftung Umwelt (DBU).

References

Akbari H, Matthews HD (2012) Global cooling updates: reflective roofs and pavements. Energ Buildings 55:2–6. doi:10.1016/j.enbuild.2012.02.055

Al-Obaidi KM, Ismail M, Malek A, Rahman A (2014) Passive cooling techniques through reflective and radiative roofs in tropical houses in Southeast Asia: a literature review. Front Archit Res 3(3):283–297

Bowler DE, Buyung-Ali L, Knight TM, Pullin AS (2010) Urban greening to cool towns and cities: a systematic review of the empirical evidence. Landsc Urban Plan 97(3):147–155

Bruse M, Fleer H (1998) Simulating surface-plant-air interactions inside urban environments with a three dimensional numerical model. Environ Model Softw 13(3–4):373–384

Coseo P, Larsen L (2014) How factors of land use/land cover, building configuration, and adjacent heat sources and sinks explain Urban Heat Islands in Chicago. Landsc Urban Plan 125:117–129. doi:10.1016/j.landurbplan.2014.02.019

Demuzere M, Orru K, Heidrich O, Olazabal E, Geneletti D, Orru H et al (2014) Mitigating and adapting to climate change: multi-functional and multi-scale assessment of green urban infrastructure. J Environ Manag *146*:107–115. doi:10.1016/j.jenvman.2014.07.025

Ebert A, Welz J, Heinrichs D, Krellenberg K, Hansjürgens B (2010) Socio-environmental change and flood risks: the case of Santiago de Chile. Erdkunde 64(4):303–313

Emmanuel R, Loconsole A (2015) Green infrastructure as an adaptation approach to tackling urban overheating in the Glasgow Clyde Valley Region, UK. Landsc Urban Plan 138:71–86. doi:10.1016/j.landurbplan.2015.02.012

Eriksen S, Aldunce P, Bahinipati CS, Martins RD, Molefe JI, Nhemachena K et al (2011) When not every response to climate change is a good one: identifying principles for sustainable adaptation. Climate Dev 3(1):7–20

Fernandez Milan B, Creutzig F (2015) Reducing urban heat wave risk in the 21st century. Curr Opin Environ Sustain 14:221–231. doi:10.1016/j.cosust.2015.08.002

Georgescu M, Morefield PE, Bierwagen B, Gand Weaver CP (2014) Urban adaptation can roll back warming of emerging megapolitan regions. Proc Natl Acad Sci U S A 111(8):2909–2914

Gill SE, Handley JF, Ennos AR, Pauleit S (2007) Adapting cities for climate change: the role of the green infrastructure. Built Environ 33(1):115–133

INE – Instituto Nacional de Estadísticas (2002) Censo de Población y Vivienda 2002. Instituto Nacional de Estadísticas, Santiago de Chile

Kates RW, Travis WR, Wilbanks TJ (2012) Transformational adaptation when incremental adaptations to climate change are insufficient. Proc Natl Acad Sci U S A 109(19):7156–7161

Klok L, Zwart S, Verhagen H, Mauri E (2012) The surface heat island of Rotterdam and its relationship with urban surface characteristics. Resour Conserv Recycl 64:23–29. doi:10.1016/j.resconrec.2012.01.009

Kraus H, Spafford A (2013) Rain gardening in the south: ecologically designed gardens for drought, deluge, and everything in between. Eno Publishers, Hillsborough

Krellenberg K, Welz J, Link F (2017) Cambio climático, vulnerabilidad urbana y adaptación a nivel municipal: Santiago de Chile y otras ciudades de América Latina. RIL Editores, Santiago de Chile

Levinson R, Akbari H (2010) Potential benefits of cool roofs on commercial buildings: conserving energy, saving money, and reducing emission of greenhouse gases and air pollutants. Energ Effic 3(1):53–109

Li H, Harvey JT, Holland TJ, Kayhanian M (2013) The use of reflective and permeable pavements as a potential practice for heat island mitigation and stormwater management. Environ Res Lett 8(1):1–14

McPhee J, Cortés G, Rojas M, Garcia L, Descalzi A, Vargas L (2014) Downscaling climate changes for Santiago: what effects can be expected? In: Krellenberg K, Hansjürgens B (eds) Climate adaptation Santiago. Springer, Berlin, pp 19–41

Measham TG, Preston BL, Smith TF, Brooke C, Gorddard R, Withycombe G et al (2011) Adapting to climate change through local municipal planning: barriers and challenges. Mitig Adapt Strateg Glob Chang 16(8):889–909

Mentens J, Raes D, Hermy M (2006) Green roofs as a tool for solving the rainwater runoff problem in the urbanized 21st century? Landsc Urban Plan 77(3):217–226

MINVU – Ministerio de Vivienda y Urbanismo (2008) Propuesta de Modificación del Plan Regulador Urbano de Santiago. Ministerio de Vivienda y Urbanismo, Santiago de Chile

Müller A, Höfer R (2014) The impacts of climate and land-use change on flood and heat hazards. In: Krellenberg K, Hansjürgens B (eds) Climate adaptation Santiago. Springer, Berlin, pp 107–126

O'Brien KL (2012) Global environmental change (2): from adaptation to deliberate transformation. Prog Hum Geogr 36(5):667–676

O'Brien KL, Eriksen S, Inderberg TH, Sygna L (2015) Climate change and development: adaptation through transformation. In: Inderberg TH, Eriksen S, O'Brien K, Sygna L (eds) Climate change adaptation and development: transforming paradigms and practices. Routledge, London, pp 273–289

Pelling M (2011) Adaptation to climate change: from resilience to transformation. Routledge, Abingdon

Perkins SE, Alexander LV, Nairn JR (2012) Increasing frequency, intensity and duration of observed global heatwaves and warm spells. Geophys Res Lett 39(20). doi:10.1029/2012GL053361

Santamouris M (2014) Cooling the cities – a review of reflective and green roof mitigation technologies to fight heat island and improve comfort in urban environments. Sol Energy 103:682–703. doi:10.1016/j.solener.2012.07.003

Smit B, Wandel J (2006) Adaptation, adaptive capacity and vulnerability. Glob Environ Chang 16(3):282–292

Stovin V (2009) The potential of green roofs to manage urban stormwater. Water Environ J 24(3):192–199

UN – United Nations (2014) World urbanization prospects: the 2014 revision. United Nations, Department of Economic and Social Affairs, Population Division

Welz J, Schwarz A, Krellenberg K (2014) Understanding hazard exposure for adaptation in a climate change context. In: Krellenberg K, Hansjürgens B (eds) Climate adaptation Santiago. Springer, Berlin, pp 127–147

Zhang J, Zhang K, Liu J, Ban-Weiss G (2016) Revisiting the climate impacts of cool roofs around the globe using an Earth system model. Environ Res Lett 11(8). doi:10.1088/1748-9326/11/8/084014

Climate Proofing of Urban Development: Regulatory Challenges and Approaches in Europe, Germany, and Beyond

Moritz Reese

1 Introduction

Climate change will increasingly give rise to environmental risks for urban agglomerations in many of the world's regions (IPPC 2014, p. 550). Riverine floods, flash floods from extreme rainfall, heat waves, aridity, landslides, and environmental degradation are natural hazards that will probably become more frequent in the wake of global warming (IPPC 2014, p. 552; for Europe: IPCC 2014, p. 1279; EEA 2016). Cities around the world need to be aware of these potential climate impacts, to thoroughly assess local risks and to adapt their development with a view to strengthening resilience – a challenge that is often referred to as "climate proofing" (CP) (Romero Lankao and Qin 2011; Birkmann and Fleischauer 2009). In its 5th Assessment Report, the IPCC has – again – analysed the multiple dimensions of this urban governance challenge, including aspects of awareness-raising, knowledge production, policy mainstreaming, stakeholder involvement, capacity building and financing, and, in line with recent literature (e.g., Lowe et al. 2009, Kehew et al. 2013, Sanchez-Rodriguez 2009; also Reese 2015 and Reese et al. 2010a, b) the Panel has particularly highlighted the key role of urban land-use and infrastructure planning, as well as the importance of the legal framework that regulates urban planning and development, respectively (IPPC 2014, p. 576 with many more references).

Taking this up, this paper aims at further exploring how CP can be fostered within the legal framework of urban development. After an overview of the impacts of climate change and respective adaptation requirements in Sect. 2, a set of general key requirements for steering urban development towards climate resilience will be presented (Sect. 3). Then, some general conceptual questions about how these

M. Reese (✉)
Department Environmental and Planning Law, Helmholtz Centre for Environmental Research-UFZ Leipzig, Leipzig, Germany
e-mail: moritz.reese@ufz.de

S. Kabisch et al. (eds.), *Urban Transformations*, Future City 10,
DOI 10.1007/978-3-319-59324-1_19

governance requirements can best be implemented in the legal framework will be discussed (Sect. 4) before taking, by way of example, a closer look at whether and how this is done in European and German law (Sect. 5). Environmental impact and risk assessment procedures, which play a key role in this regard, are analysed in terms of how they might best be geared towards climate proofing (Sect. 6), and finally, the options and possible function of a separate climate adaptation planning instrument will be briefly considered (Sect. 7). Ahead of all this, a brief review of the relevant climate impacts and needs for adaptation sets the scene (Sect. 2, below).

2 Climate Change Impacts in the Urban Context

The most significant impacts of climate change in the urban context, related to land-use, are expected to arise from an increase in extreme weather conditions and events, and can be summed up by the categories storms and flooding, and heat and aridity (EEA 2016, p. 18; EU Commission 2009). Changes in the natural environment brought on by global warming will have an impact on forestry and agriculture first and foremost, but may also play a role in settled areas. Current climate research tells us that all these impacts can be expected to occur in many parts of the world and that they constitute a gradual worsening of land-use problems that already exist (EEA 2016; IPPC 2014); at the same time, considerable uncertainty exists regarding the extent and speed of these climate change impacts, and, within a relatively short period of time, they will not be clearly distinguishable from other causes. A brief survey of the main impacts of climate change is provided in the following.

2.1 Floods: An Increased Risk of Flooding, Rising Sea Levels, and Heavy Rainfall

Many large cities are located close to major rivers, and it is likely that they will have to cope with an increasing frequency and severity of river floods as a consequence of climate change (see Kuhlicke et al., "Resilience, Adaptation and Transformation: Conceptual and Empirical Insights from Two Case Studies in Germany and Chile", in this volume). An increase in annual rainfall and a shift from the summer to the winter months has been a measurable feature in Central Europe for some time now, and it is predicted that this trend will continue (IPCC 2014, p. 1276; EEA/JRC 2008, p. 94). Increasing amounts of rainfall during the winter period, glacial melting due to global warming, less storage as snow and earlier snowmelt will probably lead to a further increase in the frequency and severity of river floods and flood-related damage in many riverine areas (IPCC 2014, p. 1280), as already clearly observed during the periods 1960–1980 and 1980–2000, when maximum levels doubled (EEA/JRC 2008, p. 96). This increase in flood risk is, however, not due, in the first instance, to climate change, but rather to housing developments and dikes encroaching on rivers, to the loss of retention areas, and to large-scale drainage of the landscape for agriculture, urban construction, and road building (EEA 2015a, p. 33; IPPC 2014, p. 1280; Wheather and Evans 2009; LAWA 1995). In this sense, climate change "merely" reinforces an existing need for action, one that has been demonstrated dramatically by the recent century floods. This need for action consists specifically in effectively containing flood risks by a combination of technical mitigation measures, on the one hand (higher dikes, polders, property protection), and prevention of causes (more space for rivers, increased water retention on plains), on the other.

A potential increase in heavy rainfall events also brings with it a greater risk of pluvial flooding in housing areas, partly due to overflowing drainage systems. There have been many examples of such flooding after heavy rainfall in the past; thus, in this respect, previous events cannot be attributed at all conclusively to climate change (DKKV and D.K.f.K. 2015, DKKV 2003). The main cause lies, instead, in a type of urban and infrastructure development that has neglected the demands of effective urban drainage. What is needed here is, on the one hand, to design more efficient systems of drainage in residential areas and, on the other, to ease the burden on them by creating retention, infiltration, and catchment areas, as well as decentralized rainwater usage. Well-planned drainage systems and a "water-sensitive" form of urban planning that combines land-related and infrastructural flood protection will be crucial for this (Moss 2011).

2.2 *Aridity*

Many regions around the world are likely to be increasingly affected by (summertime) aridity and drought, meaning that problems with water supply may occur more frequently and increase in degree, especially for large agglomerations with high water demand. In places where this trend can be expected, water supply systems and water consumption should be prepared for future shortages (see Klassert et al., "Sustainable Transformation of Urban Water Infrastructure in Amman, Jordan – Meeting Residential Water Demand in the Face of Deficient Public Supply and Alternative Private Water Markets", in this volume). This includes making provisions for remote supply, building up storage capacity, and promoting water saving technologies along with a water management cycle that converts wastewater into usable water by means of treatment.

2.3 *Summer Heat*

Throughout Europe, the summer months can be expected to bring more heat waves and an increase in daytime temperatures above 30 °C. Particularly in urban areas, this will have adverse impacts on human health, working conditions, and people's quality of life (IPCC 2014, p. 18). The extreme heat waves that occurred in the summer of 2003 in central European regions brought into sharp relief just how stressful heat can be, especially for older people and children, as well as in working environments. According to estimates, the heat wave of 2003 caused 70,000 excess deaths in 12 European countries (EEA 2015b, p. 17). By 2050, heat waves are projected to cause 150,000 excess deaths per year in the European Union alone (EEA 2015b, p. 17). As a result, many countries have implemented improved heat warning systems (IPCC 2007, No. 12.5.11) but much more can be done, especially in cities. There are a number of ways in which urban heat stress can be mitigated, including town planning measures, such as improving air flows and providing more green spaces and shady areas, along with architectural adaptations (see, for example, Gill et al. 2007, see Welz et al., "Adapting Built-Up Areas to Climate Change: Assessment of Effects and Feasibility of Adaptation Measures on Heat Hazard", in this volume). This is obviously a key challenge for urban planning.

2.4 Changes in Nature and Loss of Biodiversity

The general warming of air, water, and soil, changes in precipitation regimes, in the water cycle, and in the availability of nutrients in water and soil, will have a direct impact on species and habitats. This will partly manifest itself in the migration of many species and habitats to "cooler" regions and to higher locations, and we can also expect considerable changes to occur in biozonotic relations between habitats (IPCC 2014, p. 14). In particular, those habitats located in more southerly regions and affected by aridity will come under increasing pressure of migration, while species diversity could initially increase in northern European regions. The main type of action that needs to be taken in relation to spatial planning includes, for example, adapting existing nature reserve strategies to observed and predicted changes and establishing suitable migration corridors (for a detailed account, see Schumacher et al. 2013, Gies and Albrecht 2013). These challenges relate predominantly to rural areas, but they may also become relevant for the urban natural environment and, in particular, for the areas immediately surrounding cities and towns, and for any further expansion of housing developments.

3 Climate Proofing as an Administrative and Regulatory Challenge

The overview given in Sect. 2 of the main spatial impacts of climate change and possible adaptation requirements, despite its brevity, makes it clear that the term climate change adaptation encompasses a heterogeneous set of problems. These problems affect diverse sectors and regions; the main factor linking them together as a focus for adaptation action is their shared cause – climate change (Reese 2010,

p. 14) – and their overlap with many other drivers shaping future urban risks and vulnerabilities. However, despite the many different problems, it is abundantly clear that the task of adaptation has also involves overarching challenges, particularly in relation to urban development (Reese 2015, p. 18; Reese 2010 p. 12, 2012, p. 400). The key challenges arise from the following factors:

- the extent of people's *awareness and knowledge of climate risks* is still relatively modest;
- there is still considerable *uncertainty* about the precise impacts of climate change (IPCC 2014, p. 176; EU Commission 2013, p. 19);
- environmental conditions are changing under the influence of climate change with an unprecedented *dynamic* (IPCC 2013, p. 4);
- in the long term, profound changes may occur in regional environmental and human settlement conditions, necessitating *anticipatory adaptation*, especially with regard to long-lasting developments and infrastructures (EEA 2016, p. 8).

With regard to the governance of urban development and, in particular, the responsible planning and permitting administrations, the above mentioned challenges imply the need for:

- *clear responsibility:* Actors must be aware of the potential risks arising from climate change impacts and they must perceive CP as a part of their development responsibility.
- *thorough assessment*: The relevant risks and adaptation options must be assessed as thoroughly as possible in order to acquire the best possible knowledge basis for climate sensitive land-use decisions (IPCC 2014, p. 579; Craig 2010, p. 40).
- *long term orientation*: Planning and investment decisions that will have structural, long-term impacts require an anticipatory orientation towards long-term climate impacts (Craig 2010, p. 53). Moreover, when it is likely that certain land uses cannot be upheld, due to climate impacts in the long run, it is advisable that this is anticipated and land users are granted an adequate transition period.
- *regular review*: To ensure that ongoing adaptation is possible with respect to unforeseen developments and newly recognized needs for action, a dynamic orientation is also required, in the sense of regular monitoring – and, where necessary, adjustment – of the risk assessment and of land-use decisions (see, for example, Swart et al. 2009, p. 145; Reese 2012, pp. 216, 402)

The key task of ensuring that these "CP requirements" will be effectively fulfilled falls to the laws regulating urban development and facility approval (EEA 2016, p. 8). These laws regularly determine the tasks, competences, aims, standards, and procedures of urban planning, zoning, and permitting.

It is self-evident that fulfilment of the abovementioned CP requirements depends strongly on whether these requirements are accounted for and expressly regulated, as a part of the tasks and competences of the planning authorities (Hirokawa and Rosenbloom 2013, p. 325). Climate proofing – as outlined above – implies considerable effort and cost, and it may raise additional issues of conflict. Therefore, it cannot be expected that thorough assessment, long-term orientation, and regular review will be effectively conducted on a voluntary basis. Hence, a legal framework that is equal to the task of adaptation will have to respond to the above-mentioned challenges by means of, in particular, the following steering approaches:

- Urban planning *responsibility* for adaptation ought to be underlined by explicitly elevating climate adaptation to a management objective and a planning concern. In addition, there can and should be clear guidelines regarding the extent to which the precautionary principle is to be applied in the context of adaptation.
- To encourage a *thorough assessment* of significant climate impacts, risks, and adaptation options (climate risk assessment), a number of relevant legal obligations can be put in place. The standard of this assessment can be calibrated by including requirements related to breadth, depth, time horizon (long-term orientation), and the inclusion of specialist expertise, and it can be adjusted with particular regard to staffing and financial capabilities. Another important aspect to consider when establishing governmental tasks for assessing climate risks is that knowledge about flood, drought, weather, and landslide risks may be needed not only to enable appropriate urban planning but possibly also to enable private sector actors to adapt their individual land-use decisions to the local risk situation, in the context of a minimum level of precaution guaranteed by zoning (construction planning) law. Risk assessment thus acquires a synergetic dual function: it enables a risk-appropriate framework to be put in place by means of binding development planning while simultaneously enabling users to fully exploit this legal framework in a risk-appropriate way, by providing them with the necessary risk knowledge. It is clearly a public sector, governmental task to provide such "risk knowledge", given that it is a public good and that individual users cannot feasibly undertake the necessary assessments (observations, model simulations, scenarios) themselves.
- With regard to long-lasting structural dispositions, adequate *long-term orientation* should be ensured for both assessment and land-use decisions. The latter can be facilitated by legal entitlements to devise terminable usage rights and to fade out existing usage rights in the long run and without compensation, e.g., in areas prone to increasing flood risks.
- *Monitoring* and *reviewing* the relevant developments and existing land-use decisions can also be enforced by law. In particular, obligations to monitor and review can be linked to formal development plans and also to relevant permit decisions. In order to minimize the effort, it seems important to implement a graduated approach and to require substantial review only where the necessity is indicated by a preliminary screening.

In the following section, we analyse in more detail how planning laws and related land-use regulations can implement the above-mentioned essentials of CP by looking, in particular, at the example of German and European law. However, before focussing on this example, it is explained, in more general terms, how urban development is usually framed by planning and zoning regimes – in most jurisdictions of the world – and which general conceptual questions arise when addressing the essential requirements of CP.

4 General Challenges of Including Climate Proofing in Urban Development Law

An effective system of public land-use planning is an indispensable foundation for sound urban development. Accordingly, the large majority of countries and regions in the world have more or less distinctive urban planning laws that make it possible for public – often local authority – planners to make stipulations regarding the use of land in order to guarantee a properly functioning and well-balanced structure of use (Nolon and Salkin 2006, p. 57). Although planning systems may differ considerably as to design and effectiveness, very similar strategic issues and options arise when it comes to anchoring the main features of CP. These include:

- *Division of tasks between overall planning and sectoral protection regimes:* Public urban development planning often amounts to overall planning. In other words, it arranges and orders urban settlement areas comprehensively with regard to all relevant issues of concern. The public authorities responsible are thus tasked with assessing all these issues and finding balanced solutions within an integrative deliberation process. Because there are limits to specialization in terms of individual issues in this kind of overall planning, it may be expedient to submit relevant sectoral aspects of urban development to special plans and to allocate these to a specialized authority. Such sectoral regulations for land use, which are also relevant for the urban context, are provided for – as we show below – in the EU Floods Directive and in German nature conservation law. In strategic planning terms, the question that arises for any legal system is whether or not the above-mentioned environmental risks in urban development should be placed within the remit of a special assessment and planning regime and, if this is done, how to integrate this into the overall planning. The analysis below demonstrates that the challenge of CP is basically calling for a (further) development of risk-specific sectoral planning instruments, in order to provide specialist assessments and concepts as a solid basis for integrated, climate-proof urban development (Birkmann et al. 2010).
- *Requirements for the assessment and representation of climate impacts and risks*: In principle, an assessment of climate impacts should be required – implicitly – as a matter of course in overall urban planning, whenever these impacts may adversely affect the planned uses and impede the intended development. Wherever climate adaptation has become a legal planning objective – such as in Germany (§ 1 subpara. 5, 2nd sentence, Federal Building Code – BauGB) – this underlines the administrative requirement to conduct an assessment. However, assessment obligations that are merely implicit and abstract cannot be expected to generate much impetus for a thorough climate impact assessment. Only *specified minimum requirements* regarding the extent, depth, and form of assessments, and the way they are represented, will be capable of ensuring that adequate funding is provided for such a thorough climate impact assessment and that "inconvenient" truths are not deliberately concealed. Furthermore, only uniform legal

standards of assessment will be capable of improving the comparability of data, their transparency, and public availability, as well as the knowledge base as a whole (Kment 2010). Depending on the type of risk, the indicators concerned, the level of ambition regarding precautionary action, and available capacity, the following measures may become part of such a catalogue of requirements:

- evaluate the best available climate models and regional development scenarios,
- where necessary, produce climate models with a certain minimum resolution in keeping with the latest scientific standards,
- commission expert reports on relevant climate impacts/parameters,
- for certain climate impact risks, determine risk zones according to uniform benchmarks and create risk maps,
- name relevant uncertainties,
- consider and provide information on assessment results and risk analyses from other planning procedures,
- determine adaptation needs and options for the sectors affected,
- publish the results of the above assessments in a transparent form,
- ensure public participation,
- monitor the results of the assessments at regular intervals,
- observe and document trends in climate-sensitive environmental parameters according to uniform benchmarks.

• *Potential of Environmental Impact Assessment (EIA) schemes and/or their anchoring in a special assessment/planning procedure*: The above-mentioned requirements can be part of a formalized CP module, as a part of the obligatory urban planning exercise. However, if a plan-related EIA is already provided for, within urban planning, the question arises whether this plan-related EIA can be used and rendered fit for the purpose of a CP and, if so, how. Besides, it certainly makes sense to provide regional climate projections on a more central level and independent of local spatial planning.

In the following, we see what answers are offered by European and German law to these issues of designing regulations for climate proofing urban development, and which further-reaching developments are under discussion. This example will also be of interest to the non-German and non-European reader, because European and German law can be said to be at a comparatively advanced stage of development; as such, it also illustrates, relatively precisely, the regulatory issues and options that need to be considered by legislators around the world seeking to develop urban planning regulations that are based on risk and adequate for the task of adaptation.

5 Climate Proofing of Urban Development in German and European Law

The presentation of the German and European legal framework begins with a brief analysis of the German urban planning regime and of how it accounts for the key requirements of effective CP, in general (5.1). Subsequently, we look more specifically at how each of the relevant risks is addressed by planning law and – where relevant – by special sector regulations (5.2–5.6).

5.1 Integrated Urban Development: Urban Planning Law

Germany has a well-developed urban planning regime that is regulated on federal level by the Federal Building Code (BauGB) and basically composed of two types of municipal spatial plans:

- *Municipal land-use plan*: Every municipality is obliged to draw up a land-use plan that covers the whole territory of the municipality. This plan determines, on a medium scale, the main lines and locations of important public infrastructure and coarsely assigns municipal land for settlement, industry, agriculture, horticulture, open space, or nature protection.
- *Zoning plans*: The municipalities are also requested to draw up zoning plans on a local scale, as far as this appears necessary for ensuring sound spatial development. Zoning plans determine the admissible land uses on every plot in the planning areas. For this purpose, a large variety of possible land-use determinations are provided as planning instruments in the BauGB.

Both plans have to follow the general objectives and principles of sound urban development, as set out in Sec. 1 and 1a of the Federal Building code. These planning objectives include "sustainable development" and "a healthy urban environment." Adaptation to climate change is expressly mentioned as a planning objective in Sec. 1(5) and 1a(5) BauGB, as are the "requirements of flood protection" (Sec. 1(6) No. 12) and the targets of nature and species conservation (Sec. 1(6) No. 7a). Moreover, it is stipulated that sector plans based on nature conservation, water, and waste law must be taken into account.

With regard to the key requirements of CP, a positive statement is that climate adaptation is expressly determined as a planning objective and that the planning objectives cover – more or less explicitly – all relevant impacts and risks. Thus, the "responsibility requirement "is apparently satisfied.

As to the "assessment" requirement, the diagnosis conforms to what has been mentioned in Sect. 4 as a general regulatory precondition of effective CP: Making climate adaptation a general planning objective does not ensure that climate-related risks and adaptation options are thoroughly assessed. Instead, concrete *minimum requirements* regarding the scope, depth, and form of the assessment are needed, as specified above. The Federal Building Act does not contain specific requirements on climate risk assessment. However, it obliges the planning authorities to conduct a strategic environmental impacts assessment (SEIA), and this directs us to the question of whether the referenced regulations on SEIA are suitable for ensuring a complete and thorough CP assessment. This question will be dealt with below, once more risk-specific contributions of relevant sector regulations such as, most notably, flood protection law, have been considered.

As to urban planning law, it remains to be noted that no obligation to monitor the performance of existing plans and to formally review them within certain time frames exists in the Federal Building Code. Reviewing and updating of land-use and zoning plans is widely left to the discretion of the municipalities and, in practice, zoning plans are quite often out-dated and "outrun" by factual development. Monitoring of the environmental performance is, however, demanded as part of the obligatory SEIA procedure (Sec. 14 m of the Federal Act on Environmental Impact Assessment – UVPG) and, again, this raises the question of whether the current shape of the SEIA is sufficient to also enforce complete CP assessment and monitoring.

5.2 *Riverine Floods: The EU Floods Directive (FD) and the Federal Water Act (WHG)*

As pointed out above, the increasing risk of riverine floods is one, or the most relevant, impact of climate change, particularly also in the urban context. However, flood protection has always been an important issue for urban development and has also long been subject to regulatory efforts in Germany and, more recently, also in the EU. Special provisions for protection from and precautionary measures against the flooding of rivers and coastal regions are contained in §§ 73 ff. of Germany's Federal Water Act, along with additional regional state laws. These regulations represent the current status of a highly dynamic development of laws on flooding in the past 10 years, influenced crucially by European legislation, namely, by the stipulations of the EU Floods Directive 2007/60/EG.[1] Currently applicable laws on flooding now constitute a synthesis of, mainly, the following EU and national legal instruments:

- Risk assessment and identification of flood risk areas: According to Articles 4 and 5 FD and Sect. 73 WHG, the relevant authorities were obliged to conduct a preliminary flood risk assessment by the end of 2011 and, on that basis, to identify those areas that are prone to "significant floods". This flood risk assessment and identification of flood risk areas is to be revised in 2018 and then every 6 years thereafter.

[1] Directive of 23.10.2007 on the assessment and management of flood risks, OJEU of 6.11.2007, p. 27 ff.

- Flood hazard and flood risk maps: State authorities are also obliged to prepare flood hazard and flood risk maps for all flood-risk areas (Article 6 FD, p. 6 Sec. 74 WHG). Flood hazard maps should display the geographical areas that could become flooded due to (a) floods with a low probability (extreme event scenarios), (b) floods with a medium probability (likely return period of ≥100 years), and (c) floods with a high probability, where appropriate (Article 6.3 FD, Sec. 74.2 WHG). In addition, flood risk maps should show the potential adverse consequences associated with these flood scenarios, in accordance with more specific criteria provided by Article 6.5 of the Floods Directive. These hazard and risk maps were to be completed by December 2013 and are to be revised every 6 years thereafter (Article 14.2 FD, Sec. 74.6 WHG).
- Flood risk management plans: On the basis of the maps mentioned above, flood risk management plans must be established and duly coordinated at the level of river basin district (Art. 7 FD, Sec. 75 WHG). Appropriate objectives for the management of floods must be established in these plans, which should also stipulate appropriate measures for achieving the objectives. According to Sec. 75 WHG, the primary aim of the flood risk management plan is to mitigate the negative consequences of a flood with medium probability alongside rivers, whilst in coastal areas, extreme events should also be taken into account. The planned measures should address all aspects of flood risk management, focusing on prevention, protection, and preparedness, including flood forecasts and early warning systems. The flood risk management plans were to be finalized by 22.12.2015 and are to be revised every 6 years thereafter (Sec. 75.6 WHG).
- Flood risk prevention areas: In addition to the above-mentioned European instruments, Sec. 76(2) No. 1 WHG stipulates that risk areas likely to be flooded in a flood event with medium probability (statistically: 100 years return period) must be designated by administrative ordinance as flood risk prevention areas. Here, planning and construction of new buildings is subject to wide-ranging restrictions (set out in Sec. 78 WHG) aimed at preventing the creation of new or increased flood risks and ensuring unhindered run-off.
- Flood retention conservation areas: Areas used for flood retention purposes are to be formally designated as flood retention conservation areas (Sec. 76(2) No. 2 WHG). In these areas, construction activities are basically prohibited, and only very limited exemptions are admissible in accordance with § 78 WHG. In addition, a general obligation to protect and maintain retention areas and to provide compensation for the taking of such areas is stipulated in Sec. 77 WHG.
- Climate change adaptation as a legal objective: Sec. 6(1) No. 5 imposes the general obligation on pubic water and flood-risk managers to prevent "possible negative impacts from climate change". This stipulation makes it clear that all the instruments mentioned above are to be implemented with a view to preventing additional flood risks arising from climate change. Moreover, by stating that merely "possible" effects should be prevented, the formulation of the CP principle clearly indicates a precautionary approach, where the mere possibility of negative climate impacts may legitimize and also necessitate preventative measures.

In view of these instruments and principles, European and German flooding laws today can be seen, overall, as constituting a progressive programme of regulation that, at any rate, addresses – and largely appears to fulfil – the key requirements for adaptive land-use management and effective CP, as outlined in Sects. 3 and 4 (for further details, Reese 2011, p. 19; Jablonski 2014, p. 424). CP is expressly declared to be task for the relevant authorities and it is clearly stipulated that a precautionary approach shall be applied in the pursuit of this task. The obligation to identify risk regions, the detailed stipulations for doing so, and the obligation to map flood risks and potential for damage constitute a new kind of legal instrument of risk assessment that does justice to the demands of an anticipatory risk assessment process and to risk awareness building. This instrument of flood risk management planning additionally guarantees that measures can be developed and mutually adjusted during the planning process. Risk assessment and flood risk planning should be revised at 6-year intervals, so that the condition of dynamic orientation is also given, in potential terms.

The key problem of flood protection, however, traditionally lies in its *implementation*: all too often, in the conflict between flood risk precaution and economic interests linked to land use, the latter are given priority. This is essentially the reason why the instrument of a "flooding zone" (*Überschwemmungsgebiet* as defined in §§ 76, 78 WHG) has been introduced; it consists of a series of legal restrictions on building that are exempt from all land-use planning considerations. This definitely applies to the flooding zones defined in § 76 Section 2 No. 1 WHG, which serve to avoid any further potential for damage in the settlement area. These flooding zones should be strictly separated according to the statistical 100-year flood. The associated planning and building restrictions thus apply in all settlement areas that would be affected by a 100-year flood. Exceptions to the building restrictions are possible, but only under conditions that guarantee that the potential for damage does not increase.

5.3 Extreme Rainfall and Urban Flash Floods: Water Law and Urban Planning Law

With regard to pluvial flood risks that may result from increasingly heavy rainfall events, it also appears sensible, in principle, to conduct a risk analysis and to map the risks. For residential areas, a risk analysis would need to establish the amounts of precipitation with which the land, as well as the drains and sewers, can cope and where (and to what extent) there is a threat of flooding. The instruments of flood risk assessment and risk mapping described above, however, relate to riverine flooding only. Flooding from drain overflows is expressly exempted (see Article 2.1 EU Flood Directive 2007/60 and Sec. 72, second sentence, of Germany's Federal Water Act). Thus, a thorough, public risk assessment is not guaranteed in any binding way.

The prerequisite for avoiding flooding due to heavy rainfall events in settlement areas is – as mentioned above – a coordinated combination of efficient wastewater infrastructure and "water-sensitive" urban planning. Whereas the development of retention and runoff areas, as well as regulated surface run-off, are essentially the responsibility of land-use and urban planning, the development of wastewater infrastructures, in the narrower sense is, first and foremost, the task of wastewater laws at Federal (national) and federal state levels. Federal law delegates responsibility for wastewater disposal, including rainwater management, to the institutions provided for within federal state law (§ 56 WHG). These are tasked with developing and maintaining wastewater systems for managing graywater and rainwater. According to § 55 subsection 2 WHG, the principle that "precipitation water (…) should seep or drip into the ground locally or be channelled directly via a drain or sewer into a body of water without becoming mixed with graywater, as far as this does not contradict water regulations or other public legal regulations or water management concerns" applies.

The development of coordinated wastewater infrastructure systems, as well as seepage, retention, and catchment areas, presupposes coordinated planning, and this, in turn, can only succeed on the basis of a correspondingly integrated wastewater infrastructure planning that incorporates the urban planning components of rainwater management. This infrastructure planning needs a specific sectoral planning regime and must be assigned to the competent sectoral administrations. It can by no means be adequately covered as just one aspect of comprehensive spatial planning and zoning. There is no special provision for this kind of water infrastructure planning, however, in German laws. Although most water laws of the federal states contain the obligation to draw up local wastewater disposal strategies, there is no parallel obligation to systematically determine the urban planning requirements for sustainable, climate-proofed rainwater management and to articulate them transparently in urban plans (Wickel 2015); consequently, it has rarely been done in practice to date. In the area of urban drainage, then, there is a gap in steering provisions based on planning law in Germany, which also makes it very difficult to take effective overall planning responsibility for water-sensitive urban development, to ensure a meaningful assessment of climate change impacts and adaptation options, as well as to continuously monitor and review the climate compatibility of the water infrastructure development (Reese 2010, p. 173; Wickel 2015).

5.4 Heat Waves – Urban Planning and Nature Conservation Law

Hazards from summer heat are not subject to any specific legal risk assessment standards and sectoral planning obligations. It remains, thus, a part of the general responsibility of overall urban planning and permitting to ensure sufficient shade, air circulation, and cooling green/blue infrastructure. In this respect, the legal framework does not go beyond the general objectives and instruments of urban spatial planning and zoning presented above (5.1). As mentioned, these objectives include the aim of ensuring "a healthy urban environment" and "adaptation to climate change", and this certainly also applies to urban summer heat. However, without specific assessment and planning standards – like those established with regard to flood-risk – the key requirements of effective CP are not effectively enforced in terms of urban heat stress.

Another instrument that can play a proactive role in this regard is the so-called "landscape and green space planning", as provided in the Federal Nature Conservation Act (BNatSchG – sec. 9). By means of landscape planning, nature conservation authorities may determine "conservation targets, requirements, and measures", in order to implement the general objectives of nature conservation and landscape development within the local context. The objectives of nature conservation and landscape development include "improvement of air-quality and climate" and "conservation of green spaces in residential areas". Landscape plans are to be monitored as to their effectiveness and revised as soon as this is necessary, due to factual developments (Sec. 9(4)). Mitigating summer heat is not explicitly mentioned as an aim of the landscape plans but this can easily by subsumed as a sub-purpose of the above-mentioned more general aims. Thus, this sectoral planning instrument can be used by the relevant authorities to proactively assess the heat risks, also with view to climate warming effects, and to develop corresponding green-infrastructure concepts. However, a landscape plan has no direct external effect and must be approved and implemented by spatial planning and zoning or designation of protected areas.

In view of the key requirements of effective CP, it is clear that none of the above-mentioned instruments specifically aims at or is geared towards managing heat risk related to climate change. There are no specific standards for the assessment, mapping, monitoring, and management of climate change-related heat risk. Hence, this field is, currently, generally the responsibility of the relevant local administrations. This is probably due to the fact that increasing heat stress is still not perceived as a very serious climate change impact – this may change in the future.

5.5 *Aridity and Water Shortage: Water Law and Urban Planning Law*

Securing urban water supply in times of drought is an issue of technical infrastructure and water saving but, to a considerable degree, also a matter of urban land use and planning. The most relevant land-use related challenge is the protection and replenishment of local groundwater resources. To this end, the water administrations are empowered by the water law to designate special water protected areas and to restrict any type of land use that could negatively affect the quality or quantity of the protected water body (Sec. 51 of the Federal Water Act). Beyond this, it is clear that safe water supply for large agglomerations must rely on a coherent and resilient system of sources and reservoirs, as well as measures for rainwater management and re-use. Much of this relates directly to urban land use and, in particular, to green and blue infrastructure, and it is also obvious that complex risk assessment and planning is needed to integrate all relevant factors for safe water supply into a functioning and drought-proof system and to also develop clear requirements about urban space and infrastructure.

However, German water law does not provide a formal instrument or specific standards for the assessment and planning of (climate proof) infrastructure (Wickel 2015). Hence, this planning task and the associated responsibility of making the water supply climate proof is basically left up to the relevant administrations and water service providers. As a consequence, spatial demands of secure water supply are not systematically established and formally articulated in terms of overall urban planning and zoning. In the absence of a formal urban water supply plan, there is also no basis for regulated monitoring and review. In Germany, all this is certainly due to the fact that there is simply no serious water scarcity problem in the region. In the future, however, this might change and make it necessary to underpin the key requirements for climate proofing with adequate planning and assessment standards.

5.6 Urban Nature and Ecosystem Changes: Nature Conservation Law and Landscape Planning

The spatial adaptation requirements of nature conservation are addressed, in the first instance, by nature conservation law. Nature conservation law provides a set of instruments for designation and protection of particularly valuable sites and, beyond this, for ubiquitous management of nature and landscape. Both approaches can be used for the purpose of climate proofing urban nature as follows:

With respect to area-wide management, the first and most important instrument is the special nature offset regime, set out in Sec. 14–18 BNatSchG and Sec. 1a(3) BauGB. According to these, provision of adequate, real offset must be provided for any taking of open space and nature. This nature offset prevents, to a certain degree, a net loss in green infrastructure, and, according to the circumstances, this can also help to strengthen climate resilience and to mitigate adverse effects of climate change. However, there is no legal obligation to gear nature offset towards CP, and whether this is done is decided on by the responsible authorities.

The instrument of "landscape planning", described above in Sect. 5.3, does offer a means of steering conservation efforts – including those derived from the offset regime – towards CP. Landscape planning constitutes a regulatory basis for a specialist assessment of climate change impacts on (urban) nature and for devising adequate development concepts and measures. However, as shown in the previous section, such a proactive use of this sectoral planning instrument for the purpose of CP is not sufficiently anchored in the existing legal framework.

Site protection instruments are regularly designed to protect particularly valuable habitats and landscape elements. As a rule, site protection includes both restrictions regarding adverse developments and also proactive obligations to maintain and support the protected ecological elements. These site- related instruments can well be used as a basis to adapt the location, range, and design of conservation areas and to set the required land-use limitations, so that the conservation areas and biotope networks can be prepared for climate change and land uses that run counter to adaptation can be prohibited (Köck and Möckel 2009, p. 318). A particularly strong conservation status is provided by the European Habitats Directive 92/34/EEC for so-called "Natura 2000" areas, designated according to Article 4 of the Directive. New developments that might have negative impacts on the conservation status of such sites have to undergo a thorough impact assessment in which cumulative effects, including pressures from climate change, should, in principle, also be assessed, if there is any indication that such effects may occur.

In sum, nature conservation law offers a range of instruments for effective protection of "green urban assets", also against adverse effects from climate change. The instrument of nature and landscape planning provides a legal basis and legitimation for in-depth specialist assessments of such climate effects and also for the development of adaptation measures – not only for the sake of nature, as such, but also with a view to improving urban environment and liveability. Mirrored against the key requirements of CP, however, the performance of these conservation law instruments still appears to be relatively weak: Unlike the Federal Building Code and the Federal Water Act, the Nature Conservation Act does not even provide an explicit obligation or objective to mitigate the effects of climate change. Assessment of climate change related impacts, risks, and adaptation options are not explicitly prescribed, and the obligation to revise landscape plans is not timed: instead, it vaguely depends on whether revision is "necessary as a consequence of factual developments". Moreover, there is not even an effective obligation for urban planners to react immediately to nature conservation planning and to revise the corresponding spatial and zoning plans accordingly.

5.7 *Interim Conclusions*

The examination of the relevant legal framework reveals that the key requirements for effective CP are reflected with differing intensity, depending on the type of risk addressed. We have seen, in particular, that:

- The laws on integrated urban planning and zoning explicitly mention climate change adaptation as a primary objective of the planning exercise and that they provide sufficient empowerments to the authorities for determining the necessary land-use restrictions and conditions. However, there are no specific CP assessment standards that could ensure that climate change impacts and risks, as well as adaptation requirements and options, are thoroughly assessed, transparently presented to the public, and adequately considered in planning decisions. A timed review obligation that ensures a regular revision of the planned land-use arrangements is also lacking.

- With regard to riverine flood risk, the CP requirements are widely accounted for by special flood protection laws. There is, today, a well-developed body of law on risk assessment and management, ensuring thorough assessment of future flood risk, risk and hazard mapping, and specialist planning of prevention measures. All this is also subject to obligatory cyclic review, and, moreover, German water law imposes direct land-use and planning restrictions in settlement areas prone to a 100-year risk of flooding.
- Urban floods from heavy rainfall and sewage overflow are not included in the above flood-risk regime. Legal obligations and standards for the assessment and communication of these risks are missing and there are insufficient requirements for integrated planning of public rainwater management that could serve as a basis for adequate long-term orientation of long-lasting infrastructure developments, for the integration of such concepts in overall urban planning, and for continuous monitoring and regular revision.
- Urban heat stress is not covered by special regulations at all. It is clear that mitigating urban heat is an implicit part of the legal objective to plan towards a "healthy urban environment" and to adapt planning to the expected effects of climate change, as laid down in Sec. 1(5) and (6) and 1a(5) BauGB. However, without any specific standards for risk assessment, communication, and management and, because heat is not mentioned expressly as an issue of climate-resilient urban development, the key requirements of effective CP are only weakly enforced.

All together it has become evident how risk specific specialist assessments and planning are informing and facilitating climate-proof urban planning. Moreover, the instruments of flood prevention areas and nature conservation sites determine a fixed minimum level of prevention that urban planning can well exceed but must not fall below. However, in many regards, specific risk assessment, communication and planning standards do not exist and, in so far, the responsibility of ensuring CP remains entirely in the area of overall urban planning and zoning.

In this respect, it has been noted that the CP requirement of regular review is not reflected in urban spatial planning and zoning law. As to the requirement for thorough assessment, however, it still remains to be examined whether adequate assessment standards are provided within the instrument of plan-related "strategic environmental assessment" (SEA), as prescribed by the EU Directive 2001/42 and transposed into German law by the Law on Environmental Impact Assessment (Gesetz über die Umweltverträglichkeitsprüfung – UVPG).

6 Strategic Environmental Assessment (SEA) – An Effective Climate Proofing Instrument for Urban Planning?

In the debate about legal approaches to climate adaptation, various proposals have been made to use the project-related EIA and the plan-related SEA for climate change impact assessment and to tighten up the corresponding legal framework accordingly (see especially EU Commission 2013; Kment 2010, 67 suggesting a strategic adaptation assessment; also Reese et al. 2010a p. 345; Fischer 2013, p. 194; Meyer 2014, p. 222). There are a number of indications that the binding elements of environmental reporting, assessment of options, public participation and ongoing monitoring might easily be extended to such a planning-related impact assessment and, in this sense, might guarantee a thorough and transparent assessment of impacts and adaptation options (Reese et al. 2010a, b, p. 375; Reese 2012 p. 409). If these instruments are to be rendered fruitful for a fully effective climate change impact assessment, SEA law will need to be complemented and systematically expanded, not least in terms of the basic assessment *environment* → *humans*. The specialist literature speaks, in this regard, of the need for a "reverse environmental impact assessment" (Parejo Navaras 2015).

According to existing law, the SEA serves exclusively to assess the effects of planned uses on the environment. In urban climate adaptation, however, what is also at stake, above all, is the impact of climate-related environmental trends on land use. This applies to the risks of flooding, heavy rainfall, storms, heat and aridity. These environmental risks are not included directly as objects of assessment within the SEA, to date (Fischer 2013, p. 197). However, it would be judicious to add them, because the "environmental robustness" of anthropogenic land uses is frequently influenced by the same general conditions as their environmental impact, so that considerable synergies can be expected on this basis (Birkmann and Fleischauer 2009). For the project-related EIA, which is regulated in a legally binding way throughout Europe by the EIA Directive, such an expansion has already occurred by way of the most recent update of the Directive,[2] albeit tentatively and in a somewhat obscure location: according to Annex IV No. 5 f) of the Directive, the environmental report must, in the future, also describe the "vulnerability of the project to climate change". With regard to the SEA, such an addition has not yet been made. It may also be appropriate to demand that not only the vulnerability of plans to climate change impacts but also – as a part, as it were, of an alternative assessment in line with Article 5 Para. 1 SEA Directive 2001/42/EC or § 14 g Subsect. 1 EIA Directive – sensible adaptation options should be assessed and depicted.

The question also arises as to whether the standard of impact assessment associated with existing environmental assessment law is sufficient, throughout, for assessing climate change impacts. The German Federal Administrative Court has, at all times, sought to restrict the scope of the assessment and has limited it to providing information on the basis of available knowledge (methods, empirical evidence, predictive knowledge) about effects for humans and the environment (Reese et al. 2010a, b, p. 348). The Court has decided, with regard to environmental impact assessment, that it is "not conceived as a search procedure that serves the purpose of revealing environmental impacts that elude conventional means of knowledge acquisition."[3] The environmental assessment, it states, is, above all, not an instrument

[2] Directive 2014/52/EU of the European Parliament and European Council of 16.4.2014.

[3] Official Journal of the German Administrative Court (BVerwGE) 100, 238, 248.

for further developing the state-of-the-art of science.[4] Accordingly, then, demands cannot be made, in the context of the SEA, to commission, for example, regional climate models or scientific studies on the local impacts of climate change.

It does, at least, seem worth considering whether it would make sense for the SEA to include more detailed assessments of climate change impacts. What should also be considered, fundamentally, is how to appropriately distribute the assessment and research needs prompted by the effects of climate change between the public and the private sector and how public assessment tasks should be distributed efficiently between administrative departments and levels. With regard to overall spatial planning, attention should be given to not burdening it with assessment and evaluation tasks that can be undertaken more efficiently by sectoral assessments or by research and assessment services unrelated to planning. Within the jurisdiction of EU law, it should be recalled that the most significant areas of spatial climate change adaptation have already been largely addressed *de lege lata* by legal instruments of assessment and communication and that there is, therefore, relatively little pressure on overall planning in this respect – including with regard to the SEA. The main issue that will need to be addressed in those areas where there are, as yet, no appropriate assessment and planning instruments is whether the relevant laws also need to be tightened up. The important example of the – lack of - wastewater management planning - has already been presented above (see Sect. 5.3).

7 Special "Climate Adaptation Plans" as a Regulatory Option

Finally, with regard to effectively tackling assessment and research tasks related to climate adaptation, consideration should be given to whether it may be expedient to institutionalize basic assessments of regional climatic changes and their impacts as part of an overarching climate adaptation planning and to distribute them – on an appropriate scale – between central, regional, and local administrative levels (Reese et al. 2010b, p. 395; Reese 2010, p. 410; Fischer 2013, pp. 304 ff.; Mayer 2014 p. 222). The idea of establishing an independent "adaptation planning" instrument is advocated by some authors (e.g. Meyer 2014, pp. 211 ff.), whilst others are skeptical particularly with regard to potential duplications of existing plans and administrative procedures (Fischer 2013, p. 306). Of course, any kind of independent adaptation planning would have to be clearly distinct from the urban and sectoral planning instruments described above.

Over recent years, in Germany, a wide variety of climate adaptation strategies have been devised voluntarily at all levels and, very recently, the federal state of North Rhine-Westphalia has even, for the first time, placed its state government under legal obligation to produce an adaptation strategy as part of a climate protection

[4] Official Journal of the German Administrative Court (BVerwGE) 100, 238, 248; 100, 370, 377.

plan.[5] If, therefore, a legally constituted climate adaptation plan does appear to be attainable politically, it is worth considering combining this plan with an assessment, that can be used – not least as a foundation for regional and urban planning – to promote research on and modelling of regional climatic changes.

8 Conclusion

The above analysis leads to the following observations and theses, which require further consideration and debate:

- To ensure that urban development incorporates adaptation to climate change, land-use regulation and planning regimes need to be oriented towards modern approaches to risk regulation. Specifically, this means that binding standards and procedures should be included to ensure that thorough risk assessments or projections are considered, that monitoring occurs, and that spatial and infrastructure planning is subject to regular review.
- It appears reasonable to assign the assessment and management of particularly relevant risks (flood, nature, water) to specialized administrations and sectoral planning instruments, in order to ensure that expert evaluations feed into adequate sectoral concepts as preparatory instruments for climate proof urban planning and zoning.
- EU and German laws provide examples of highly advanced regulation in the field of river flood risk management including, in particular, obligations and standards for risk assessment, risk mapping, and management planning. Nature conservation provides a special planning instrument (nature and landscape planning) that can well be used to assess climate impacts and develop adaptation concepts and measures for urban nature. However, proactive use, for the purpose of CP, is not expressly demanded by the law. With regard to the other climate change related impacts, special assessment and planning standards are not provided in existing law, and the task of effective CP lies mainly within overall urban planning and zoning. In so far, the questions arises as to whether and to what extent a thorough assessment of climate risks and adaptation options can be ensured by the general instrument of (strategic) environmental impact assessment.
- Environmental impact assessment procedures are regularly applied in many countries across the globe and, often, also in the context of urban spatial planning and zoning. EIA schemes can, in principle, be used and up-graded to include not only the impacts of human land use on the environment but also – vice versa – the risks to which human land use is exposed and which may increase due

[5] See § 6 Sect. 4 No. 6 Climate Protection Act of North-Rhine Westphalia: This Act obliges the Government of North-Rhine-Westphalia to draw up, by 2013, and revise every 5 years, a Climate Protection Plan. This plan must also contain sector-specific strategies and measures aimed at mitigating and adapting to the adverse impacts of climate change.

to climate change. This applies, at least, to the plan-related strategic environmental assessment as regulated by EU-Directive 2001/42 and an appropriate amendment to this directive is suggested.
- Thorough climate impact assessment often requires a sound knowledge of global, regional, and local climate developments and impacts – knowledge that cannot reasonably be acquired by local actors and municipalities in the context of an individual project or planning process. Such global assessments and research tasks can, however, be assigned by law to higher, supra-regional levels of government, e.g., as part of a regional or national climate change programme.
- Monitoring and regular reviewing of plans and developments is another crucial element of risk-oriented governance that can be readily enforced by adaptation-based environmental and planning laws. In German law, this is the case with riverine flood protection but there are no obligatory review schemes for overall spatial planning. In this regard, further discussion about how to make spatial planning more sensitive and adaptive to an increasingly dynamic environment is required.

References

Birkmann J, Fleischauer M (2009) Anpassungsstrategien der Raumentwicklung an den Klimawandel: "Climate Proofing" - Konturen eines neuen Instruments. Raumforsch Raumordn 67(2):114

Birkmann J, Garschagen M, Kraas F et al (2010) Adaptive urban governance: new challenges for the second generation for urban adaptation strategies to climate change. Sustain Sci 5:185–206

Craig RK (2010) Stationarity is dead - long live transformation: five principles for climate adaptation law. Harvard Envrion Law Rev 34:9–75

DKKV – Deutsches Komitee für Katastrophenvorsorge (2003) Hochwasservorsorge in Deutschland - Lernen aus der Katastrophe 2002 im Elbegebiet. Deutsches Komitee für Katastrophenvorsorge e.V, Bonn

DKKV – Deutsches Komitee für Katastrophenvorsorge, D.K.f.K (2015) Das Hochwasser im Juni 2013: Bewährungsprobe für das Hochwasserrisikomanagement in Deutschland. DKKV-Schriftenreihe Nr. 53. DKKV, Bonn

EEA – European Environmental Agency (2015a) Exploring nature-based solutions — The role of green infrastructure in mitigating the impacts of weather- and climate change-related natural hazards. Technical Report No. 12/2015. EEA, Copenhagen

EEA – European Environmental Agency (2015b) EEA Signals 2015 – Living in a changing climate. http://www.eea.europa.eu/publications/signals-2015. Accessed 15 Feb 2017

EEA – European Environmental Agency (2016) Urban adaptation to climate change in Europe 2016 - Transforming cities in a changing climate, EA Report No. 12/2016. http://www.eea.europa.eu/publications/urban-adaptation-2016. Accessed 15 Feb 2017

EEA/JRC - European Environmental Agency/Joint Research Council (2008) Impacts of Europe's Changing Climate – 2008 Indicator Based Assessment, EEA Report No. 4/2008. EEA, Copenhagen

EU Commission (2009) Regions 2020 – The climate change challenge for European Regions, Background document of the Directorate General for Regional Policy to Commission Staff Working Document Sec(2008)2868. EU Comission, Brussels

EU Commission (2013) Guidance on integrating climate change and biodiversity into strategic environmental assessment. http://ec.europa.eu/environment/eia/pdf/SEA%20Guidance.pdf. Accessed on 15 Nov 2016

Fischer C (2013) Grundlagen und Grundstrukturen eines Klimawandelanpassungsrechts. Mohr Siebeck, Tübingen

Gies M, Albrecht J (2013) Legal study and recommendations for climate change adaption. IÖR, Dresden

Gill SE, Handley JF, Ennos AR, Pauleit S (2007) Adapting cities for climate change: the role of the green infrastructure. Built Environ 33(1):115–133

Hirokawa KH, Rosenbloom J (2013) In: Verschuuren J (ed) Research handbook on climate change adaptation law. Edward Elgar, Cheltenham/Northampton

IPCC – Intergovernmental Panel on Climate Change (2013) Summary for Policymakers. In: Climate change 2013: the physical science basis. Contribution of working group I to the fifth assessment report of the Intergovernmental Panel on Climate Change. Cambridge University Press, Cambridge/New York

IPCC – Intergovernmental Panel on Climate Change (2014) Climate change 2014: impacts, adaptation, and vulnerability. Part A: global and sectoral aspects. Contribution of working group II to the fifth assessment report of the Intergovernmental Panel on Climate Change. Cambridge University Press, Cambridge/New York

Jablonski S (2014) Hochwasserschutzrecht. Nomos, Baden-Baden

Kehew RB, Kolisa M, Rollo C, Callejas A, Alber G, Ricci L (2013) Formulating and implementing climate change laws and policies in the Philippines, Mexico (Chiapas), and South Africa: a local government perspective. Local Environ 18(6):723–737

Kment M (2010) Anpassung an den Klimawandel – Internationaler Rahmen, europäische Strategische Adaptionsprüfung und Fortentwicklung des nationalen Verwaltungsrechts – (Habilitationsvortrag). JuristenZeitung:62–72

Köck W, Möckel S (2009) Naturschutzrecht im Zeichen des Klimawandels – Vorläufige Bewertung und weiterer Forschungsbedarf. Natur und Recht 31(5):318–325

LAWA – Bund-Länder-Arbeitsgemeinschaft Wasser (1995) Leitlinien für einen zukunftsweisenden Hochwasserschutz – im Auftrag der Umweltministerkonferenz. http://lawa.de/documents/Leitlinien_d59.pdf. Accessed on 15 Nov 2016

Lowe A, Foster J, Winkelman S (2009) Ask the climate question: adapting to climate change impacts in urban regions. Center for Clean Air Policy (CCAP), Washington, DC

Meyer K (2014) Adaptionsplanung – wie die Raumordnung auf die Herausforderung Klimawandel reagieren kann. Nomos, Baden-Baden

Moss T (2011) Planung technischer Infrastruktur für die Raumentwicklung: Ansprüche und Herausforderungen in Deutschland. In: Tietz H-P, Hühner T (eds) Zukunftsfähige Infrastruktur und Raumentwicklung. Handlungserfordernisse für Ver- und Entsorgungssysteme. Akademie für Raumforschung und Landesplanung, Hannover, pp 73–94

Nolon JR, Salkin PE (2006) Land-use in a nutshell, 5th edn. West Group, St. Paul

Parejo Navaras T (2015) Reverse environmental impact assessment analysis for the adaptation of projects, plans and programs to the effects of climate change in the EU. Columbia Public Law Research Paper Nr. 14-445. https://papers.ssrn.com/sol3/papers.cfm?abstract_id=2565856. Accessed 15 Feb 2017

Reese M (2010) Die Herausforderungen der Klimaanpassung an Bürger, Politik und Recht. In: Reese M, Bovet J, Möckel S, Köck S (eds) Rechtlicher Handlungsbedarf für die Anpassung an die Folgen des Klimawandels – UBA-Berichte 1/10. Erich-Schmidt, Berlin, pp 3–12

Reese M (2011) Das neue Recht des Hochwasserschutzes vor den Herausforderungen des Klimawandels. Natur und Recht 33(1):19–28

Reese M (2012) Klimaanpassung im Umwelt- und Planungsrecht – Konzeptionelle Herausforderungen und Optionen. Verwaltungsarchiv 2012:399–420

Reese M (2015) Klimaanpassung im Raumplanungsrecht. Zeitschrift für Umweltrecht 26(1):16–28

Reese M, Bovet J, Möckel S, Köck W (2010a) Rechtlicher Handlungsbedarf für die Anpassung an die Folgen des Klimawandels – UBA-Berichte 1/10. Erich-Schmidt, Berlin

Reese M, Köck W, Möckel S (2010b) Räumliche Gesamtplanung. In: Reese M, Bovet J, Möckel S, Köck S (eds) Rechtlicher Handlungsbedarf für die Anpassung an die Folgen des Klimawandels – UBA-Berichte 1/10. Erich-Schmidt, Berlin, pp 336–402

Romero Lankao P, Qin H (2011) Conceptualizing urban vulnerability to global climate and environmental change. Curr Opin Environ Sustain 3(3):142–149

Sánchez-Rodríguez R (2009) Learning to adapt to climate change in urban areas. A review of recent contributions. Curr Opin Environ Sustain 1(2):201–206

Schumacher J, Schumacher A, Krüsemann E, Rebsch S, Becker R, Niederstadt F et al (2013) Natuschutzrecht im Klimawandel – Juristische Konzepte für naturschutzfachliche Anpassungsstrategien. Springer, Berlin

Swart R, Biesbroeck R, Binnerup S, Carter T, Cowan C, Henrichs T et al (2009) Europe adapts to climate change comparing national adaptation strategies – PEER report no 1. Partnership for European Environmental Research, Helsinki

Wheater H, Evans E (2009) Land use, water management and future flood risk. Land Use Policy 26(1):251–264

Wickel M (2015) Rechtliche Pflichten, Verfahren und Anreize zur besseren Vernetzung von Siedlungswasserwirtschaft und Stadtentwicklung. In: Gawel E (ed) Die Governance der Wasserinfrastruktur, Band 2: Nachhaltigkeitsinstitutionen zur Steuerung von Wasserinfrastruktursystemen. Speyer & Peters, Berlin, pp 399–459

Decision Support on Flood Management in Complex Urban Settings. Is Risk Assessment the Right Approach or Do We Need Decision Heuristics?

Volker Meyer

1 Managing Flood Hazards in Urban Areas – from Hazard Protection to Risk Management

Natural hazards, such as floods, droughts, heat waves, avalanches, etc., have severe societal impacts. This is not only, but in particular, the case in cities, where people and assets concentrate, and hence, the values at risk and potential damages accumulate (Rink et al. 2015). As a consequence of urban growth and climate change, these risks are likely to increase in many cities in the future, confronting urban decision makers with the question of whether their current portfolio of risk management options is sufficient or if more radical urban transformations are necessary to mitigate the societal impacts of natural hazards.

For flood risk mitigation, decision makers can choose from a range of potential options, from structural, hazard-oriented measures, such as dikes or reservoirs, to more vulnerability- or exposure-oriented measures, such as warning systems or even land-use regulations or restrictions. The main question is how to choose among all these options (including variants and combinations) in order to come to an optimal or, at least, a satisficing solution, considering all of the measures' positive and negative effects, including the costs of the various options, their positive effects in terms of risk reduction and also the residual, remaining risks. Considering all these aspects in the decision-making process poses a great challenge to decision makers, in particular in an urban environment that is not static, but under transformation and, hence, highly dynamic and shaped by uncertainty (Kabisch and Kuhlicke 2014).

A variety of approaches and decision rules can provide (and have provided in the past) support for decision makers. However, different decision rules may lead to

V. Meyer (✉)
Department of Economics, Helmholtz Centre for Environmental Research-UFZ, Leipzig, Germany
e-mail: volker.meyer@ufz.de

S. Kabisch et al. (eds.), *Urban Transformations*, Future City 10,
DOI 10.1007/978-3-319-59324-1_20

different recommendations. Therefore, the question is which decision rule is best suited to support flood management in urban areas.

In recent decades, risk assessment and management has become the dominating scientific paradigm in flood management (Merz and Thieken 2004; Tung 2005; Schanze 2006). Additionally, in practice, it often replaced – at least in part – the earlier concept of hazard protection (Klijn et al. 2008).

Hazard protection aims mainly at mitigating the consequences of floods by controlling the hazard. For example, flood protection aims at limiting and controlling flood hazards by providing a pre-defined safety-standard by means of dikes, barriers, dams, etc. The first part of the decision process typically consists of identifying flood mitigation measures that provide this pre-defined safety standard (e.g., for river floods, protection against floods up to a 100-year event, i.e., a flood that occurs, statistically, once in 100 years). From an economic point of view, the decision rule for this flood protection approach is provided by cost-effectiveness analysis: The flood mitigation measure that achieves this safety-standard target at lowest costs is selected.

However, this decision rule and its underlying concept obviously have some drawbacks (Meyer et al. 2012). Firstly, the safety-standard approach only indirectly addresses the main target, i.e., the reduction of damage. For example, for one and the same safety level, damage would be reduced to a greater extent in urban areas, where people and values at risk accumulate, than in rural areas. Hence, it would be more efficient to provide higher safety standards for cities than for rural areas, but this is not considered by the safety-standard approach. Secondly, the safety-standard approach favours hazard-related measures. In the case of flood protection, the safety standard is a hydraulic target, which can be achieved mainly with structural measures, such as dikes, dams, barriers, etc. Such measures may be characterized by certain drawbacks; for example, they may create an illusion of complete safety or could have considerable negative environmental impacts. Exposure- or vulnerability-related measures, such as land-use regulations or warning systems, which are less inclined to cause these effects, do not (or only indirectly) address the hydraulic target and are therefore discriminated by this decision rule. Such measures may reduce damage in a more efficient way, but it is difficult for them to achieve the hydraulic safety target. Consequently, the hazard protection decision rule may lead to sub-optimal decisions, with 1) insufficient protection of urban areas and 2) an insufficient share of non-structural measures.

The *risk management* approach tries to address these shortcomings. In combination with cost-benefit analyses, the approach primarily aims at efficiency, minimising the societal costs of natural hazards, including the damage costs, as well as the costs of risk mitigation measures themselves (Plate 2000; Meyer et al. 2013; Kreibich et al. 2014; Kabisch and Kuhlicke 2014). In order to estimate the damage costs, it is necessary to assess the risk, i.e., the expected annual damage due to flooding.

According to the definition of Knight (1921), risk is a function of probability and consequences. The economic risk due to flooding can be expressed in terms of annual average damage, i.e., the product of flood probabilities and estimated flood damage for the entire range of possible events (Merz and Thieken 2004; Samuels et al. 2009; Meyer et al. 2009; Hallegatte 2014).

Damage itself is a function of hazard intensity, exposure, and vulnerability (IPCC 2012; Fuchs et al. 2011). Hazard intensity refers to characteristics of the hazardous phenomenon itself (e.g., flood, earthquake, drought), exposure refers to the location and number of people or economic assets in hazard-prone areas, and vulnerability refers to their susceptibility to suffer damage and loss, for example due to unsafe housing and living conditions or merely due to the physical susceptibility of buildings, in terms of flood depth/damage relationships (IPCC 2012; Fuchs et al. 2011). All three elements can be altered by adaptation (i.e., explicit risk reduction measures).

By combining these definitions, risk can be expressed as:

Risk = f (hazard probability, hazard intensity, exposure, vulnerability, adaptation)

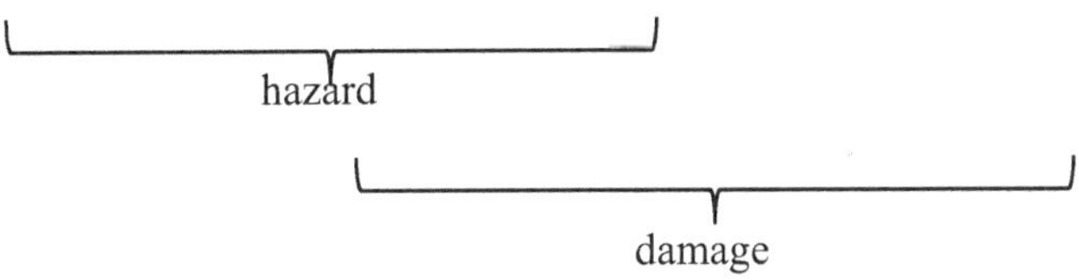

The risk-reduction effect (the annual average damage avoided), as the main benefit of risk management options, is then compared to the costs of these measures, in order to identify efficient measures by means of cost-benefit analyses (CBA).

In the context of CBA, efficiency is measured by using the net present value (NPV) as the major output criterion. The NPV is defined as the sum of discounted benefits minus the sum of discounted costs over the lifetime of an option or project (Hanley and Spash 1993):

$$NPV = \sum_{t=0}^{n} B_t (1+i)^{-t} - \sum_{t=0}^{n} C_t (1+i)^{-t}$$

A NPV > 0 indicates a positive impact of the option on social welfare.

Ideally, the NPV should be maximized in order to meet the Pareto optimum. Pareto optimality (or economic efficiency) is defined as an allocation of resources such that no further reallocation is possible that would create gains in production or consumption for some persons without simultaneously imposing losses to others (Young 2005). Generally, it is assumed that, with an increasing level of investments in risk reduction, benefits increase with a decreasing rate, while costs increase with an increasing rate (Young 2005; Weck-Hannemann and Thöni 2006; see Fig. 1).[1] If this is the case, the Pareto optimal level of risk reduction (R*) would be where marginal benefits equal marginal costs (ibid.).

[1] Admittedly, empirical evidence from the UK has shown that, at the national level, costs of flood protection could also increase with an even higher declining rate than the benefits (Pearce and Smale 2005). In this case, the NPV of flood protection projects increases as investments increase.

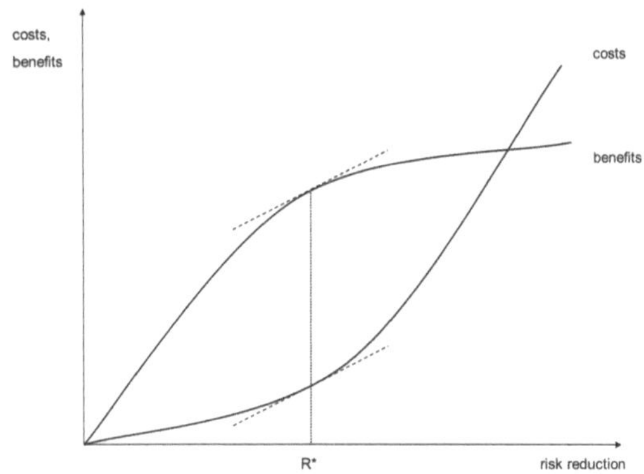

Fig. 1 The Pareto optimum of risk reduction (Source: Based on Young 2005; Weck-Hannemann and Thöni 2006)

This risk assessment/CBA approach currently predominates in the scientific literature on natural hazards management (e.g., Schanze 2006; Meyer et al. 2013; de Moel et al. 2015) and is, furthermore, reflected in practice in many countries (e.g., MAFF 1999; FOE 2010; Klijn et al. 2008) as well as in European legislation (the "Floods Directive", European Parliament and the Council 2007). However, in practice, hybrid approaches between hazard protection and risk management are often adopted. For instance, in the federal State of Saxony, Germany, flood-safety standards are still prescribed, but combined with efficiency-oriented considerations: Firstly, the safety standards are differentiated with regard to land-use classes, in order to reflect that high protections standards tend to be efficient only if protected values accumulate (i.e., in particular in cities and villages), whereas, for agricultural land, only lower protection standards are likely to be efficient (LTV 2003). Secondly, measures are prioritized (among other criteria) by their efficiency. In other words, efficiency considerations play a prominent role, but not regarding the question *whether* a measure is implemented but *when* it is conducted (SMUL 2005).

2 Drawbacks of the Risk-Management Approach

However, the risk assessment and management approach also has its drawbacks. According to Knight (1921), risk assessment and management requires a quantification of probabilities and consequences. In contrast, the term uncertainty defines decision situations in which potential outcomes might be known but probabilities cannot be quantified (ibid.). Mousavi and Gigerenzer (2014) add the category of fundamental uncertainty, in which some of the alternatives and outcomes, in addition to probabilities, are unknown (also referred to as deep uncertainty; see Groves and Lempert 2007).

For flood hazards, it has to be acknowledged that the input information used for risk assessments often contains very substantial uncertainties (Grelot et al. 2008; Meyer et al. 2013; Saint-Geours et al. 2015). For example, in flood risk assessment, all input components may be quantified, i.e., hazard probabilities and characteristics, exposure in terms of number, location, and value of assets at risk, as well as susceptibility/vulnerability in terms of depth-damage functions. But these quantifications are associated with substantial uncertainty bounds, leading to an overall risk estimate that is also highly uncertain (Handmer 2003; Merz and Thieken 2005).

In addition, indirect damage, i.e., costs induced by the direct impacts, e.g., by interruption of business and traffic networks, as well as intangible, non-market damages, such as environmental and health effects, are often not considered in these assessments (Meyer et al. 2009, 2013). There are methods available to quantify these types of damages (Przyluski and Hallegatte 2011; Markantonis et al. 2012 for an overview), but their assessment is associated with even higher uncertainties than for direct damages (Meyer et al. 2013). Furthermore, the future development of risk is highly uncertain due to a) climate change, leading to changes in the hazard probabilities and b) socio-economic developments, leading to changes in the exposed elements at risk (Bouwer 2013; Hallegatte 2014). With regard to the challenges of climate change, Groves and Lempert (2007) state that "there is widespread agreement that this traditional optimum expected utility approach [i.e. the risk assessment/CBA approach, comment by the author], at least in its most basic form, is insufficient to address decision challenges with the characteristics of climate change". This is even more the case if we consider decisions in urban areas, in which predictions on future developments are even more uncertain, due to their complexity and highly dynamic nature (Kabisch and Kuhlicke 2014).

3 Decision Support Approaches that Consider Uncertainties

Several approaches have emerged that try to tackle the problem of uncertainties within the framework of risk assessment, such as Robust Decision Making (Lempert et al. 2003; Groves and Lempert 2007), Real Option Analysis (e.g., Woodward et al. 2014) and Dynamic Adaptation Pathways (Haasnoot et al. 2013).

Robust Decision Making (RDM) aims at considering uncertainties in all input components. RDM uses computer simulation models to create large ensembles of plausible future states that are used to identify candidate robust strategies and systematically assess their performance (Groves and Lempert 2007). By conducting multiple model runs (e.g., of the risk assessment CBA), this approach is able to map and identify options that are relatively robust, i.e., perform well under all possible future scenarios. Hence, the approach does not aim at identifying optimal but, rather, robust decisions.

Real Option Analysis also starts from the basis of risk assessment CBA, but it adds flexibility, i.e., the option to postpone the decision and to wait for more certainty (e.g., for the effects of climate change on flood probabilities, Woodward et al. 2014).

The Dynamic Adaptation Pathway approach is more related to a safety-standard approach, but it adds even more flexibility: If a predefined target (e.g., a protection target) will no longer be met in the future (e.g., due to climate change), it adds the possibility of switching to another adaptation measure (Haasnoot et al. 2013). In addition, the most cost-effective pathway can be identified, but the target itself is not defined by efficiency, as in the risk assessment/CBA approach described above.

However, the question remains whether risk-based decision approaches are best suited to tackle flood management, considering the related uncertainties, which could be classified as "fundamental" (Mousavi and Gigerenzer 2014) or "deep" uncertainties (Groves and Lempert 2007).

In this sense, the Heuristic Decision Making approach (Gigerenzer and Brighton 2009; Gigerenzer and Gaissmaier 2011; Mousavi and Gigerenzer 2014) seems to be a promising alternative to risk assessment and management for decision support under uncertainty. Heuristics are simple decision rules that ignore part of the information, with the goal of making decisions not only more quickly and frugally, but, under some conditions, also more accurately than more complex methods (Gigerenzer and Gaissmaier 2011).

Typical examples for such heuristics are, e.g., Tallying and Take-the-Best. Tallying means ignoring weights and weighting all cues (or decision criteria) equally. It entails simply counting the number of cues favouring one alternative, in comparison to others. Take-the-Best, on the other hand, ignores some of the cues. The decision rule here is to (a) search through cues in order of validity, (b) stop searching as soon as a cue is discriminated, and (c) choose the alternative this cue favours (see Gigerenzer and Brighton 2009 for a description of various heuristics).

Research on heuristics originated in cognitive science, trying to find out how human minds make decisions (Tversky and Kahneman 1974). As Gigerenzer and Brighton (2009) point out, the traditional view on heuristics was that many human decisions are based on such heuristics; however, they were often seen as second-best and partly irrational: People use heuristics only because of the cognitive limitations of human minds. Simon (1955, 1979) stated that people "satisfice", i.e., search for "good-enough" solutions to meet their aspiration level, rather than maximize their utility and search for optimal solutions.

This point of view is based on the so-called accuracy-effort trade-off: "Information and computation cost time and effort; therefore, minds rely on simple heuristics that are less accurate than strategies that use more information and computation" (Gigerenzer and Brighton 2009). From this point of view, the use of a simple decision rule would only make sense if its benefits, i.e., the reduced costs of finding a decision (e.g., avoiding the costs of a detailed flood risk assessment) would outweigh its costs, i.e., the marginal social costs of a sub-optimal decision, compared to the optimal solution (e.g., the unavoided flood damage in the case of a sub-optimal flood management option).

However, and this is the main point raised by Gigerenzer and Brighton (2009), heuristics are not always second best. They provide empirical evidence for so-called less-is-more effects, which means that more information or computation can decrease accuracy (ibid.). For instance, in a study on 20 datasets from psychological, biological, sociological, and economic inference tasks, a traditional multiple regression model was compared to the Tallying and Take-the-Best heuristic. Of course, multiple regression performed better in fitting the sample, but both the

Tallying and Take-the-Best heuristic performed better in terms of predictive accuracy, e.g., predicting the rest of the population in cross-validation (ibid.).

According to Gigerenzer and Brighton (2009, Gigerenzer and Gaissmaier 2011), such less-is-more effects can be explained by the bias-variance dilemma: When models are made more complex, the bias of the model is reduced but, in turn, the variance of the model can increase, due to sampling error, leading to lower overall predictive accuracy. Gigerenzer and Brighton (2009) also clearly state that this is, of course, not always the case, but depends on the decision environment. Heuristics perform well, in particular, when decisions have to be made 1) based on small sample sizes and/or 2) under conditions of high uncertainty. Therefore, the main question is not whether heuristics perform better than more complex decision rules, but what would be the best decision rule for a certain decision environment, i.e., which decision rule is "ecologically rational".

4 Are Heuristics Ecologically Rational for Flood Management Decisions? Hypotheses and Research Outline

Coming back to the problems flood risk assessment and management are facing, and relating them to the approach of decision heuristics outlined above, the author offers two hypotheses:

Firstly, regarding the high uncertainties in flood risk management, it is very likely that such less-is-more effects occur, i.e., that simple decision rules lead to the same or even better results (in terms of predictive accuracy) than complex risk assessment/cost-benefit models.

The rationale for this first hypothesis is the following: As outlined above, assessments – in particular in urban areas – have to deal with substantial uncertainties in most parts of the flood risk assessment model chain; these range from natural uncertainties of extreme values statistics, through model uncertainties in hydrological and economic models, to the uncertainties related to climate and socio-economic change scenarios, of which the last-mentioned, in particular, are high in urban environments. Furthermore, models often have to rely on small samples, such as, e.g., flood depth/damage functions, which are often based on few observations or little expert knowledge. In addition, it is likely that there are some basic input components of flood CBA (or cues, in the terminology of decision heuristics) which 1) can be assessed with a relatively high level of certainty and 2) explain a substantial part of the CBA outcome (e.g., investment costs, number of buildings saved from flooding). In such a decision environment, it is likely that less-is-more effects occur and that that heuristics may perform as well as, or even better, than complex risk assessment/CBA models – but this has not been tested, to date.

But how can the predictive accuracy of different decision rules in flood management be measured? One option would be to review past decisions on flood management measures, to check whether different decision rules would have led to different decision outcomes, and to test how the alternative solutions would have performed for past events. However, due to the infrequent nature of flood events, this performance test would be based on only very few events and would therefore have a relatively random outcome. A second option would be to test the outcome of different decision rules against a larger set of random, simulated events. However, these random events would also follow assumed probability distributions of events, and their impacts would need to be estimated by means of existing damage-evaluation models. Therefore, the outcome would probably be more or less the same as that estimated with the risk assessment model chain. This leads to a third option: Existing flood risk/CBA data could be used, but with uncertainty bounds in all the input components (Grelot et al. 2008; Saint-Geours et al. 2015). A ranking based on the full set of information would be assumed to be the benchmark (the best possible estimate). To simulate real world uncertainty, only a sample of all the input components would then be used to test different decision rules, i.e., the traditional risk assessment/CBA approach against various other decision rules (including different decision heuristics, such as Tallying and Take-the-Best). By varying sample sizes, number of input components, and uncertainty bounds, which decision rule performs best under which conditions could be analysed. On this basis, at least an impression about the performance of different decision rules in real world decision-making situations for flood management could be achieved.

The second hypothesis is that decision heuristics are closer to real world decision- making practice than complex risk assessment/CBA models. Whilst flood risk assessment/CBA constitutes the core of the formal decision rules in flood management practice in many countries, it is often embedded in a wider decision framework. For example, in England and Wales (MAFF 1999), in France (CGDD 2014) and in Saxony, Germany (SMUL 2005), the CBA is embedded in a multi-criteria framework. This allows decision makers to consider also social and environmental criteria that have not be included in the CBA framework, because their monetization – or even quantification – is difficult (Meyer et al. 2013). This means that, in practice, the strict logic of expected utility optimization is already complemented with other formal decision rules, often also taking into account equity-based considerations (Johnson et al. 2007; see Gawel and Kuhlicke, "Efficiency-Equity-Trade-Off as a Challenge for Shaping Urban Transformations", in this volume). Furthermore, theoretical considerations from Public Choice theory (Gawel et al. 2016) and interviews with decision makers (Meyer et al. 2012) suggest that informal decision criteria often play an important role in the decision-making process, e.g., regional balance of measures, pressure or resistance from various interest groups, media and voters, budget considerations, etc.

In addition to the normative evaluation of decision heuristics, i.e., the comparison of different decision rules with regard to their predictive accuracy, a descriptive assessment should also be conducted (Mushavi and Gigerenzer 2014). Whereas the normative evaluation outlined above searches for the best decision rule in a given decision environment, the descriptive evaluation seeks to understand how people actually make decisions in given situations and to describe their underlying decision heuristics.

From the author's point of view, it would be very interesting to combine both perspectives on decision heuristics in flood management, in order to:

1. analyse which explicit and implicit decision rules are actually applied in flood management and to find out how decision rules should be designed to match the normative objectives and practical needs of decision makers;
2. investigate which decision rules perform best in terms of predictive accuracy in which decision environment; and
3. combine both in order to identify decision heuristics that perform well in terms of both perspectives, i.e., are practically viable and sufficiently accurate.

The results of such an analysis could also have important practical implications for decision support in urban areas, for flood management, but potentially also for other decision problems related to urban transformation. If the first hypothesis were to be confirmed, good or even better decisions in flood management could be made with less time and effort. This would be particularly relevant for decision support at the urban level, where reliable quantitative estimates of future risks are almost impossible, due to the complexity and dynamics of urban transformations. But it would also create an added value for regions where decision makers do not have the resources to conduct detailed risk assessments, e.g., in developing countries. Finally, decision rules that are more transparent and comprehensible for decision makers and even for laypersons could be formulated. In this sense, the decision heuristics approach seems to be very promising for decision support in flood management in urban areas, but this will require further research.

Acknowledgments I would like to thank Pauline Brémond, Katrin Erdlenbruch, Frédéric Grelot, Reimund Schwarze and Oliver Gebhardt for discussing the topic with me and/or comments on earlier versions. The work on this topic was supported by the Prix Gay-Lussac Humboldt, funded by the French Ministry for Higher Education and Research, in conjunction with the Académie des Sciences, Institut de France.

References

Bouwer LM (2013) Projections of future extreme weather losses under changes in climate and exposure. Risk Anal 33(5):915–930

CGDD - Commissariat général au développement durable (2014) Analyse multicritères des projets de prévention des inondations: Guide méthodologique, + annexes. Ministère de l'Environnement, de l'Énergie et de la Mer

De Moel H, Jongman B, Kreibich H, Merz B, Penning-Rowsell E, Ward PJ (2015) Flood risk assessments at different spatial scales. Mitig Adapt Strateg Glob Chang 20(6):865–890

European Parliament and the Council (Ed) (2007) Directive 2007/60/EC of the European Parliament and of the Council of 23 October 2007 on the assessment and management of flood risks. Off J Eur Union L 288:27–34

Fuchs S, Kuhlicke C, Meyer V (2011) Editorial for the special issue: vulnerability to natural hazards – the challenge of integration. Nat Hazards 58(2):609–619

Gawel E, Lehmann P, Strunz S, Heuson C (2016) A public choice framework for climate adaptation - barriers to efficient adaptation and lessons learned from German flood disasters, UFZ Discussion Papers 3/2016. Helmholtz Centre for Environmental Research – UFZ, Leipzig

Gigerenzer G, Brighton H (2009) Homo heuristicus: why biased minds make better inferences. Top Cogn Sci 1(1):107–143

Gigerenzer G, Gaissmaier W (2011) Heuristic decision making. Annu Rev Psychol 62(1):451–482

Grelot F, Bailly J-S, Blanc C, Erdlenbruch K, Mériaux P, Saint-Geours N, Tourment R (2008) Sensibilité d'une analyse coût-bénéfice - enseignements pour l'évaluation de projets d'atténuation des inondations. Ingénieries EAT 14:95–108

Groves DG, Lempert RJ (2007) A new analytic method for finding policy-relevant scenarios. Glob Environ Chang 17(1):73–85

Haasnoot M, Kwakkel JH, Walker WE, ter Maat J (2013) Dynamic adaptive policy pathways: a method for crafting robust decisions for a deeply uncertain world. Glob Environ Chang 23(2):485–498

Hallegatte S (2014) Natural disasters and climate change. An economic perspective. Springer, Cham

Handmer J (2003) The chimera of precision: inherent uncertainties in disaster loss assessment. Aust J Emerg Manag 18(2):88–97

Hanley N, Spash CL (1993) Cost-benefit analysis and the environment, vol 499. Edward Elgar, Cheltenham

IPCC - Intergovernmental Panel on Climate Change (2012) Summary for policymakers. In: Field CB, Barros V, Stocker TF, Qin D, Dokken DJ, Ebi KL et al (eds) Managing the risks of extreme events and disasters to advance climate change adaptation. A special report of Working Groups I and II of the Intergovernmental Panel on Climate Change. Cambridge University Press, Cambridge/New York, pp 3–21

Johnson C, Penning-Rowsell E, Parker D (2007) Natural and imposed injustices: the challenges in implementing 'fair'flood risk management policy in England. Geogr J 173(4):374–390

Kabisch S, Kuhlicke C (2014) Urban transformations and the idea of resource efficiency, quality of life and resilience. Built Environ 40(4):497–507

Klijn F, Samuels P, Van Os A (2008) Towards flood risk management in the EU: state of affairs with examples from various European countries. Int J River Basin Manag 6(4):307–321

Knight FH (1921) Risk, uncertainty, and profit. Hart, Schaffner & Marx, Boston

Kreibich H, van den Bergh JC, Bouwer LM, Bubeck P, Ciavola P, Green C et al (2014) Costing natural hazards. Nat Clim Chang 4(5):303–306

Lempert RJ, Popper SW, Bankes SC (2003) Shaping the next one hundred years: new methods for quantitative, long-term policy analysis. RAND, Santa Monica

LTV Landestalsperrenverwaltung des Freistaates Sachsen (2003) Erstellung von Hochwasserschutzkonzepten für Fließgewässer. Empfehlungen für die Ermittlung des Gefährdungs- und Schadenpotenzials bei Hochwasserereignissen sowie für die Festlegung von Schutzzielen (unpublished)

MAFF - Ministry of Agriculture, Forestry and Fisheries (1999) Flood and coastal defence project appraisal guidance. MAFF, London

Markantonis V, Meyer V, Schwarze R (2012) Valuating the intangible effects of natural hazards, review and analysis of the costing methods. Nat Hazards Earth Syst Sci 12:1633–1640

Merz B, Thieken AH (2004) Flood risk analysis: concepts and challenges. Österreichische Wasser- und Abfallwirtschaft 56(3–4):27–34

Merz B, Thieken AH (2005) Separating natural and epistemic uncertainty in flood frequency analysis. J Hydrol 309(1–4):114–132

Meyer V, Scheuer S, Haase D (2009) A multicriteria approach for flood risk mapping exemplified at the Mulde river, Germany. Nat Hazards 48(1):17–39

Meyer V, Priest S, Kuhlicke C (2012) Economic evaluation of structural and non-structural flood risk management measures: examples from the Mulde River. Nat Hazards 62(2):301–324

Meyer V, Becker N, Markantonis V, Schwarze R, van den Bergh J, Bouwer L et al (2013) Assessing the costs of natural hazards - state-of-the-art and knowledge gaps. Nat Hazards Earth Syst Sci 13:1351–1373. doi:10.5194/nhess-13-1351-2013.

Mousavi S, Gigerenzer G (2014) Risk, uncertainty, and heuristics. J Bus Res 67(8):1671–1678

Pearce DW, Smale R (2005) Appraising flood control investments in the UK. In: Brouwer R, Pearce D (eds) Cost-benefit analysis and water resources management. Edward Elgar, Cheltenham, pp 71–92

Plate EJ (2000) Risikoanalyse im Hochwasser- und Küstenschutz. In: Franzius Institut für Wasserbau und Küsteningenieurwesen (ed) Risikomanagement im Küstenraum. Beiträge zum internationalen Workshop 30./31. März, Universität Hannover, pp. 1–14

Przyluski V, Hallegatte S (2011) Indirect costs of natural hazards, CONHAZ Report, www.ufz.de/index.php?en=35939. Accessed 2 Nov 2016

Rink D, Banzhaf E, Kabisch S, Krellenberg K (2015) Von der "Großen Transformation" zu urbanen Transformationen. Gaia 24(1):21–25

Saint-Geours N, Grelot F, Bailly JS, Lavergne C (2015) Ranking sources of uncertainty in flood damage modelling: a case study on the cost-benefit analysis of a flood mitigation project in the Orb Delta, France. J Flood Risk Manage 8(2):161–176. doi:10.1111/jfr3.12068

Samuels P, Gouldby B, Klijn F, Messner F, van Os A, Sayers P, et al. (2009) Language of risk—project definitions. Floodsite project report T32-04-01, second edition. www. floodsite.net/html/partner_area/project_docs/T32_04_01_FLOODsite_Language_of_Risk_D32_2_v5_2_P1.pdf. Accessed 2 Nov 2016

Schanze J (2006) Flood risk management - a basic framework. In: Schanze J, Zeman E, Marsalek J (eds) Flood risk management - hazards, vulnerability and mitigation measures. Springer, Heidelberg, pp 149–167

Simon HA (1955) A behavioral model of rational choice. Q J Econ 69(1):99–118

Simon HA (1979) Rational decision making in business organizations. Am Econ Rev 69(4):493–513

SMUL - Sächsisches Staatsministerium für Umwelt und Landwirtschaft (2005) Ergebnisse der landesweiten Priorisierung von Hochwasserschutzmaßnahmen. Sächsisches Staatsministerium für Umwelt und Landwirtschaft. Dresden, Germany. [online]. http://www.umwelt.sachsen.de/umwelt/download/wasser/051206_HwskMaListe_GU_HswskRang_051206.pdf. Accessed 23 Feb 2017

Tung YK (2005) Flood defense systems design by risk-based approaches. Water Int 30(1):50–57

Tversky A, Kahneman D (1974) Judgment under uncertainty: heuristics and biases. Science 185(4157):1124–1130

Weck-Hannemann H, Thöni M (2006) Kosten-Nutzen-Analyse als Entscheidungsgrundlage im Naturgefahrenmanagement: Verfahren, Vorzüge, Vorbehalte, alpS working paper series. a. Z. f. Naturgefahrenmanagement, Innsbruck

Woodward M, Kapelan Z, Gouldby B (2014) Adaptive flood risk management under climate change uncertainty using real options and optimization. Risk Anal 34(1):75–92

Young RA (2005) Economic criteria for water allocation and valuation. In: Brouwer R, Pearce D (eds) Cost-benefit analysis and water resources management. Edward Elgar, Cheltenham, pp 13–45

Reflections

Sigrun Kabisch, Florian Koch, Erik Gawel, Annegret Haase, Sonja Knapp, Kerstin Krellenberg, and Andreas Zehnsdorf

According to the expectations and targets that we formulated in the introduction of this volume, this final chapter represents a critical reflection of the acchivements. We want to sketch out the lessons we learned within the procedure of producing the volume, and formulate some "take-home messages" that might be helpful for other scholars and practitioners working on the issue of urban transformations towards sustainable urban development that call for appropriate research perspectives in particular regarding interdisciplinarity.

One starting point of our research documented in this volume was the observation that the debate on the pathways and fates of urban development has been always confronted with the challenge of how to organize and provide living conditions that enable a dignified life for all people. Cities crystallize into societal hotspots, incorporating a broad spectrum of challenges that became visible and perceptible for the

S. Kabisch • F. Koch • A. Haase • K. Krellenberg
Department of Urban and Environmental Sociology, Helmholtz Centre for Environmental Research-UFZ, Leipzig, Germany
e-mail: sigrun.kabisch@ufz.de; florian.koch@ufz.de; annegret.haase@ufz.de; kerstin.krellenberg@ufz.de

E. Gawel
Department of Economics, Helmholtz Centre for Environmental Research-UFZ, Leipzig, Germany
e-mail: erik.gawel@ufz.de

S. Knapp
Department Community Ecology, Helmholtz Centre for Environmental Research-UFZ, Halle (Saale), Germany
e-mail: sonja.knapp@ufz.de

A. Zehnsdorf
Centre for Environmental Biotechnology, Helmholtz Centre for Environmental Research-UFZ, Leipzig, Germany
e-mail: andreas.zehnsdorf@ufz.de

S. Kabisch et al. (eds.), *Urban Transformations*, Future City 10,
DOI 10.1007/978-3-319-59324-1

local population. Thus, the urban realm is considered as an ideal research environment to discuss and investigate conditions and requirements of transformations, and to elaborate and test practice-oriented knowledge that may have major impacts on the improvement of human living conditions. We define urban transformations as fundamental, multi-dimensional, non-linear alterations of resource use, population developments, and governance modes; we also established values and behaviors as baselines for further conceptual discussions. Coming from a socio-environmental background, we focus on resource-efficiency, quality of life, and resilience as major dimensions of our perspective on urban transformations, acknowledging, especially, the interlinkages and interdependencies between these dimensions.

Under the following five headlines, we summarize our general results and insights.

The Acknowledgement of a Variety of Urban Transformation Approaches

The debate about transformation covers a number of origins and understandings and always depends on the respective thematic genesis. Facing a terminological imbroglio in the scientific literature, we provided, in our introductory chapter, one basic definition of our understanding of urban transformations. Most of the contributions in this volume are in line with this definition. Some contributions critically evaluate and further adapte their conceptual approach, according to the respective research question, e.g., through the use of a transition approach. We consider our initial understanding as contours of a conceptual and methodological framework that allows further supplements to achieve a robust and consistent definition for meaningful dialogues. This will enrich the debate and provide evidence for the dynamics of the object of research.

Thus, urban transformations can differ markedly in terms of form, character, and process; these differences can be described according to their locations on spectra that range from incremental to radical changes, from top-down to bottom-up initiatives, from fast to slow implementations, and from steered to uncontrolled developments.

We do not identify any endpoint for urban transformations; instead, from our scientific point of view, we specify urban transformations as a research mandate.

The Concept of Urban Transformations as an Umbrella Term

The concept of urban transformations has proved to be a key concept that enables an intensive exchange between a variety of disciplinary approaches and backgrounds. As an umbrella term, it provides enough openness to facilitate exchanges between experts from different disciplines dedicated to the solution of selected scientific problems. It helps to analyze complex natural and/or societal changes that can be poorly understood by one or a few disciplines only. Therefore, the concept needs to include the theories and methods of a variety of disciplines. This stimulates new insights for dealing with conceptual approaches and methodological design. Based on a common understanding concerning the urban transformations challenges, we identified, shared, and tailored research targets. To elaborate appropriate comprehensive solutions and recommendations for policy and practice, excellent in-depth studies have to be part of the research design. The key concept of urban transformations interlinks these studies to obtain additional knowledge and solutions. Substantial disciplinary and interdisciplinary knowledge and expertise are the decisive prerequisites for tackling challenging, often complex questions of high societal relevance, such as the United Nations Sustainable Development Goals.

The Inter- and Transdisciplinary Benefits

The collaboration of scientists from different disciplines and backgrounds under the roof of a common umbrella term, such as urban transformations, allows for interdisciplinary studies of different shape, scope, and depth. It fosters an intensive stimulation of new ideas by avoiding sectoral silo thinking. New interactions among disciplines concerning the discussion and development of concepts, subjects, and methodologies are encouraged to fertilize each other. Thus, it focuses on thematic priorities and identifies new research questions. These interrelations can create more efficient problem solutions and raise the level of understanding of urban complexity. Successful interdisciplinary work has to build on disciplinary expertise. Interdisciplinary exchange enriches the thematic debate, opens new scientific perspectives and strategic dialogues, and provides new proposals for project constellations.

Research dedicated to local challenges that require urgent solutions delivers highly acknowledged results if representatives of the respective local institution are part of the research team from the very beginning of the project. This is understood as transdisciplinary research, which can be seen as crucial for ensuring impact and relevance. While the plea for more inter- and transdisciplinarity is not new and has been continuously repeated by different urban and environmental scholars, it needs to be acknowledged that inter- and transdisciplinarity remains challenging: Adequate resources to realize this kind of research, the willingness to leave the beaten track of disciplinary research, as well as the need to provide suitable knowledge and solutions according to academic evaluations standards are all essential. According to our research experiences, the concept of urban transformations has proven to offer an adequate environment for inter- and transdisciplinary research.

The Special Feature of Our Urban Transformation Approach

The normative perspective of our urban transformations concept towards sustainability, as an overarching direction of change along the three dimensions – resource efficiency, resilience, and quality of life – distinguishes our concept from others. The "translation" of the term transformation by the three dimensions allows for a more precise and manageable relationship between conceptual thinking and concrete empirical analyses. The major question about how to achieve sustainable urban development through resource efficiency, resilience, and quality of life acknowledges different ways, according to the particular context. It includes synergies in development strategies, trade-offs and conflicts between diverging plans, big steps forward as well as regressions in terms of non-linearity. This refers to the local requirements without neglecting the general context.

The Merits of Our Approach

The merits of our approach can be specified as content-related, methodology-related, and governance-related.

The content of our research program encompasses a socio-environmental approach towards sustainable urban development, executed by a variety of transformation pathways. We have elaborated the three major dimensions of resource efficiency, quality of life, and resilience and their interrelatedness with respect to both the general sustainability goal and a number of selected challenges that characterize urban diversity.

In terms of methodology, we have demonstrated the benefits of well-established research designs that combine theoretical framings with empirical case study approaches and consider different temporal and spatial scales. In addition, the new combination of different disciplinary measurements and surveys yielded evidence for deeper and more convincing insights into complex issues. Thus, the analytical dimension focuses on understanding how change or stagnation is occurring, interpreted, framed, shaped, negotiated, and implemented.

We stress the importance of governance in bringing socio-technical and socio-environmental innovations into practice. This transdisciplinary research-practice interface pays attention to the decision-making processes and the stakeholder involvement, as well as the gaps between political statements and discussions on the ground. We shed light on the institutional framings that hinder the application of socio-technical solutions. And we place new technological solutions in a societal context, illuminating limits and obstacles to implementing theses novelties. Driver constellation, political power, financial interests as well as legal framings are addressed.

A Final Note

This volume demonstrates the results of a 3-year research process on urban transformations. Being based on the exchange and the collaboration between a plurality of natural, social and technical scientists, this type of research process can be considered itself as a living lab in academic collaboration. It demonstrates what multi- and interdisciplinary research, understood broadly and including a large number of very different disciplinary backgrounds, could achieve in terms of novelty, mutual understanding, and cross-fertilization, but it also reveals both the potentials and the limitations of this ambitious approach. During the research and publication process, the many contributors had the opportunity, but also faced the challenge, to repeatedly reflect on their own research background by constantly exchanging with researchers from other disciplines and generating new knowledge and research approaches in an inter- and transdisciplinary circle. Even though not all questions could be answered and many new questions have emerged, the volume provides novel insights into how urban transformations can be thought about, conceptualized, and investigated.

Index

S. Kabisch et al. (eds.), *Urban Transformations*, Future City 10,
DOI 10.1007/978-3-319-59324-1

Kolibri 360 GmbH
Ferdinand-Jühlke-Straße 7
99095 Erfurt, Deutschland
produktsicherheit@kolibri360.de